ENCYCLOPEDIA OF THE OCEANS

ENCYCLOPEDIA OF THE
OCEANS

Dorrik Stow

OXFORD
UNIVERSITY PRESS

This edition published by

OXFORD
UNIVERSITY PRESS

Great Clarendon Street, Oxford OX2 6DP

Oxford University Press is a department of the
University of Oxford. It furthers the University's objective
of excellence in research, scholarship, and education, by
publishing worldwide in:

Oxford   New York

Auckland   Bangkok   Buenos Aires   Cape Town   Chennai
Dar es Salaam   Delhi   Hong Kong   Istanbul   Karachi
Kolkota   Kuala Lumpur   Madrid   Melbourne   Mexico City
Mumbai   Nairobi   São Paulo   Shanghai   Singapore   Taipei
Tokyo   Toronto

Oxford is a registered trademark of Oxford University Press
in the UK and in certain other countries

British Library Cataloguing-in-Publication Data
Data available

ISBN 0-19-860687-7

Printed in China

10 9 8 7 6 5 4 3 2 1

AN ANDROMEDA BOOK

The Brown Reference Group plc
(incorporating Andromeda Oxford Limited)
8 Chapel Place
Rivington Street
London EC2A 3DQ

© 2004 The Brown Reference Group plc
The moral rights of the authors have been asserted
Database right The Brown Reference Group plc

Project and Art Director   **Ayala Kingsley**
Editor   **Tony Allan**
Cartographic Manager   **Richard Watts**
Cartographic Editor   **Tim Williams**
Picture Management   **Helen Simm, Claire Turner**
Picture Researcher   **Alison Floyd**
Production   **Alastair Gourlay, Clive Sparling**
Indexer   **Ann Barrett**

CONTENTS

PREFACE

"It may be that the oceans are the last, best hope of Earth."

—Dr. Roger Revelle, Director, Scripps Institution of Oceanography, 1950-64

As science-fiction writer Arthur C. Clarke once said, our brilliantly blue planet really should be called "Ocean," for liquid water covers more than 70 percent of its surface. Life on Earth originated in the oceans over 3.5 billion years ago, eventually establishing our oxygen-rich atmosphere, while today, the oceans regulate our climate and stabilize the conditions for life. Although people have explored this watery realm for centuries, scientific understanding of the oceans is still in its infancy. There is a lot we don't know about the oceans' structure, processes, and inhabitants. Much of the knowledge we do have has been skillfully compiled in the *Encyclopedia of the Oceans*. This beautifully illustrated reference volume will introduce you to a fascinating and complex world, and help you better understand the role that the oceans play in our lives. You will also learn how the oceans—and ocean sciences—are being transformed as the twenty-first century begins.

Human beings have always modified their environment to suit their needs. But over the last half century, such alterations have assumed global proportions as we divert and dam major rivers, cut down great swathes of forest, release pollutants into the atmosphere that damage the vital ozone layer, and continue to pump out the vast quantities of greenhouse gases that are beginning to affect our climate. So far, the oceans have met us with resilience, but our exploitation of their resources and disregard for their well-being cannot continue unchecked.

Humans are changing the Earth, so ocean and earth sciences are, out of necessity, changing too. In addition to studying past eons, scientists must now also work to understand the next hundred years, and make useful forecasts of what the future may have in store. The fusion of ocean and earth science, modern biology, and information technology will lead to a continuity of awareness of Earth's systems and their interactions with the human environment. Such integration will strengthen our disaster responses and our ability to manage and sustain our environment and cope with climate change. The upcoming century of ocean exploration will be as eventful and innovative as the last. And it is sure to be even more significant for the future of mankind.

Charles F. Kennel

Director, Scripps Institution of Oceanography at the University of California, San Diego

THE OCEAN FRONTIER

From the dawn of history the ocean has served as a hunting ground for fishermen and a highway for transport and trade. It has offered adventure and challenge to explorers and scientists, as well as providing the battlefield for a long pageant of human conflicts. The sheer immensity of the ocean's reach and its unfathomable depths have been a source of wonderment for our thinkers, and its shorelines a playground for our children, while its beauty and vitality have moved us all.

Earth is an ocean planet, and the oceans are the last great frontier for human endeavor. It is often stated, truly, that we know less about the deep-sea floor than we do about the surface of the Moon. Indeed, we have walked on the dusty surface of our nearest neighbor in space and mapped its every crater, but we are very far from being able to set foot in the deep ocean. The ocean challenge is now firmly before us and it is likely that, by the end of the century, our children's children will be as familiar with the mysteries of the deep seas as we are now with their shoreline.

But environmental problems remain preeminent. Jacques Cousteau, a pioneer of ocean science and its popularization, once wrote of the ocean that it was "a threatened oasis. In our solar system, the Earth is the only planet with an appreciable supply of liquid water. This rare gift is essential for life and, consequently, as the only intelligent and conscious species, mankind should consider the protection of the water system—rivers, lakes, seas, and oceans—the first condition for survival."

There seems little doubt, however, that, in order to support the world's rapidly expanding population, scientific and technical ingenuity will be directed with increasing urgency toward the ocean's great wealth. As we continue to harvest its bountiful resources, great uncertainty remains over our ability to conserve, manage, and protect rather than to pillage and destroy.

The sea has proved at times a dangerous ally, threatening our well-being as well as bringing gifts. History's annals are full of accounts of its power to wreak havoc with people's lives. Times have changed little in this respect. Tropical cyclones can intensify at sea to such an extent that high winds and heavy rainfall can wreak major destruction and cause widespread flooding where they cross the shoreline. The populous coasts of South and Southeast Asia are always at risk during the hot summer months, and each year the coastal states of the southern USA and the whole of the Caribbean region brace themselves for a new hurricane season. Scientists now can monitor, and sometimes predict, natural hazards, but the ability of governments and local populations to mitigate their effects lags far behind.

The big ideas

For 3,000 years at least, philosophers and, later, scientists have been drawn to the ocean and have pondered on its shores, but only in the last century has the science of oceanography begun systematically to unravel its mysteries. Already these studies have produced some of the most momentous discoveries of recent times and have led, directly or indirectly, to four seminal ideas that shape our thinking today.

First, they have shown that life on Earth had its origins some 3.5 billion years ago in the primordial ocean that covered the surface of the planet. Countless legions of organisms have lived and perished in the long process of evolution since those first tiny cells stirred. The living world we know is one of constant evolution and change, of great profusion and profound frailty, of extinction and opportunism.

Secondly, the Earth is in continual motion, albeit at a rate barely measurable except in geological time. Ocean crust is continually created and destroyed. Mountains rise from the sea, slowly erode, and are once again returned as dust to the ocean basins, only to rise again. Neither the land nor sea is immutable.

A third, allied notion is that patterns of change are forever cyclic, although the timescales vary and the past is never exactly repeated. The cycle of rocks, of sea level, and of climate are all inextricably linked to the oceans and to life.

Finally, the natural environment—ocean, earth, and atmosphere—that we share with other species is the single most important variable for our survival. It can and has been altered by our own activities and must, therefore, be carefully managed. Many of the living and non-living resources we extract from the environment are non-renewable.

Hokusai's *Great Wave off Kanagawa*, one of the 19th-century Japanese master's most celebrated color prints, captures something of the sea's destructive power as well as its restless beauty.

These fundamental concepts not only exert profound influence as paradigms in natural science, but have also reached into popular consciousness. Evolution, motion, cyclicity, and change have an everyday reality beyond science that helps us to realize our place in the world. The care of the natural environment, the management of its resources, and the appreciation of its influence on global climate, are an increasingly challenging element of our lives and our relationship with the oceans.

Research and discovery

As scientists continue to probe the oceans and delve into the seafloor sediments and rocks on which they lie, so new discoveries are continually being made. Most of these add relatively little to our understanding, but others seem more important, although their true significance has yet to be determined.

Each new square meter of continental-shelf sediment that is sifted by biological oceanographers can be expected to yield one or two new species, slowly adding to the compendium of marine organisms. But drilling several kilometers into the sediment below the seafloor has revealed an entirely new world of tiny microbial organisms, living in the minute pore spaces between grains of rock. This completely unexpected discovery is profoundly significant in terms of where life can exist, both on this planet and elsewhere in the Solar System. Equally remarkable are the vibrant communities only recently found surrounding metal-charged hot-water vents in the deep ocean. Their existence has already challenged our thinking about the origin of life on Earth.

A growing array of scientific satellites constantly orbits the planet at lightning speeds. Sophisticated information technology systems housed in remote-sensing laboratories around the world can barely cope with the volume of data they return to Earth. Oceanographers can now track the origin and propagation of the rogue waves that are so perilous to shipping or map the high concentrations of plankton that will attract large schools of fish. Such information can be relayed almost instantaneously to the world's fishing fleets.

These eyes in space have given us the ability to detect very subtle but ocean-wide mounds of water on what we once thought of as a flat sea surface. They can record sea-surface temperatures that supply information on the spawning of tropical cyclones and the imminent development of unusual ocean–climate conditions. Remarkably, they can also penetrate below the surface, yielding a new view of ocean-floor topography. The distribution of submarine mountains and valleys, slopes and scarps, submarine fans and mud volcanoes presents new data and challenges for geological oceanographers. Deeply towed instruments greatly refine the scale of observation, while remotely operated submarine vehicles can now roam freely beneath

permanently ice-covered seas. The list of discoveries is as long as it is exciting, yet even so it is dwarfed by the enormous challenge of what we now know we do not know.

Future challenge

We experienced an exponential advance in science and technology through the 20th century, witnessed, for example, by the progress of the space program, which began with the ambitious dream that the human species was ready at last to explore outer space. The first landing on the Moon was followed by unmanned flights to Mars and missions of reconnaissance to Jupiter, Saturn, and the outer planets. A few of these small metallic ambassadors from Earth are even now hurtling out of the Solar System altogether. This great project can be heralded as a crowning achievement for the planet's life, which has been totally earthbound for 3.5 billion years and limited to the ocean for the first 3 billion of these. It is also a remarkable luxury when a third of the human race is starving and a further third is living in poverty, and when 70 percent have neither adequate medical care nor clean drinking water.

But space exploration has brought back unexpected treasures—not just the first prized fragments of Moon rock or, important as it is, the knowledge that water and life do not abound on the other planets of the Solar System, but, quite simply a cosmic view of the Earth itself. Images taken in orbit, or from the surface of the Moon, showed us, for the first time, the world as a whole—the white, swirling clouds in the atmosphere and the blue expanse of ocean that covers 70 percent of its surface. Frontiers between countries, for which so many men and women have fought and died through the years, are neither tangible nor visible. We are all passengers on the same small, isolated blue planet, third out from the Sun. Oceans and continents, mountains and rivers are all teeming with a rare and beautiful life. The Earth and her life systems are an ecological unity.

Ocean space is the last great frontier on the planet. Our scientific and technical capabilities have at last enabled us to penetrate the vast expanse beneath the waves. But they have also given us the ability to radically alter the global ecosystem. At the dawn of a new century we are witnessing major changes to the atmosphere, biosphere, and hydrosphere, all brought about unintentionally by the activities of an ever-increasing human population. Seen primarily as an economic entity, our view of the world is shaped by human greed, rather than human need.

As we enter the third millennium, therefore, we face the most important challenge ever—that of ensuring careful stewardship and sound management of the planet, not only for ourselves but for the whole rich diversity of life whose world we have the privilege to share.

THE LURE OF THE SEA

There seems to be a basic human urge with respect to the sea that began in the shadowy mists of pre-history and that still continues today. Then as now, the sea was alternately savage and powerful, serene and beautiful. For our distant ancestors its far horizons must have seemed another world, mysterious and boundlessly challenging. It is this lure of the unknown that has intrigued humankind since earliest times—and that ocean scientists have only just begun to explore.

The sea played a central role in many early cultures. The ancient Greeks traveled and fought in sleek galleys like this one, powered by oar and sail and steered by a couple of paddle-like rudders.

Scientists now believe that the beginnings of human evolution took place in East Africa, and that it was perhaps as long as 7 million years ago that our ancestral lineage separated from that of the great apes. The early waves of migration out of Africa by *Homo erectus* took place over 1.5 million years ago and were entirely overland. The evidence to date suggests that the first migration by sea took place 700,000 years ago, after which, in time, all corners of the ocean planet were opened up.

There is much debate, however, about just why the human species felt the need to take up the challenge of long sea crossings at a time when only primitive stone tools were in use and when navigation was largely a matter of chance. Perhaps the human mind has always been restless and inquisitive, the human spirit stirred by adventure and romance. Maybe the initial drive was for food and resources, or for battle and territory. The answer is probably all of these, in part, and somehow more.

Migration and adventure

It was a remarkable discovery, made as recently as 1998, that established the presence of *Homo erectus* on Flores Island, somewhere near the middle of the Indonesian archipelago, around 700,000 years ago. The entire prehistory of sea voyages had subsequently to be rewritten, for the next well-established marine crossings were not made until after the first wave of *Homo sapiens* migrations out of Africa, through southern Asia and to Southeast Asia. There is clear evidence for the colonization of Australia around 50,000 years ago, which must have involved deliberate migration across the Banda and Timor seas from Indonesia. Between this date and the *Homo erectus* one yawns a 650,000-year gap, which will almost certainly be slowly filled as marine archaeologists redirect their sights and attention.

More remarkable still is the story of how the Polynesian people came to colonize over 10,000 islands scattered across 26 million sq km (10 million sq mi) of Pacific Ocean. Starting out from the Philippines and New Guinea roughly contemporaneously with the first sea voyages to Australia, they spread slowly throughout the mid-Pacific. By around 500 BCE the Marshall,

Solomon, and Fiji islands had all been populated. But the easternmost destinations of the Polynesian peoples—the Marquesas chain and Easter Island—lie almost half way around the globe from the Philippines. And a quite astounding feat of navigation must have been required to locate Hawaii, one of the most remote island groups anywhere in the world and consequently among the last to be discovered. Their settlement, which must have involved an epic voyage through the equatorial doldrums and then beneath the strange stars of the Northern Hemisphere, was a crowning achievement even for such able seafarers.

We do not know what prompted these migrations or what lured the Polynesians to leave the safety of one island group and push at the frontiers of exploration toward an unknown destiny. Certainly they became master sailors who designed and built great double-hulled sailing ships, some capable of transporting up to 100 people at a time. They mapped out the location of their island homes using stick charts made from bamboo and shells, much like those still used in Micronesia today. They learned to navigate by the stars, but also used each subtle nuance of the marine realm—the direction of rhythmic or refracted waves, the breath of winds, the color of sunrise and sunset. They noted the type and clustering of marine life, the flight of birds, the phases of the Moon, and the state of the sea. Little of this knowledge was ever written down; instead, it was passed down the generations by word of mouth, carefully incorporated into the rituals of a seafaring tradition.

Trade and influence

Throughout history and across the world, different peoples have reached out to the sea for a multitude of reasons. The Egyptians perfected the art of shipbuilding for river and coastal trade around 6,000 years ago, but it was the Phoenicians who became the true maritime masters of the Mediterranean region. They explored and traded to every corner of the Mediterranean Sea, sailed into the Atlantic as far as southwest England, and even circumnavigated Africa. At about the same time that Phoenician dominance began to wane in the west, the Chinese were drawn to the sea, plying their trade, language, and culture through the Far East. By around 400 BCE they had invented the magnetic compass and had become highly accomplished sailors.

On the other side of the world, while the Egyptians were first sailing the River Nile and well before the pyramids had been built, the Ciboney Indians migrated from their South American homeland. Using lightweight balsawood rafts, perhaps with large palm fronds for sails, they were the first to colonize the Caribbean islands. From about 5,000 years ago, a second wave of people, the Arawaks, set out from the Orinoco Delta region in dugout canoes and rafts. Slowly they, too, spread out across the

great library of Alexandria in Egypt—became the first to calculate its circumference to within 8 percent of its actual value. He also invented the concept of latitude and longitude, a system then further refined by the astronomer Hipparchus and the geographer Ptolemy. The Greek legacy was further advanced in the 1st millennium CE by the Arab people, whose trade and influence expanded even beyond that of the Greek world.

Inspiration and change

The same urge to chart the unknown and understand its mysteries inspires scientists today. Yet we cannot pretend that it is only science that drives us on; there has always been something more. Somewhere deep within the core of the human soul there is a beat that matches the rhythm of the sea.

Myths and legends are one way of coping with what we cannot explain, and mythology concerning the sea is often dominant among seafaring peoples. For the ancient Greeks, the Cretan King Minos was the first ruler of the seas; later, Poseidon became a powerful divinity, while Aphrodite, goddess of love and beauty, was born out of the foam. The Polynesian people believed the world was created by their sea-god, Tagaroa, while the islands they inhabited were wrenched from the bottom of the ocean by a famed hero, Maui, who used the magical jawbone of an ancestor as a hook.

From the tranquillity of Canaletto's views of the port of Venice to Turner's wild seascapes and the Impressionists' mastery of light playing on water, the sea has inspired artists through the ages. Great composers, too, have captured facets of the ocean world in their music—the crash of surf and the rustle of pebbles, the swash and backwash of waves on a sandy beach. For poets, the ocean theme is as alive and ever-changing as the sea itself. Whether in Homer's epic *Odyssey*, which portrays a lengthy sea voyage and the multitude of adventures that befall its hero, or in Basho's minimalist haiku, the lure of the sea is ever-present.

It is present, too, in our own lives. The sand and the sea, rock pools and sand dunes, are some of our most cherished playgrounds. Lazy seaside vacations or active water sports, sea fishing, sailing, and cruises draw us in droves to the ocean. Some individuals seek an even closer relationship; by battling the elements in round-the-world sailing races, they are exposed to both the loneliness and fury of the untamed ocean.

Lesser and then the Greater Antilles, largely replacing the Ciboney peoples. Their combination of agricultural acumen and maritime culture was to survive through several more millennia, as the regional population expanded to maybe 6 million people.

It was the descendants of these first island Arawaks who greeted Christopher Columbus and his companions when they arrived in the Bahamas in 1492. Within a century of the European discovery of America, however, the native population had all but vanished, principally as a result of virulent Old World diseases against which they had no immunity. Voyaging by sea can transfer disease and warfare just as it does culture and trade.

But it was the ancient Greeks who exhibited the first truly scientific interest in the sea, thereby laying the foundations of modern oceanography. The explorer Pytheas journeyed as far afield as Norway and Iceland, and developed a method for determining latitude from the North Star. Aristotle, the father of natural philosophy, described in great detail what had so far been learned of natural history, including the first-known treatise on marine biology. The Greeks had already concluded that the Earth was (more or less) spherical before Eratosthenes—then in charge of the

Note on dates

Historical dates are usually given as BCE (Before Common Era) and CE (Common Era). Earlier dates are sometimes abbreviated to mya (million years ago).

GREAT VOYAGES OF DISCOVERY

Large areas of the world were quite unknown to the people of medieval Europe. In the Dark Ages that followed the fall of the Roman Empire, knowledge of the broader world and of science in general had retrenched; even the concept of a flat Earth had gained prominence once more. Yet the early 1400s marked the start of a century of amazing sea voyages. This expansion, which formed part of the Renaissance period in Europe, has become known—with a distinctly Eurocentric bias—as the Age of Discovery.

In a wider perspective, of course, other peoples had populated nearly every corner of the world apart from Antarctica many centuries before. Some, like the Polynesians, still flourished in their island homes scattered across the Pacific Ocean.

In the latter part of the 1st millennium CE, a race of feared warriors rose to prominence in the North Atlantic, raiding at will from their base in Scandinavia. These were the Vikings (a name meaning "pirate"), who would arrive suddenly in swift warships and fight their way inland, robbing towns and villages as they went. But the Vikings were also fine sailors whose longships could safely traverse the open ocean. They colonized Iceland and Greenland, and visited Newfoundland in North America by 1000 CE, 500 years before its rediscovery by the Spanish. They also ventured to the Mediterranean and Black Seas and may have sailed around the southern tip of Africa into the Indian Ocean.

As far as we know, there was only limited contact between the Viking and Arab worlds. Arab sailors were masters of the Indian Ocean at the time. They had a good understanding of the monsoon climate and its effects on sea conditions, and took the idea of the magnetic compass from the Orient for better navigation.

The Chinese, who already had a long seafaring tradition, became ever more skilled and adventurous in this era. Chinese expansion reached a zenith under the Ming Dynasty in the early 15th century, when a magnificent fleet of 317 ships and 37,000 men was commanded by Admiral Zheng He. The largest vessel was a nine-masted treasure ship 134m (440ft) long, which was built to carry luxury materials and artworks to distant lands across the Indian and Pacific oceans. The intention seems to have been to give away the treasures in order to demonstrate the wealth and superiority of Chinese civilization. Although this phase of expansion was relatively short-lived, some of the technical advances the Chinese developed, such as a central rudder, watertight compartments, and sophisticated sails that could be operated from the deck, filtered through into European designs.

European expansion

In hindsight, the European voyages of discovery have often been glamorized as expeditions undertaken for adventure or to advance knowledge. The truth was far more mercenary. The journeys were largely inspired by the search for an alternative channel for the lucrative trade between east and west, the existing overland route having suffered a fatal blow when the Turks captured Constantinople in 1453.

Prince Henry of Portugal was an early visionary who saw clearly that ocean exploration could solve the trade-route problem and yield great wealth in the process. He established a center for the study of marine science and navigation, and insisted that all captains under his patronage should compile detailed charts wherever they sailed. From 1451 to 1470 he championed and funded many expeditions to the South Atlantic, and opened up new commerce with West Africa. But it was not until a storm blew the navigator Bartolomeu Dias round the Cape of Good Hope in 1487 that the route around the southern tip of Africa was finally rediscovered. Other ships then sailed around the Cape; in 1498, Vasco de Gama reached India by this route.

Meanwhile, an Italian mariner named Christopher Columbus had become convinced that he could reach the promised lands of the East by sailing westward around the globe. He persuaded the King of Spain to pay for an expedition that set out from Cádiz in 1492. But Columbus mistakenly calculated the Earth to be only about half its actual size. Unsurprisingly, therefore, he mistook his first landfall on a small Caribbean island for his goal of India

Christopher Columbus exorcises a waterspout in this fanciful artwork illustrating his pioneering voyage of 1492, which opened the way for the European settlement of America.

short-lived, for barely six weeks later Magellan was killed on the Philippine island of Mactan, caught up in a dispute among local tribes. One of his captains, Sebastián del Cano, struggled on for a further 16 months, eventually reaching Seville in September 1522 with only one ship and 18 men remaining.

Piracy and pillage

The circumnavigation of the globe was a crowning achievement of the age, but unfortunately its accomplishment marked the beginning of a far less pleasant period of global history—a time when the European powers set out to exploit the lands they had discovered. Exploration continued through the 16th century and was not without importance to the future of ocean science; it led to the charting of new waters and new territories, to the drafting of improved maps, and to a greater knowledge of the world in general. But a prime motive was still the search for new and simpler trade routes to the Orient. Both Jacques Cartier of France and Martin Frobisher from England set out to find a northwest passage to China around Canada. Cartier sailed up the St. Lawrence estuary as far as the future Montreal; Frobisher pressed further north, eventually foundering against the Canadian Arctic islands. A Dutch sea captain named Willem Barents sought a route to the East by sailing northeast past Norway, locating Svalbard and what is now the Barents Sea on his way to northern Siberia. Trapped in the floating ice of the Arctic Ocean, he managed to build a shelter and survive until the ice melted once more. At about this time, too, Gerardus Mercator constructed a map projection of the world, principally for navigational purposes, which is still in general use today.

However, dominance at sea coupled with growing wealth led to excessive greed; the European maritime nations became the new pirates of the high seas. Their mariners were quite unscrupulous in their subjugation of native peoples, and wholly acquisitive in their rape of other nations' natural resources and the theft of their treasures. Objects of priceless archaeological value were looted and lost, or melted down for coin. The civilizations they encountered were destroyed and enslaved, or perished rapidly through the introduction of exotic diseases.

But the piracy was glamorized and the exploitation condoned at the highest levels. As monarchs became richer, their roving mariners were knighted or otherwise honored. The Spanish conquistadors who defeated the Aztec and Inca peoples of Central and South America were feted in their home land. Sir Francis Drake was undoubtedly a remarkably skilled mariner, sailing around the world from 1577 to 1580 and captaining many transatlantic crossings. Yet even if he was a hero of Elizabethan England, he was also a glorified pirate; like so many other sailors of this era, his legacy was not to ocean science.

or Japan, believing he had strayed either too far north or south to encounter the wealthy cities he had expected. Despite three further trips to the New World, he died convinced he had discovered islands off the Asian coast .

Other explorers quickly followed. John Cabot sailed from England to Newfoundland in 1497, also seeking a new route to the East. In 1499, another Italian, Amerigo Vespucci, reached the South American coast, sailing as far south as the Amazon River; America was to be named after him. The errors of the early explorers were eventually rectified, and the new continent of America was represented for the first time on the world map published in 1507 by the cartographer Martin Waldseemüller.

Circumnavigating the world

However, the prize of being the first person to sail around the world still waited to be claimed—and the Waldseemüller map indicated a possible route, around the tip of South America. Taking up the challenge, the Portuguese navigator Ferdinand Magellan set off in 1519 with a flotilla of five ships and 230 seamen. Almost a year after his departure and with only three ships left, he finally located the narrow passage around South America's southern extremity that now bears his name.

The next months were desperate ones for Magellan. The Pacific crossing was completely unknown and proved to be much longer than anticipated. Half-starved and racked by disease, his crew eventually made landfall on the island of Guam in March 1521. Their relief must have been enormous, but it was

PIONEERS OF OCEAN SCIENCE

The challenge and excitement of marine exploration was reawakened in the 18th century. The pioneers of ocean science embraced the oceans with the same sense of courage and adventure displayed by the navigators of the Age of Discovery, but now with scientific discovery at the top of the agenda. Tremendous advances in our understanding of the seas were made over the next 200 years. This work is still in progress today, although with a level of technical sophistication quite undreamed of in earlier times.

The move to the oceans came partly as fallout from the scientific revolution that began when certain individuals of genius, Galileo and Newton among them, carved new pathways for the physical sciences. It stemmed in part, too, from political and economic expansion, as a new phase of empire-building had just begun. People recognized clearly that scientific and technical acumen, including mastery of the seas, yielded great advantages in gaining and holding territory and power.

Birth of scientific oceanography

It is both convenient and fitting to tie in the birth of modern scientific oceanography with the departure of HMS *Endeavour* under the command of Captain James Cook from Plymouth, England, in 1768. There were, however, several important precursors who had already helped establish the discipline. Robert Boyle had investigated the relationship between temperature, salinity, pressure, and depth in the sea. The Italian diplomat and scientist Luigi Marsigli had compiled his *Histoire Physique de la Mer*, the first book to deal entirely with the ocean. The Swiss mathematician Leonhard Euler had published an important work on the lunar forces responsible for tidal movement, and Benjamin Franklin had charted the Gulf Stream.

Yet the challenge of accurately determining position at sea remained to be solved; without this information, any long voyage remained hazardous. Whereas latitude was easy to calculate from the stars, no simple method was available for longitude, so in 1714 the British government offered a £20,000 prize (equivalent to an amazing US$14 million in modern terms) for the first person to discover one. The secret turned out to lie in devising a means of timekeeping that was reliable over long periods and in all ocean conditions. The quest became a lifetime obsession for the clockmaker John Harrison, whose radical new design featured a spring escapement rather than a pendulum. His No. 4 Chronometer was first tested at sea in 1861, and found to be only five seconds slow after a complete crossing of the Atlantic Ocean. This achievement eventually earned Harrison his due reward when he was 80 years old.

Captain Cook made three long and remarkable exploratory voyages between 1768 and 1780, mainly through the Pacific and Antarctic oceans. For two of these, on HMS *Resolution* and *Discovery*, he had the advantage of a Harrison-style chronometer. His navigation was outstanding and his charting of many Pacific islands, including New Zealand, was invaluable. He sailed to latitude 71°S, where he encountered icebergs and pack ice, although never actually sighted Antarctica. Later he sailed equally far north into the Bering Sea. Cook took scientists from the Royal Society on his various voyages, but his own observations mark him as an exceptional scientist in his own right.

A wooden diving bell was devised in 1775 by the Scottish inventor Charles Spalding. Supplied with barrels of air, the bell could be raised and lowered by its occupants by means of a tackle and weight.

He gathered careful and extensive data on geography, bathymetry, geology, biota, currents, tides, and water temperatures.

As empire-building progressed through the 19th century, so enthusiasm for ocean science blossomed. Valuable observations were made on expeditions from Russia, France, Holland, and the USA as well as from Britain. President Thomas Jefferson established the US Coast and Geodetic Survey in 1807 and mandated coastal charts of the entire United States. The British explorer Sir John Ross ventured into the heart of the Arctic Ocean, while his nephew Sir James Clark Ross led a later expedition to the Antarctic; both pioneered new methods of depth soundings and deep-sea sampling for sediments, biota, and water. Lt. Charles Wilkes led the US Exploring Expedition on a 4-year circumnavigation of the world, starting in 1838; the team's exploits ranged from ascending the volcano Mauna Loa on Hawaii to charting a large sector of the east Antarctic coast.

The voyage of the Beagle

Amid all the activity, the work of one person stands out—Charles Darwin. At the age of 22 Darwin joined the survey ship HMS *Beagle*, sailing on a mission to map the South American coastline, as the ship's naturalist. In the course of an epic voyage from 1831 to 1836, his studies included important work on barnacle biology, an explanation of how atolls and other types of coral reefs develop around volcanic islands, and many detailed observations of both living and fossil species.

Darwin's observations on fossils and on variations within living species—particularly the subtle differences between closely related island species such as the Galápagos finches—led to his seminal work *On the Origin of Species*, first published in 1859, expounding how organisms evolve over many generations under the influence of natural selection. The theory was bitterly contested at the time because it apparently conflicted with the biblical story of Creation, but it is currently the dominant scientific paradigm, accepted throughout the world.

Darwin's work may have thrown the Church into turmoil, but it provided a tremendous stimulus for marine biologists everywhere. The focus of research initially remained near the surface and in coastal waters, partly because they were easier to study but also thanks to Sir Edward Forbes's influential book, *Distribution of Marine Life*, published in 1854. Despite several earlier discoveries in deep trawls and dredgings, Forbes argued that sea life could not exist in the so-called "azoic" zone below about 600m (2,000ft). Coastal laboratories were set up to study the physiology and life histories of the many new species that were being discovered. The important international oceanographic centers at Woods Hole (Massachusetts), Scripps (California), and Naples all began life as marine biological stations.

The concept of an azoic zone, however, was doomed to a short life. In 1868, two British biologists, W.B. Carpenter and C. Wyville Thomson, persuaded the Admiralty to allow them the use of its survey ships so they could dredge in deep water. They soon obtained incontrovertible evidence of a thriving deep-water community previously unknown to science.

The Challenger expedition

The same two scientists were soon to set up one of the most significant voyages in the history of ocean science, even coining the word "oceanography" to describe the purpose of the enterprise. The round-the-world voyage of HMS *Challenger* between 1872 and 1876 was an unqualified success. The expedition—a purely scientific mission with multidisciplinary objectives—covered 127,600km (79,300mi) through all the world's oceans apart from the Arctic. In its course the scientists aboard took hundreds of deep soundings, grabs, trawls, and dredgings, measured salinity, temperature, and water density, and brought back many tons of sediment and biological and sea-water samples for further, detailed analysis.

The true wealth of the expedition's findings only began to unfold when Sir John Murray published the 50-volume *Challenger Report* between 1880 and 1895, following Thomson's early death. In all, 4,717 new marine species were described, and the distribution of seafloor sediments was documented for the first time. Old concepts, such as that of the azoic zone, were buried, and new paradigms emerged, some of which held sway for many years before they, too, were overtaken by new evidence. For example, the *Challenger* data suggested that the deep-sea floor was covered by nothing more than a shallow layer of biogenic ooze and red clay; almost 100 years were to pass before the erroneous nature of this concept was eventually conceded.

The instrumentation on board *Challenger* did not allow for the construction of a detailed picture of ocean circulation. Even so, the measurements taken proved its existence beyond doubt, thus confirming the pioneering work on physical oceanography undertaken by the American Matthew Maury, which was first published in his book *The Physical Geography of the Seas* in 1855.

The expedition was the crowning achievement of the pioneering era, establishing oceanography as a fashionable science of enormous practical and economic significance. Oceanographic institutes opened, consolidated, or expanded in many different countries. The *Challenger* expedition made the case for voyages dedicated to scientific ends and encouraged the development of more sophisticated and standardized instrumentation. It forced the pace of ocean science, emphasizing the necessity for a multidisciplinary approach, and helped spur the international cooperation that has become an essential part of oceanography today.

MODERN OCEANOGRAPHIC RESEARCH

Oceanographic research today is as adventurous as ever—perhaps even more so. Our growing knowledge has only served to open windows on the unknown, revealing fresh scientific questions. But the technology at our disposal makes it easier to find answers and ensures that exploring the ocean realm remains immensely rewarding.

The advent of satellite remote sensing and powerful computer modeling of complex ocean phenomena has committed some scientists entirely to the laboratory or to work in front of a computer screen. However, oceanography is still very much a field science—and the field area is often remote and quite inaccessible. The practical difficulties involved are easily matched by the personal hardships endured by individual researchers. Scientists and their students may become very wet, cold, and seasick, and are often hugely frustrated when sophisticated equipment fails to work or, worse still, is lost at sea. Oceanography can be a remote and lonely science, but the thrill of discovery and the undeniable beauty of the ocean world provide an ample reward for our engagement in the ocean quest.

Motivation and money

Ever since the *Challenger* expedition in the late 19th century, which cost the British treasury an estimated US $200,000, a staggering sum of money for the time, individuals and governments have had to be realistic about the cost of the new science. Oceanography is "big science," necessarily so because of the size of the canvas on which it is undertaken, the nature of the equipment involved, the sea-going vessels that are required, and the scale of the questions being tackled. It has always demanded a cooperative approach involving the backing of large institutions or governments.

The advances made in the early part of the 20th century were generally achieved by major national expeditions, in some cases backed by rich patronage at the highest level. Prince Albert of Monaco and King Carlos of Portugal, for example, both sponsored important oceanographic missions at the turn of the century. These pioneering ventures were followed by the British *Discovery* expedition to the Antarctic Ocean, the German *Meteor* and Swedish Deep-Sea research cruises primarily in the Atlantic Ocean, and the work of the purpose-built US research vessels *Atlantis* and *E.W. Scripps*. Many other countries became involved, so that by the middle of the century there were already some 300 institutions devoted to marine research around the world, from Russia and Japan to Australia and Argentina.

Oceanographic progress has always depended on a mix of politics, economics, and science. The first transatlantic telegraph cable was laid as early as 1858, and now parts of the ocean floor are crisscrossed with communication cables, as well as with oil and gas pipelines and other installations. The need for accurate seafloor surveys to facilitate this work has long been an economic driver. Oil exploration took its first tentative steps into the offshore realm in 1924 and has since moved into progressively deeper waters. Today companies are routinely drilling on the continental slope in water depths of over 2,500m (8,000ft). The technological and environmental challenges posed by oil prospecting has provided a huge incentive for deep-sea research.

The potential wealth to be won from the oceans' bountiful living resources has likewise stimulated much research, especially in biological oceanography. The whaling industry was already well established in the remote South Atlantic island of South Georgia and around Antarctica in the early 1900s, and the British *Discovery* mission, although its aims were laudably scientific, was funded largely to support this lucrative business. Understanding fish ecology, the causes of diminishing stocks, and the often wide fluctuations in fish numbers and distribution has taxed the best efforts of ocean scientists. The challenge is even greater today, when a burgeoning world population requires ever more food resources, when massive factory fleets ply the high seas, and when the "blue revolution" in mariculture is all set to fulfill its promise.

Sadly, ocean science has also benefited from the greed of empire and the needs of military strategists. The Second World War in particular, and the Cold War that ran its course through the ensuing five decades, both demanded as never before an intimate knowledge of the seafloor. Echosounding replaced the

Resembling the ornate forms of some plankton, this robotic submersible heads for deep water. In the future, remotely operated research vehicles could spend months at a time exploring the seabed.

cumbersome and time-consuming single depth soundings by which oceanographers had previously built up an imperfect and patchy view of ocean-floor topography, and it soon became routine on all navy and research vessels. More recent developments using sidescan sonar and sophisticated computer software yield very precise high-resolution maps of large swaths of seafloor, while satellite altimetry has recently produced complete coverage of all the world's oceans at a lower resolution.

Although the age of empire has drawn to a close, its legacy of worldwide sea trade has steadily expanded. The scientific need to understand global water circulation or the generation of freak waves, to track storms and predict hurricanes, becomes all the greater the more the seas are used. Today tourism has become the world's largest multinational industry, and the market is still growing fast. A huge part of the trade centers on coastal or island destinations and on maritime leisure activities, passenger ferries, and luxury cruises. Tourism also creates some of the most serious environmental problems for the oceans as well as for the world as a whole, posing a significant and immediate challenge for oceanographers. As yet, however, there has been minimal direct investment in marine science from the tourist industry.

The scientific quest

Whatever their source of funding, oceanographers strive to address the major scientific questions of the day—and for the future. What is the nature and shape of the ocean floor and how does it interact with the waters above? What are the sources, sinks, and fluxes of sediment to and within the oceans? Can we tap the oceans' long-term memory by drilling deep into the sedimentary record? How are the oceans stirred and what happens to the continuous supply of chemical salts from land and from submarine volcanoes and hot springs? Can we understand and model the links between ocean and climate—and what might these tell us about global warming? What new biological discoveries and fresh organisms will deep-sea exploration next reveal? These are only a few of the principal scientific challenges that face us today: ask any oceanographer and he or she will doubtless mention a dozen more, none of them easy to answer and all requiring largescale funding. But the rewards are of undoubted scientific, economic, and societal importance.

The task of addressing these questions falls to research oceanographers and their students at many hundreds of universities and institutes around the world. Typically, they work together in interdisciplinary teams, often involving cooperation between several institutes and countries. Because the problems are so large and the oceans are a truly global system, an increasing number of international and regional projects have been spawned in the past few decades—the World Ocean Circulation Experiment, the Global Geosphere–Biosphere Program, Global Ocean Ecosystem Dynamics, the European Marine Science and Technology programs, the Tropical Ocean and Global Atmosphere Program, and many others. Some of these, and the numerous exciting findings they have generated, are discussed in the concluding chapters of this book.

Deep-sea drilling

The largest, most costly, and longest-lasting international program in ocean science began in 1968 as the Deep-Sea Drilling Project, under the joint direction of a series of oceanographic institutes in the USA. They commissioned a state-of-the-art research vessel, the *Glomar Challenger*, capable of drilling more than 1,000m (3,300ft) below the seafloor in water depths of over 6km (3.5mi); the ship was also equipped to maintain its position over a single drill site for weeks at a time using satellite-controlled dynamic positioning. Many goals were set for this ambitious new quest, not the least being to test the then-radical hypothesis of seafloor spreading.

Fifteen years later, after voyaging more than 600,000km (375,000mi) and drilling a total of 1,112 holes throughout the world's oceans, the *Glomar Challenger* was eventually decommissioned. Five other nations had joined the project by that time, and scientists from dozens of different countries had sailed aboard this most prestigious of research vessels. Not only had it helped to entrench the theory of plate tectonics and seafloor spreading as a fundamental paradigm in the earth and ocean sciences, but an unprecedented wealth of scientific discovery had been achieved. The overall project was so successful that it was immediately replaced by the Ocean Drilling Program, using a bigger and still more versatile ship, the *JOIDES Resolution*, supported by many more nations. At the time of writing, a further 20 years later, the scientific momentum rolls ever onward—a new Integrated Ocean Drilling Program is being phased in, this time operated jointly from Japan and the United States. It will be a multiplatform operation, initially employing two ships including the recently launched *Chikyu*, with many European and Asian nations directly involved in jointly funding and planning the scientific mission.

Thus the pace of discovery achieved by oceanography in the 20th century is being carried forward with renewed vigor and excitement into the third millennium. It is vitally important that these endeavors, coupled with the many other oceanographic investigations in progress now and in the future, should be directed toward the needs of society and of the world as a whole. Whatever the results may be, we must ensure that they are used for the benefits of the majority of humankind rather than to serve narrow economic greed or military ambition.

Ocean Systems

The physical nature of the ocean is the sum of a complex interplay of cycles. Some of these are immediately obvious to anyone standing on the seashore, and critical to sailors—the pulse of the breaking waves, the rise and fall of the tides, the great surface currents. Some are on a grand scale, others intricate—the crucial relationships of the air-ocean interface that regulate our weather and climate, or the finely balanced chemistry of seawater that is essential to life. And some are imperceptible—the hidden currents of the deep ocean, the cycle of the rocks from land to ocean to the Earth's interior and back, and the slow creep of the continents over the eons.

Part of a recent lava flow on Kilauea, Hawaii. The giant Mauna Loa volcano and the smaller Kilauea, on its eastern flank, make up the southern half of the island. Together they form the world's tallest mountain—reaching 4km (2.5mi) above sea level, but more than 10km (6mi) from its gigantic base on the Pacific seafloor— built up over the last few million years by the accumulation of thousands of lava flows. Volcanic activity is almost as old as the Earth itself, and has been responsible, over the long course of geological history, for the formation of ocean crust, the generation of ocean water, and the evolution of our atmosphere.

PLATES ON THE MOVE

CHAPTER 1

Drifting continents and spreading seas

Out of the cloud of dust and gas that condensed to create the Solar System, on the edge of the spiral galaxy we call the Milky Way, the Earth was formed around 4.5 billion years ago. As the third planet out from the Sun, it was neither so close and hot that the surface was covered in a cocktail of poisonous vapors, nor so distant that only frozen wastes prevailed. In fact, it is the only planet we know with an appreciable supply of liquid water on its surface. From hot and tumultuous beginnings, the Earth gradually cooled; as the molten materials separated out, a thin surface crust developed, and slowly the continents and ocean basins were formed. Rain fell and fell, filling up the basins to form the seas and oceans of eons past.

For as long as we can trace back into that distant past, both continents and oceans have been moving, changing, and evolving, at a rate perceptible only in geological time. New crust is continually being formed beneath the oceans; ocean basins grow, only to contract and be again destroyed, leaving ancient ocean water covering a crust many times its junior. Islands rise from the sea, continents grow and divide. The crust and deep interior of the Earth act as a unified system, driven by forces of convection that result from the continued cooling of the planet's core. These processes, known as plate tectonics, form mountain chains and deep-sea trenches, and explain the nature and distribution of earthquakes and volcanoes. This is the great unifying paradigm of the earth and ocean sciences, developed over the past half-century.

21

The Earth is unique within the Solar System, its liquid oceans and protective atmosphere having survived the tumultuous early years of planetary history. Without water, life as we know it could not exist.

The Ocean Planet

KEY TERMS

BIG BANG THEORY

ACCRETION

PLANETESIMAL

SILICATE

CORE

MANTLE

CRUST

CRUSTAL DIFFERENTIATION

BASALT

LITHOSPHERE

ASTHENOSPHERE

ATMOSPHERE

DEGASSING

Our search for the origin of life, the universe, and everything goes back to the earliest recorded mythologies. In today's world, science seems to offer the most promising avenue of inquiry into such mysteries, even though the farther back in time we attempt a scientific explanation for the natural evolution of the universe, the less reliable our cognizance becomes. Nevertheless, it is important to establish as accurate a scenario as possible, in order to better understand the development and current nature of the Earth.

Currently, the most widely accepted explanation of the origin of the universe is the Big Bang theory. This derives from astronomical observations that the universe appears to be everywhere expanding, with every galaxy racing away from every other at incredible velocities. Rewinding the clock backward to time zero, it would appear that at what astronomers take as the beginning of time, between 12 and 15 billion years ago, all energy and matter were compressed into an inconceivably dense point, and the universe began in a cosmic explosion, before which moment there was no time, nor space, nor matter, in the sense in which we understand these terms.

Multiple thunderstorms over the Pacific (*right*). Four billion years ago, the oceans formed from an almost constant deluge of storms in the primordial atmosphere.

Birth of the Solar System The next few billion years witnessed the expansion and thinning of the universe, coupled with the formation of pockets of concentrated matter from which galaxies were created, each with their own

Planets of the Solar System

Planet	Diameter (km)	Mass (Earth = 1)	Average density (kg/m³)	Interior	Surface	Atmosphere/water
MERCURY	4,880	0.06	5430	Iron core, not known if molten or solid	Heavily cratered	Tenuous atmosphere, mostly H, He
VENUS	12,104	0.81	5250	Probably molten	Lava plains, low hills, some volcanoes	Dense, hot atmosphere, mostly CO_2
EARTH	12,756	1.00	5520	Solid inner core, molten outer core, rocky mantle	Tectonic plates, volcanoes, some craters, weathering	Atmosphere mostly N_2, O_2; liquid surface water
MARS	6,794	0.11	3950	Probably solid core, molten rocky mantle	Some craters, weathering, dormant volcanoes	Atmosphere mostly CO_2, frozen water at poles, liquid water in the past
JUPITER	142,984	17.8	1330	Terrestrial (solid) core, liquid metallic hydrogen inner shell, liquid hydrogen mantle	"Ocean" of liquid hydrogen	Mostly H, He; dynamic cloud belts; hurricane features; some icy moons
SATURN	120,536	95.16	690	Similar to Jupiter, larger core	No solid surface	Mostly H, He; uniform belt and zone structure
URANUS	51,118	14.54	1290	Terrestrial core, liquid water shell, liquid hydrogen/helium mantle	No solid surface	H, He, some CH_4 in higher clouds; weak belt and zone system
NEPTUNE	49,532	17.15	1640	Similar to Uranus	Similar to Uranus	Similar to Uranus
PLUTO	2,300	0.002	2030	Probably rocky	Rock and frozen methane	Mostly N_2

continued ever since. This solar energy is the principal, though not the only, source of energy on Earth, fueling the atmosphere and the oceans and giving rise to and sustaining life. Interestingly, the same type of energy, used to destructive effect, powers nuclear bombs.

Grains of dust in the enveloping space around the Sun collided and fused together to form solid bodies, or "planetesimals," ranging from a few meters to thousands of kilometers in diameter. The larger planetesimals finally swept up the smaller ones to form the planets of the Solar System in their present orbits. The four inner planets—Mercury, Venus, Earth, and Mars—are relatively small and made up of dense materials: rocks and metals. Most of the volatile materials were swept to the cold outer reaches of the Solar System, forming the giant planets—Jupiter, Saturn, Uranus, and Neptune, made up of ice and gases. Beyond them lies tiny Pluto, a strange, frozen mixture of methane, water, and rock, while more than three times farther away is the newly discovered planetoid "Sedna" —possibly the most distant object in the Solar System.

Evolving Earth

Somehow, the Earth evolved from a ball of cosmic dust into a habitable planet with clearly differentiated continents and oceans, and an oxygen-rich atmosphere. The first 500 million years of that evolution resemble a lost manuscript telling a tale about which we can only speculate. But there is little doubt that it was a violent story of constant bombardment and intense heat. Astronomers now believe that the Moon originated during this early, tumultuous phase from a cataclysmic impact between Earth and a Mars-sized proto-planet. The impact showered debris into orbit, some of which aggregated to form the Moon. The oldest lunar rocks brought back to Earth are dated at 4.46 billion years, which we can assume was close to the time of impact.

Two other very important events occurred at this stage, with long-lasting significance for the Earth. The first was that the impact is believed to have knocked the Earth askew from a vertical spin axis to its present inclination of about 23°, and may also have had further influence on the nature of the planet's orbit.

billions of separate stars. Somewhere in the far corner of the universe, a spiral galaxy came into existence, not unlike many others that we can now observe with powerful telescopes. Just over 4.5 billion years ago, a medium-sized star—our own Sun—and a series of planets revolving around it formed along the outer edge of one of the spiral arms.

The Solar System formed from the action of gravity on a rotating cloud of gas and fine dust, known as a nebula. Similar nebulae in outer space far beyond the Solar System are mostly composed of the gases hydrogen and helium, the two elements that make up over 99 percent of our Sun, and a range of dust-sized particles chemically similar to materials found on Earth and the other planets. The diffuse, slowly rotating cloud, originally subspherical in shape, contracted under the influence of gravity to become a rapidly spinning, flattened disk, with material concentrated at the center to form a proto-sun.

Under the unceasing pull of gravity, this proto-sun became dense and hot. Temperatures rose to millions of degrees, at which point nuclear fusion—the fusing together of hydrogen atoms under intense heat and pressure to form helium— began. Nuclear fusion releases enormous amounts of energy, and it has

Core **1**

Mantle **2**

Crust **3**

As the Earth was formed, the intense heat generated by continued bombardment and by the decay of radioactive substances created a semi-molten planet. The heaviest elements, iron and nickel, sank to form the planet's core (**1**). These and other metals combined with oxygen and silicon to form the rocks of the mantle (**2**). Finally, the lightest silicates floated to the surface and the crust began to differentiate (**3**).

Mars is known to possess perennially frozen water at the poles (as seen in this view of the northern ice cap) as well as a much greater seasonal cover of "dry" (carbon dioxide) ice. There is also good evidence for the action of running water in the past. As water is the key to life on Earth, there is intense interest in our nearest neighbor in space.

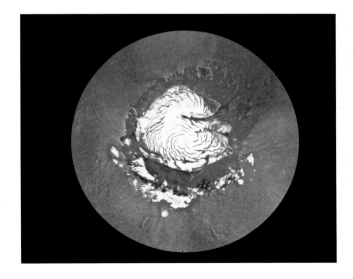

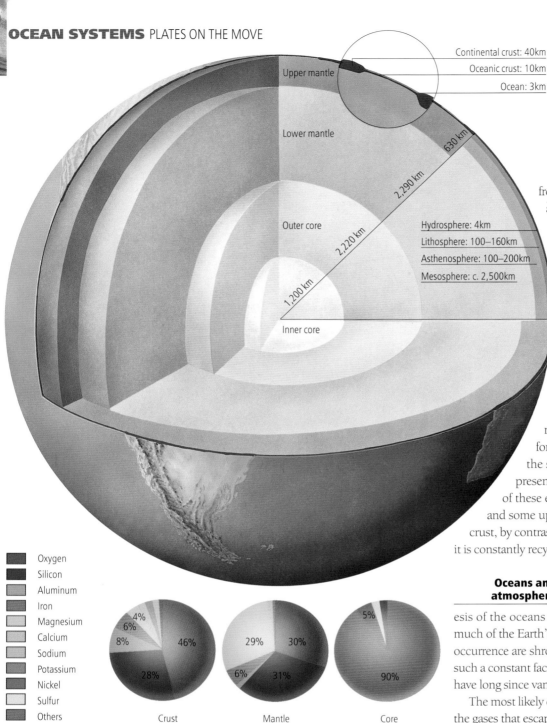

Continental crust: 40km
Oceanic crust: 10km
Ocean: 3km

Upper mantle

Lower mantle

630 km

2,290 km

Outer core

2,220 km

Hydrosphere: 4km
Lithosphere: 100–160km
Asthenosphere: 100–200km
Mesosphere: c. 2,500km

1,200 km

Inner core

Oxygen
Silicon
Aluminum
Iron
Magnesium
Calcium
Sodium
Potassium
Nickel
Sulfur
Others

4%
6%
8%
46%
28%

Crust

29%
30%
6%
31%

Mantle

5%
90%

Core

The Earth's interior consists of a series of concentric layers—core, mantle, and crust, each with a different chemical composition. An alternative division into asthenosphere, lithosphere, mesosphere, and inner and outer core, is based on physical properties such as density.

from the molten interior as the Earth was riven and shaken by repeated, violent eruptions. Continental crust, made up of still lighter materials—granites, metamorphic rocks, and sediments—took much longer to develop. Repeated melting and cooling allowed these progressively lighter materials to separate, gradually massing into small "islands"—the cores of primitive continents. Aggressive weathering by torrential rains led to the break-up of these rocks and their redistribution as sediments along the margins of those first, fragmentary landmasses. Countless cycles were repeated, and slowly larger continents formed, their lower density raising them above the surrounding basins. About 7 percent of the present-day continental crust consists of remnants of these early continents, all of them over 2.5 billion, and some up to 3.8 billion, years old. The oldest oceanic crust, by contrast, is little more than 180 million years old, as it is constantly recycled through the mantle.

Oceans and atmosphere Most important of all to our story, and to the origin of life itself, was the genesis of the oceans that make our blue planet so unique. Like much of the Earth's early history, the details of this momentous occurrence are shrouded in the mists of time. Change has been such a constant factor in Earth's evolution that most of the clues have long since vanished.

The most likely origin of both oceans and atmosphere is from the gases that escaped to the surface during differentiation.

More than 100 elements occur naturally on Earth, but over 99 percent of its mass is made up of just 10 of them. Initially the planet's chemical make-up was relatively homogeneous, but early in its history the different elements separated to give the core, mantle, and crust markedly different compositions.

This change has had a pronounced effect on the Earth's climate ever since. The second was that the heat generated by the impact almost certainly resulted in large-scale melting. Perhaps as much as 50–65 percent of the Earth's total mass melted, and the interior was softened sufficiently to allow differentiation of its components. Heavy metals, principally iron and nickel, sank toward the center, while lighter materials—oxygen, silicon, aluminum—floated to the surface. As the Earth cooled and solidified, it became a layered planet, with a central core, an intermediate mantle, and a thin outer crust, each with very different chemical compositions.

The next crucial phase in the Earth's development was the differentiation of its outer rind into oceanic and continental crust. The composition of these two is very different, as was their formation. Oceanic crust, the more primeval, is made up of dark volcanic rocks, now mainly basalts, which floated to the surface

Pohutu Geyser, New Zealand, is a typical feature of volcanic regions of Earth that are currently active. Geysers are hot-water fountains that intermittently spout steam, water, and other gases with great force, in some cases to heights in excess of 60m (200ft). Some of the dissolved chemicals precipitate as mineral deposits around the vent region as the water cools, while other gases escape into the atmosphere.

Gases emanating from volcanoes today include, among others, water vapor, carbon dioxide, hydrogen, nitrogen, and sulfur dioxide. These are released by the melting of the rocks and minerals in which they are bound. This process, known as degassing, must also have occurred 4 billion years ago, when the first volcanoes began to erupt over a still scorched Earth. Initially these hot gases contributed to the primordial atmosphere—a noxious concoction without the free oxygen that supports most life today. Small amounts of oxygen were released by the action of sunlight on water vapor, but it was not until the appearance of photosynthetic organisms that significant quantities were produced—with great impact on the evolution of life.

Comets such as Hale–Bopp, seen here, may have contributed water and gases to the early atmosphere and ocean.

Because the Earth is relatively large, the force of its gravity was sufficient to retain an atmosphere. As the planet slowly cooled, the water vapor condensed, and violent storms and endless rainfall lashed its surface, trickling and racing to fill the first ocean basins—with fresh water. Some water was also brought in as a component of extraterrestrial bodies, such as comets, during that early period of intense bombardment.

Carbon dioxide dissolved out of the atmosphere would have made the first oceans rather acidic. These acid waters then reacted with the rocks, releasing neutralizing compounds of calcium and magnesium. The composition of the oceans also changed from largely fresh water to the saline water of today, through further complex interaction between the Earth's rocks, waters, atmosphere, and lifeforms. What challenges scientists is to understand the system of delicate balances that now maintains the oceans and atmosphere in their relatively steady state.

Plate tectonics is an elegant concept that has revolutionized our understanding of the earth–ocean system. Though the movement of the plates is imperceptible to human eyes, the effects are often violent and immediate.

PLATES ON THE MOVE
Plate Tectonics and Time

KEY TERMS

PLATE TECTONICS

CONTINENTAL DRIFT

SPREADING CENTER

MANTLE PLUME

CONVECTION CELL

SUBDUCTION

LAW OF SUPERPOSITION

STRATIGRAPHY

MILANKOVITCH CYCLE

ISOTOPE

RADIOACTIVE DECAY

T he normal process of science is punctuated by momentous upheavals in thought and the overturning of long-held paradigms. A major scientific revolution occurs only rarely in any one discipline, heralding a great leap forward in understanding and ushering in a new period of intense research that serves to refine and advance the new consensus. Einstein's theory of relativity, Darwin's theory of evolution, and the more recent discovery of the structure of DNA by Francis Crick and James D. Watson are all examples of this process in action. They are some of the seminal ideas and discoveries that have gone on to shape science and the world.

Earth scientists were fortunate to witness the process of just such a revolution in the course of the 20th century, culminating in the paradigm of plate tectonics that gained general acceptance in the early 1960s. This view provides a single unifying theory that can account at the same time for the nature and for many diverse attributes of both oceans and continents, for mountain-building and sedimentary basins, for the morphology of oceans and the mechanism of continental drift, for the location and types of volcanoes and earthquakes across the world, and for much else besides.

Earlier in the century, a quieter revolution had transformed our understanding of geological time, pushing back the limits from the earlier, religious contention that the world began in 4004 BCE. Without the expanded timeframe the new thinking provided, earth scientists' current views on the origin and evolution of the Earth, and on the processes of plate tectonics that have been played out ever since, would have been untenable.

J agged peaks of the Patagonian Andes *(right)*, a region of continuing uplift. The world's greatest mountain ranges form where two tectonic plates collide.

J ava, Indonesia, as viewed from the Space Shuttle, is part of a broad island archipelago that has witnessed some of the world's most violent volcanic eruptions and devastating earthquakes. These occur as the great Indo–Australian plate plunges beneath Southeast Asia.

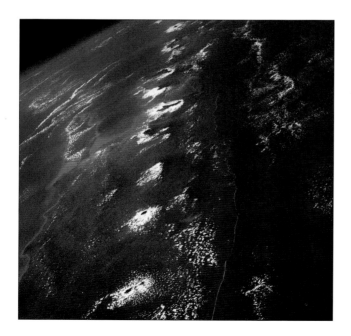

Jigsaw puzzle T he plate tectonic concept is elegant and simple, like many of the great theories that stand the test of time. It holds that the outer layer of the Earth, between 100 and 160km (60–100mi) thick, is made up of a series of large, rigid plates that are in constant motion with respect to one another, jostling and colliding. Plate boundaries are irregular and interlocking, as if in a giant, spherical jigsaw puzzle. Rates of plate motion are imperceptibly slow, being measured in just centimeters per year—roughly the same as the rate of growth of human fingernails.

Not only do the plates move, but their size and shape change constantly as new material is continually added and old material is consumed. This remarkable metamorphosis takes place at plate boundaries, the giant cracks and healed sutures that scar Earth's outer crust, and along which the majority of tectonic activity occurs—earthquakes and volcanic eruptions, hot springs, and heat loss from the Earth's interior. At divergent plate boundaries (spreading centers) new crust is being added, while at convergent margins older crust is consumed or deformed. At transform boundaries, the gigantic, rigid plates simply slide past one another, relatively smoothly but with indescribable force.

An important realization that emerged from the development of plate tectonic theory was that the plates themselves are, in fact, considerably thicker than either the oceanic or continental crust. This crustal material is more like a wrinkled outer rind,

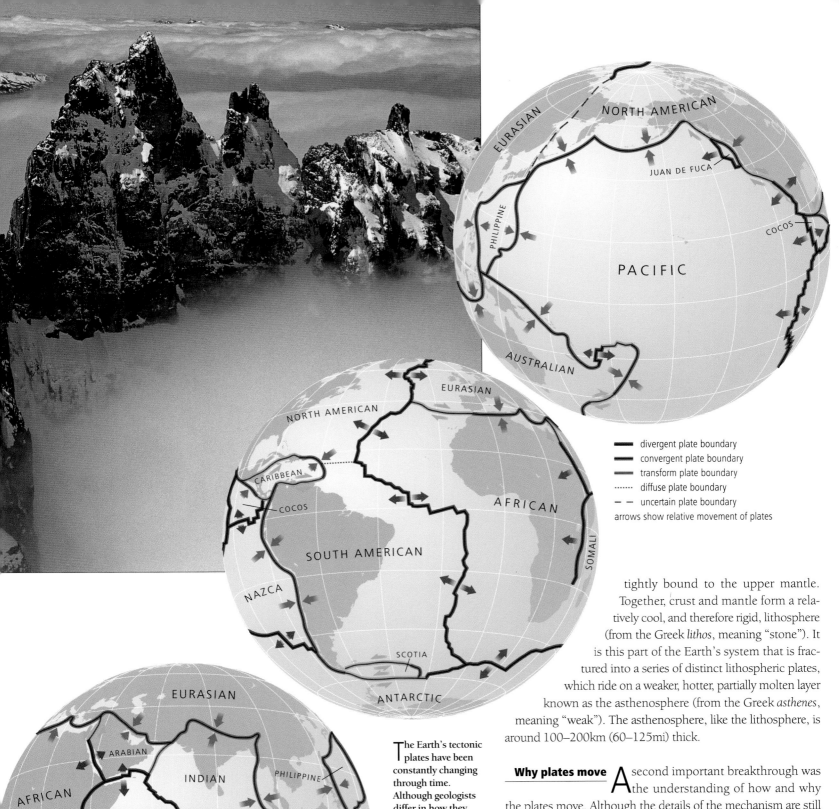

- ▬▬ divergent plate boundary
- ▬▬ convergent plate boundary
- ▬▬ transform plate boundary
- ⋯⋯ diffuse plate boundary
- – – uncertain plate boundary

arrows show relative movement of plates

The Earth's tectonic plates have been constantly changing through time. Although geologists differ in how they classify the plates, there are perhaps eight major ones—North America, South America, Eurasia, Africa, India, Pacific, Nazca, and Antarctic—and several smaller ones, as well as a number of microplates (not shown). The boundaries between some of the plates are diffuse, incomplete, or uncertain.

tightly bound to the upper mantle. Together, crust and mantle form a relatively cool, and therefore rigid, lithosphere (from the Greek *lithos*, meaning "stone"). It is this part of the Earth's system that is fractured into a series of distinct lithospheric plates, which ride on a weaker, hotter, partially molten layer known as the asthenosphere (from the Greek *asthenes*, meaning "weak"). The asthenosphere, like the lithosphere, is around 100–200km (60–125mi) thick.

Why plates move A second important breakthrough was the understanding of how and why the plates move. Although the details of the mechanism are still being debated and refined by geophysicists, a broad consensus accepts that it is in the main part due to large convection currents within the mantle. Convection is the process, most familiar to us in liquids and gases, whereby hot, less dense material rises and cooler, denser material sinks. It is, essentially, the process that drives the great conveyor belt of ocean circulation and the rapid turmoil of winds in the atmosphere, as much as it does the cooling of coffee in a mug or the rising of smoke up a chimney. At conditions of extremely high temperatures and pressures, however, even the "solid" rocks of the outer Earth can behave as an extremely viscous "fluid" that creeps or flows, and thereby allows convection to occur.

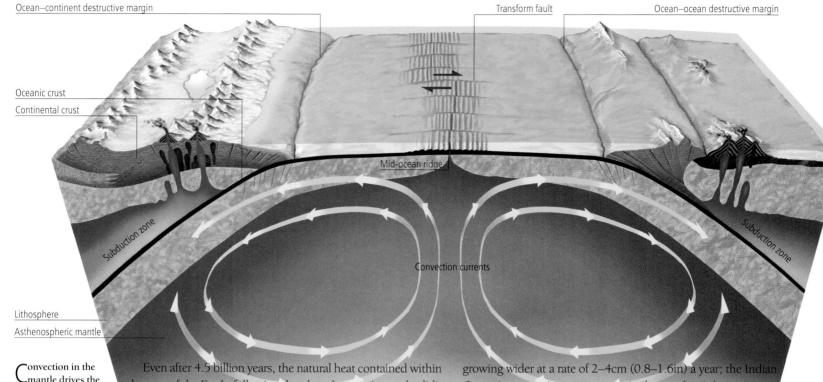

Ocean–continent destructive margin

Transform fault

Ocean–ocean destructive margin

Oceanic crust
Continental crust

Mid-ocean ridge

Subduction zone

Subduction zone

Convection currents

Lithosphere
Asthenospheric mantle

Convection in the mantle drives the slow movement of rigid lithospheric plates across the Earth's surface. Mantle upwelling at mid-ocean ridges forces plates apart as new ocean crust is created. As this crust moves away from the spreading center (divergent margin) it cools and sinks, eventually plunging back into the mantle in the subduction zones where two plates meet (convergent, or destructive, margins). Earthquakes and volcanism are associated with both types of plate boundary.

Even after 4.5 billion years, the natural heat contained within the core of the Earth, following the planet's accretion and solidi-fication, is still the principal heat engine driving convection in the mantle. Some heat is also derived from the decay of naturally occurring radioactive elements such as uranium and thorium. Rising plumes of hot mantle occur beneath divergent plate boundaries, building new crust and lithosphere at the mid-ocean ridges. The new matter cools and slowly subsides as it spreads away, eventually sinking back into the mantle at conver-gent boundaries. It is this slow process of mantle convection, played out over millions of years, that moves the lithospheric plates. Oceans and continents are like passive rafts on the litho-sphere, forever moving and changing as the surface manifesta-tion of deeper-seated convection.

Several methods now exist for calculating the rate of plate motion, although the methods become less and less accurate the further back in time we look. Most reliable for measuring present-day plate movement are direct observations using satel-lites and laser technology. These show that the Atlantic Ocean is

growing wider at a rate of 2–4cm (0.8–1.6in) a year; the Indian Ocean is attempting to grow at a similar rate, but is being severe-ly hampered by surrounding plate collisions. The fastest spread-ing center is the East Pacific Rise, along which ocean crust is being created at rates of around 17cm (7in) per year. In the Pacific Ocean as a whole, however, the plates involved are being even more rapidly consumed at convergent plate boundaries, in a process known as subduction. The net result is, in fact, a reduction in size of the Pacific to the benefit of the other oceans.

Slow beat of time Earth scientists are obsessed with time—both with the relative sequence of events in ocean and earth history that we seek to chart, and with the absolute age of rocks, fossils, ocean waters, and almost everything else we observe. To measure the rates at which natu-ral processes operate, we have devised a series of different timescales (or stratigraphies) that help in our quest to describe, understand, and measure natural Earth phenomena.

Recognition of the Law of Superposition came early in the history of geology. This holds that sedimentary layers are laid down sequentially one on top of another, so that in a normal succession of sedimentary rocks (in other words, one that has not been overturned by subsequent earth movements), the old-est rocks will lie at the base and the younger ones at the top of the pile. A timescale can thus be constructed, layer by layer, for each sedimentary rock succession exposed on land or in bore-holes drilled into the seafloor. In addition, the systematic record of evolutionary changes in the Earth's flora and fauna, drawn up by paleontologists through painstaking research on fossils, has led to the development, and constant refinement, of a timescale based upon the fossil content of rocks that can be readily corre-lated from one part of the globe to another.

The most recent and carefully refined of the relative time-scales oceanographers use has been derived from recognition of

Mt. Semeru towers over 3km (2mi) above central Java, Indonesia, almost dwarfing its neighbor, Mt Bromo. Both peaks are part of a string of active volca-noes formed where molten magma forces itself upward through weathered crust over the Java–Sumatra subduction zone.

cyclic changes in Earth's climate (the Milankovitch cycles) and of the effects that these have on a whole variety of natural events. Variations in global sea level, in the extent of ice cover, and even in the amounts of ice-rafted debris incorporated into sediments from melting icebergs, can all be linked back to climate change. Sophisticated instruments now permit measurement of very subtle changes in the ratios of chemical isotopes (principally oxygen and carbon) that have been incorporated into the shells of living organisms through time.

Absolute dating None of these relative dating techniques would, however, have enabled us to assign absolute ages and to state with confidence—as we now can—that the Earth formed 4.5 billion years ago, that the oldest ocean crust is 180 million years old, that the last dinosaurs roamed the world 65 million years ago, and so on.

That possibility came with the discovery of radioactivity by the French scientist Antoine-Henri Becquerel in 1896, and then in the early 20th century with the measurement of radioactive

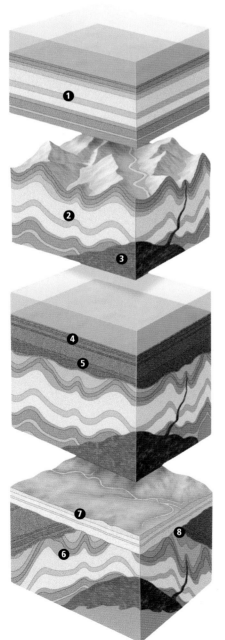

The relationships between rocks of different types allow earth scientists to construct a relative (lithostratigraphic) timescale. The basic principle is that of superposition. Sedimentary rocks are laid down sequentially, with the oldest at the base and the youngest at the top (1). However, strata do not remain undisturbed over the passage of time. Tectonic forces, heat, and pressure lead to deformation, metamorphism, and uplift (2), the deeper parts intruded by a granite mass (3). Later subsidence leads to further deposition (4) over the eroded surface—preserved as an unconformity (5). In a further phase of uplift the rocks are tilted (6) and a new sedimentary sequence (7) laid down, producing an angular unconformity (8).

decay of naturally occurring elements in rocks to obtain their absolute ages. This principle is based on the discovery that some elements, or particular forms of an element known as isotopes, are unstable and emit sub-atomic alpha particles in order to transform themselves into a different, more stable isotope. So uranium-235 decays to produce lead-207, thorium-232 to produce lead-208, and so on. What is so remarkable and fortuitous about this process is that each radioactive isotope decays to its stable "daughter" isotope at an exactly constant rate that is unique to that element. This fact provides scientists with a series of radiometric clocks, some of which tick extremely slowly, while others advance at rates that are geologically rapid. For example, half the uranium-235 in any given rock will change to lead-207 in 713 million years (this period is known as its half-life), whereas the half-life of carbon-14 is just 5,730 years. The former can be used to date the oldest rocks on earth, the latter to date organic material no more than 75,000 years old.

Radiometric dating, however precise, does have limitations. It works well with igneous rocks in which the constituent minerals were formed at about the same time, but much less well with sedimentary rocks, which often contain minerals derived from rocks of different ages. All the same, its discovery, which made possible the calibration of the stratigraphic timescale, heralded a major breakthrough in oceanography—a quieter revolution than that of plate tectonics, perhaps, but a crucial one.

These layers of sedimentary rock at Gros Morne, Newfoundland, Canada, have been tilted to near vertical by an ancient mountain-building episode that took place over 400 million years ago. Fossil remains within some of the layers allow us to place them within a biostratigraphic timescale.

The greatest mountain range on Earth lies in silence and pitch darkness over 2km (1.2mi) beneath the ocean's surface. This hidden ridge is the great assembly line for the production of ocean crust.

Backbone of the Oceans

KEY TERMS

MID-OCEAN RIDGE

MAGMA CHAMBER

DIKE

PILLOW LAVA

GABBRO

MAGNETIC REVERSAL

HYDROTHERMAL VENT

BLACK SMOKER

OPHIOLITE

MOHOROVIČIC DISCONTINUITY

One of the most remarkable features of the oceans is an almost continuous range of submarine mountains that stretches for tens of thousands of kilometers across the face of the Earth. From the Arctic Ocean, close to the North Pole, a great, jagged chain runs down the central axes of both the North and South Atlantic oceans. This range links with another that encircles the Antarctic continent, from which long spurs protrude into both the Pacific and Indian oceans. In all cases, the mountains rise up to 3km (1.9mi) above the adjacent ocean basins to within 2.5km (1.6mi) of the ocean surface and, at their broadest, boast a width of over 1,000km (620mi). This is the global mid-ocean ridge system—the backbone of the oceans.

Yet none of the peaks has been scaled, none of the ridges traversed. Only a few places have been viewed directly, from diving submersibles. Other sections have been the subject of detailed remote surveys from scientific ships, but the vast remainder lies unexplored and unknown. Yet each visit made, remote video survey undertaken, or sample recovered, each set of instruments laid down to record, each borehole that is drilled into these dark volcanic rocks, yields more and more fascinating data. New discoveries are made that challenge our understanding of the entire earth–ocean system.

Assembly line Mid-ocean ridges occur at divergent plate boundaries. They are the sites at which new oceanic crust is continually being created, as outpourings of a dark volcanic rock called basalt. The entire ridge system is essentially a continuous belt of active volcanoes, along which the Earth's internal heat is dissipated. Shallow-focus earthquakes abound as volcanic material is forced up to the surface, new crust is accreted, and the seafloor of yesterday is

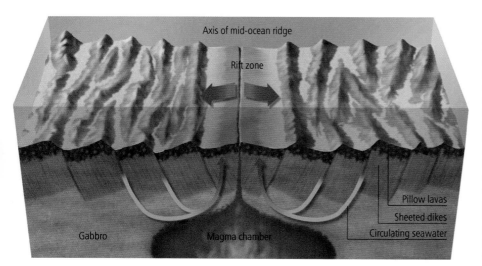

Axis of mid-ocean ridge

Rift zone

Pillow lavas

Sheeted dikes

Circulating seawater

Gabbro

Magma chamber

New oceanic crust is formed as basaltic lava extrudes onto the seafloor at mid-ocean ridges, over 2km (1.2mi) below the ocean surface, along the East Pacific Rise. At the very great pressures and icy temperatures of the deep ocean, lava oozes out slowly and rapidly cools to form globular masses known as pillow lavas.

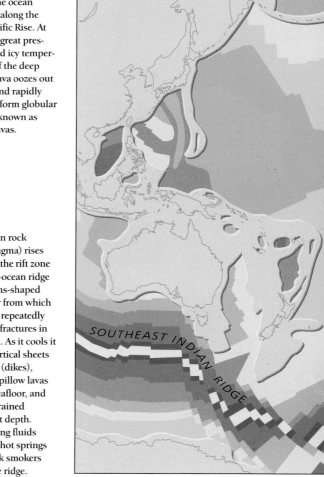

Molten rock (magma) rises beneath the rift zone of a mid-ocean ridge into a lens-shaped chamber from which it erupts repeatedly through fractures in the crust. As it cools it forms vertical sheets of basalt (dikes), fields of pillow lavas on the seafloor, and coarse-grained gabbro at depth. Circulating fluids result in hot springs and black smokers along the ridge.

SOUTHEAST INDIAN RIDGE

pushed aside. The rate of growth is slow—only about 3.5sq km (1.4sq mi) of new ocean crust is added to the Earth each year—and rates of spreading about the axis are no more than a few centimeters annually. Over geological time, however, these imperceptible amounts gradually accrue so that whole new oceans are born and grow, ultimately to face destruction in deep, narrow trenches—the Earth's great recycling plants.

Hot mantle material wells up beneath the axis of the ridge, driven by the gigantic convection cells of the Earth's interior. A hot mush of mineral crystals and molten basalt creates a flattened magma chamber only a few kilometers beneath the seafloor, from which sheets and pipes of magma (dikes) force their way through fractures in the weakened crust above and erupt repeatedly into the icy stillness of the deep ocean. Eruption beneath 2.5km (1.6mi) of water is not a violent process, but more like squeezing toothpaste out of a tube. The creation of new seafloor in this way has been captured on film, the dark, oozing basalt forming a series of small, rounded domes known as pillow lavas. The glassy rims of these pillows are testament to very rapid chilling, as lava whose temperature exceeds 1,000°C (1,800°F) meets near-freezing water.

As more and more dikes and pillow lavas make their appearance, the new crust is gradually forced apart, to begin its slow journey across the ocean on a "conveyor belt" of cooling lithosphere. The elevation of the ridge crest is due to its heat (and hence lower density), but as magma cools and moves away from the crest, the crust slowly contracts and subsides. Older crust, farther from the ridge, forms the floor of progressively deeper ocean, accumulating thick deposits of sediment from surface waters and an influx of material from the continental margins.

Measuring the age of ocean crust proved revelatory when, in the early 1960s, the realization that it grows systematically older away from ridge crests provided the remarkable, final proof of plate tectonic theory. It seems that the Earth's magnetic field interacts with cooling magma in a way that turns the seafloor, in effect, into a magnetic tape recorder. Iron-rich minerals crystallizing out from the hot lava become individually magnetized, orienting toward magnetic north, and then setting firm as the lava solidifies. The present direction of the magnetic field is referred to as "normal" and the opposite direction as "reverse". During the geologic past, Earth's magnetic field has switched back and forth between normal and reverse with an erratic periodicity, at

In the map below, showing the age of Earth's ocean floor, each colored band represents a different timespan during which a particular area of crustal material formed. Ages have been determined from a combination of magneto- and biostratigraphy, and show symmetrical crustal spreading on either side of the mid-ocean ridges. The oldest crust formed nearly 200 million years ago.

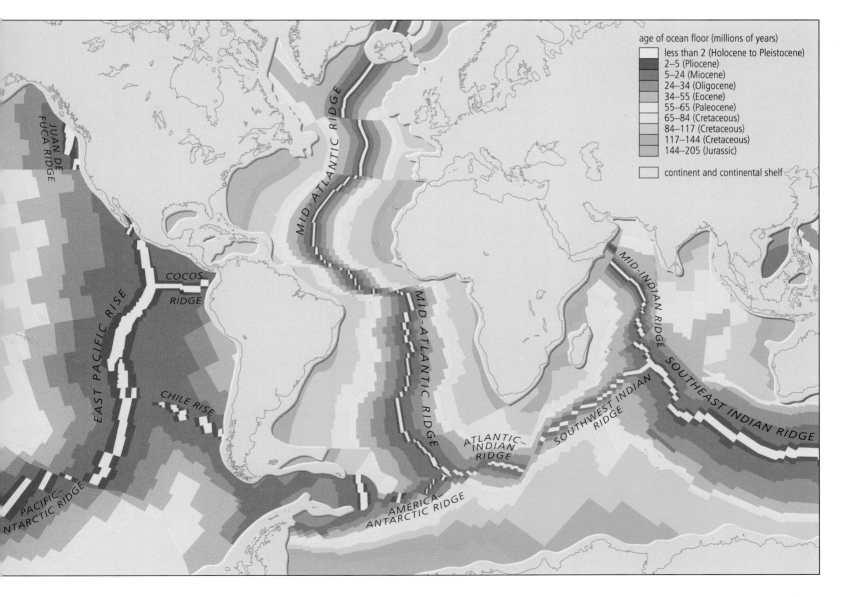

age of ocean floor (millions of years)

- less than 2 (Holocene to Pleistocene)
- 2–5 (Pliocene)
- 5–24 (Miocene)
- 24–34 (Oligocene)
- 34–55 (Eocene)
- 55–65 (Paleocene)
- 65–84 (Cretaceous)
- 84–117 (Cretaceous)
- 117–144 (Cretaceous)
- 144–205 (Jurassic)

continent and continental shelf

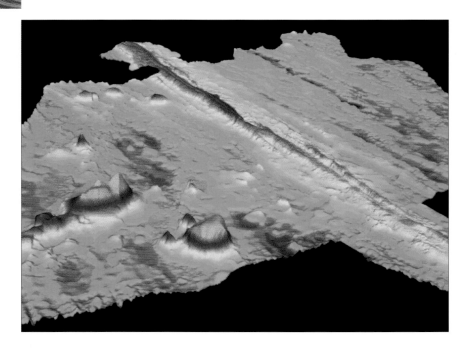

A false-color image of the seafloor on either side of the East Pacific Rise shows tensional stresses related to spreading. Faults have opened up parallel to the central ridge, creating an off-axis, steplike topography. Off-axis volcanic seamounts can be seen at left.

The Mid-Atlantic Ridge emerges from the ocean to cut across the heart of Iceland (right), providing a rare opportunity to view the process of deep-ocean floor formation on land.

A black smoker in the Pacific belches forth hot toxic metals that condense as mineral precipitates around the vent. Despite the high temperatures, pitch blackness, and metal-enriched waters, such vents are home to a unique array of organisms.

intervals ranging from many thousands to millions of years. Towing sensitive magnetometers behind research ships allows us to measure the direction of magnetization of crustal rocks. The pattern revealed is one of long, narrow bands of normal and reverse magnetic polarity, with almost perfect symmetry on either side of the ridge crest.

The ages of magnetic reversals have been carefully worked out from radiometric dating of magnetized lavas on land, so it becomes possible to assign ages to the bands of magnetized rocks on the seafloor. A complete map of the age of ocean crust can be constructed, showing that the oldest crust was formed around 180 million years ago and lies furthest from the present-day spreading center—the ridge crest. The map also allows us to calculate the rate of spreading.

Deep-ocean vents Spectacular discoveries made within the past 25 years have revealed plumes of hot, mineral-laden water spouting from hydrothermal vents at many places along the mid-ocean ridge system. Cold seawater

percolates into the many cracks and fissures associated with seafloor spreading, penetrating several kilometers into newly formed ocean crust, and leaching out a rich variety of minerals. Eventually it encounters hot magma and is forced back up to the surface, emerging on the seafloor in the form of hot springs. Dissolved minerals rapidly precipitate as these often superheated waters (reaching 300°–400°C/550°–750°F) come into contact with near-freezing temperatures at the ocean bottom. The spurting and billowing clouds that result are known as black smokers. Their precipitates can produce massive ore bodies rich in iron, copper, zinc, and a host of other metals, either as stacks of chimneys and other encrustations on the seabed, or as pools of metal-rich sediments nestling into irregular hollows along the mid-ocean ridge. Despite high temperatures, pitch blackness, and a cocktail of toxic chemicals, an extraordinary community of creatures thrives in this hostile environment.

No sooner were active black smokers discovered on the ocean floor than their fossil counterparts were found preserved on land, in the distinctive sequences of rocks known to geologists as as ophiolites, and recognized as being slices of former ocean crust. The structure of ocean lithosphere is relatively simple and uniform. Deep-sea and metal-rich sediments, pillow lavas, sheeted dikes, and layered gabbro (a coarse-grained igneous rock) form the crust, which directly overlies the dense, ultramafic (low in silica) rocks of the upper mantle. This crust–mantle boundary is termed the Mohoroviçic discontinuity (or Moho) in honor of the Croatian seismologist who discovered it in 1909, after observing a sharp increase in seismic wave velocity at that depth, as a result of the sudden change in density.

When subjected to the enormous forces involved in plate collision, fragments of ocean lithosphere can become detached and take up a new position on continental crust. Such ophiolites with their distinctively oceanic characteristics, in some instances somewhat disturbed, mixed, and altered, are familiar constituents of many orogenic (mountain-forming) belts stretched across the continents. Pillow lavas that were once the floor of a two-billion-year-old ocean form part of a much-fragmented ophiolite suite exposed in northern Quebec. Much more complete ophiolites are known from the island of Cyprus—an 80-million-year-old fragment of the now vanished Tethys Ocean—and, younger still, from many parts of the Japanese island arc. In such places, it is possible to actually walk on the floor of a vanished ocean or stand astride the famous Moho discontinuity between crust and mantle rocks.

A ridge revealed Another focus for those scientists who study the deep seafloor is Iceland. This volcanic island, lying astride the Mid-Atlantic Ridge, has been built up over millions of years from vast outpourings of oceanic lava, so that the steep scarps of the ridge itself are now exposed on land. Walking along the Thingvellir graben near Reykjavik—in Viking times, the seat of an open-air parliament—is equivalent to taking a stroll down the narrow, axial valley that runs along all slow-spreading mid-ocean ridges. The fast-spreading East Pacific Rise, by contrast, has a peaked crest.

Iceland has built up directly above a hotspot, where molten material rises from the mantle in a narrow plume that extends for as much as 600km (370mi) under the Earth's surface. Although hotspots are only sometimes found at mid-ocean ridges, this one began around 63 million years ago beneath ruptures in the continental crust that formerly joined Greenland with northern Europe. For 10 million years it spewed out lava, covering thousands of square kilometers with molten rock. The remnants of these great lava fields are found today as far apart as eastern Greenland, the Faeroe Isles, the Inner Hebrides, and

Northern Ireland. Other volcanic islands that may be associated with hotspots along the Mid-Atlantic Ridge are Ascension Island, St. Helena, and Tristan da Cunha.

Eventually the stretched continental crust began to rift, and an incipient seaway opened up between the Arctic Ocean and the southern part of the North Atlantic, which was already a respectably wide ocean. The process of seafloor spreading continued over the next 50 million years, as the present-day Atlantic took shape. The hotspot beneath Iceland is still there, evidenced by the island's 35 active volcanic centers and fields of hot springs.

When plates collide, they enter the great subduction factories beneath deep-sea trenches to be re-melted and recycled. Volcanoes, earthquakes, and the uplift of whole mountain ranges are the results.

PLATES ON THE MOVE

Collision and Slippage

If new crust and lithosphere are continually being added to the seafloor along the entire production-line of the mid-ocean ridge system, then either the Earth is steadily expanding or an equal volume of material must be destroyed. We now know the latter hypothesis is true. Destruction of ocean crust takes place in subduction zones—the world's largest recycling plants—located along convergent plate margins. These regions of great complexity and turmoil are characterized by violent volcanic eruptions and devastating earthquakes, by the profound depths of oceanic trenches where plates plunge to destruction, by young, jagged mountain ranges that are still rising today, and by a complicated geology of exotic rock formations and chaotic mélanges.

Three principal types of plate boundary at convergent (destructive) margins are shown *(right)*.
1. Ocean–continent collision, where ocean is destroyed and a mountain range formed.
2. Ocean–ocean collision, with the subduction of one plate and creation of a volcanic island-arc on the other.
3. Continent–continent collision, resulting in major mountain uplift, crustal deformation, and thickening.

Subduction and recycling Where the edges of oceanic plates collide with continents, they are subducted—literally, "forced under." Deep trenches appear in the seabed, whose dark, mysterious floors, lying over 10km (6mi) down, have mostly never been seen. It is into these gigantic gashes and beyond that the heavier oceanic lithosphere is thrust downward beneath the lighter continental material to be re-melted and recycled. The downgoing slab of material is put under a strain so enormous that it can only be released by periodic and violent earthquakes. The focal points at which these earthquakes occur follow the sloping plane of the descending plate, their magnitude gradually diminishing with depth but reaching down to several hundreds of kilometers below the ocean floor. A new technique, seismic tomography, can be used

to create images of relative temperature within the mantle. Quite remarkably, these images have shown remnants of cool lithospheric material penetrating almost as far as the mantle–core boundary, at a depth of 2,900km (1,800mi).

In the upper parts of a subduction zone, enormous frictional resistance between plate and crust causes melting on the surface of the descending plate. The wet sediments and rocks carried down on the plate surface, as well as the rocks of the deep lithosphere, begin to melt at depths of 50–100km (30–60mi), where temperatures rise to 1,200°–1,500°C (2,200°–2,750°F), creating localized pockets or chambers of molten rock. Because this magma has been formed from sediments and continental lithosphere, it is more silicic (quartz-rich) and so less dense than oceanic crust. These lighter materials rise as igneous intrusions into the lower crust, where they cool and crystallize as granites—the principal rock type of the continents. Some of the molten material forces its way right to the surface, bursting out as pressure is released in the form of the highly explosive volcanic eruptions associated with subduction zones.

The entire rim of the Pacific Ocean is marked by subduction zones, and is therefore plagued with the largest number of earthquakes and volcanoes found anywhere in the world—hence the nickname of the "Ring of Fire" given to the lands on its rim. The great plates that make up this ocean are constantly being

Figure 1 labels: Trench; Volcanic belt; Oceanic crust; Continental crust; Subducting oceanic lithosphere; Sediment wedge; Melting ocean crust

Figure 2 labels: Trench; Volcanic island arc; Forearc basin; Back-arc basin; Subducting oceanic lithosphere; Overriding ocean plate; Ocean sediments; Arc-derived sediments; Deformed ocean sediments

Figure 3 labels: Fold mountains; Ophiolites; Thrust sheets; Continental crust; Continental crust; Igneous intrusion; Subducting oceanic lithosphere; Overriding continental plate; Oceanic crust

Deep-ocean trenches

Trench	Depth (km)	Average width (km)	Length (km)
ALEUTIAN	7.7	50	3,700
MIDDLE AMERICA	6.7	40	2,800
PUERTO RICO	8.4	120	1,550
PERU–CHILE	8.1	100	5,900
TONGA	10.8	55	1,400
JAVA (Sunda)	7.5	80	4,500
PHILIPPINE	10.5	60	1,400
MARIANA	11.0	70	2,550
JAPAN	7.5	100	800
KURIL	8.4	120	2,200

recycled beneath the continents. The Nazca plate has been submerging beneath South America for at least 180 million years, and the Andes range, the longest terrestrial mountain chain on Earth, is the result. Further north, the small Juan de Fuca and Cocos plates are the only remnants of the once mighty Farallon plate, which formed the floor of much of the ancestral Pacific Ocean. Once again, 200 million years of ocean floor have been subducted. Seismic tomography reveals the cooler lithospheric slab still evident deep within the mantle, somewhere beneath the earthquake-prone Caribbean.

Along large stretches of the western and northern margins of the Pacific, subduction has created a string of volcanoes that rise as islands above the sea's surface. Groups of volcanic islands, such as Japan, the Aleutians, and the Philippines, occur about 100km (60mi) on the landward side of the oceanic trenches to which they are linked. They form great curvilinear features called island arcs. The concave side of an island arc faces the belt of sea separating it from the nearest continent, while the convex side looks toward the open ocean. Over millions of years, once distinct islands have amalgamated and landmasses have steadily grown until they begin to resemble small fragments of continent. The area between an island arc and the mainland, known as the back-arc basin, typically develops a small spreading center of its own as new volcanic material forces the crust apart.

Some of the lesser-known and more isolated volcanic peaks of the Kuril, Izu–Bonin, and Mariana arcs, however, face a running battle for survival above the relentless pounding of waves, the surge of storms, and the freak tsunami that so typify the altogether mis-named Pacific Ocean.

Seen here from the Space Shuttle, Hokkaido (Japan) and the Kuril Islands are both parts of great island arc systems within the Pacific Ring of Fire. They have grown up through time as the result of volcanic outpourings above subduction zones in the northwestern Pacific.

Mountain building Over time, the ratio of ocean crust to continental crust (and thus the extent of the oceans) has steadily decreased as the continents have grown. Although parts of the "Ring of Fire" are the result of the collision of oceanic plates, to the east the subduction takes place beneath the margins of the continents, where new continental crust is formed. The spreading of the Pacific Ocean ridges pushes the plates beneath the western edge of the Americas, causing great mountain chains to rise up in their wake. As the subducting oceanic plates slide beneath the continent, immense stresses

The Ring of Fire is a belt of intense volcanic and seismic activity encircling most of the Pacific Ocean (*below*) along the plate boundaries. It represents the Earth's response to plate destruction and material recycling through subduction zones.

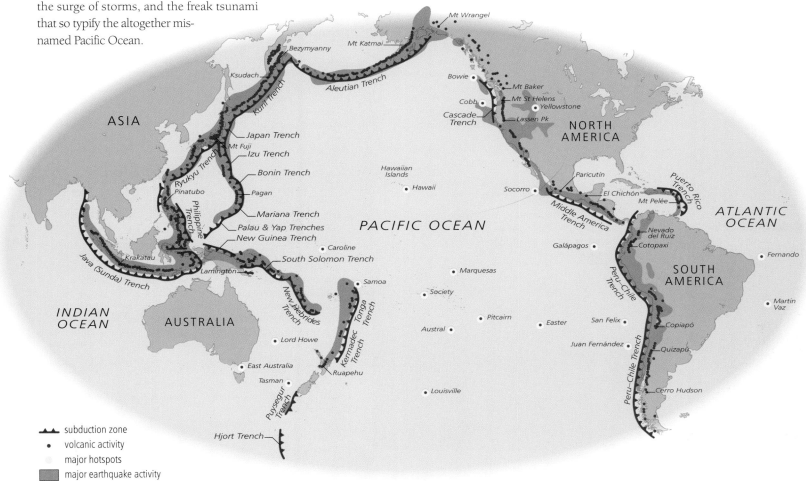

- ᴧᴧ subduction zone
- • volcanic activity
- ○ major hotspots
- ▨ major earthquake activity

Cutting its way down the western seaboard of California, the San Andreas Fault is part of a complex network of faults and associated seismic activity. The fractures mark the zone of intense shear as the Pacific and North American plates grind past one another. This region is almost entirely made up of exotic terranes—odd pieces of island arc and continental fragments welded together and bounded by faults.

occur. Magma from melting subducted material rises through the continental crust to form a chain of volcanoes—the core of a new mountain range. As weathering and erosion progressively denude the rising mountains, most of the eroded sediment cascades down in tumultuous rivers, feeds into the ocean, and is then plastered against the continental margin. Little escapes beyond into true oceanic depths, and even this material is slowly carried back toward the continent as the downgoing plate continues to subduct. Huge slices of sediment, together with

parts of the ocean crust itself, are scraped off onto the continent. In the process they are crumpled and folded, broken apart by faults, and changed into metamorphic rocks by extreme pressures and temperatures. Such chaotic assemblages of mixed rock types, accreted to the continents in this way, are known as mélanges. In addition, giant rafts or blocks of oceanic and continental material, alien to the local area and often of unknown derivation, are carried along on the descending plate and become fused into the mountain belt. These are called exotic terranes.

The Western Cordillera of North America—a broad belt of mountains extending inland from the Pacific coast to the Rockies—is largely composed of exotic terranes, each one separated from its neighbor by a fault. Changing patterns of subduction in the area have added to the complexity of the geology, producing fault zones such as the San Andreas Fault in California, where two plates grind uncomfortably past each other in opposite directions.

The Andes range, which runs along the entire western margin of South America, is still in the process of being created and has many active volcanoes, although the shallow inclination of the subducting Nazca plate has caused the center of igneous activity to shift increasingly farther inland. The rising magma produces gigantic granite batholiths ("deep rocks") beneath the surface and explosive eruptions of lava. The deep granite roots of the mountain chain are still buoyant, so that the Andes today are rising at rates of over 50cm (20in) a year.

Mount Fuji: peaceful slopes, violent past

Towering, majestic, with perfect snow-covered slopes and a prolonged dormancy that belies its violent past, Fujiyama, as the mountain is known in Japan, dominates the landscape of central Honshu, the main Japanese island. The highest and best-known of over 70 volcanoes in Japan, it last erupted nearly 300 years ago, in 1707. Fujiyama is a mature, composite volcano, currently located at a unique intersection of the Honshu and Japan island arcs. These arcs, and the numerous volcanoes associated with them, result from the subduction of the Philippine and Pacific plates beneath the Eurasian and North American plates respectively. But the origins and history of Fujiyama are altogether more enigmatic. We know that the Pacific plate is also plunging westward beneath the Philippine Sea, creating some of the deepest trenches in the world and, in the Izu–Bonin Arc, lying due south of Honshu, some of the most remote and beautiful volcanic islands. This island arc is slowly but inexorably moving north and colliding with Japan. Over the past few millions of years a series of Izu–Bonin volcanoes have impacted and become accreted onto the mainland. Through this weakened crust, Fujiyama began to erupt one million years ago. Next in line is the O-shima volcanic isle, now lurking in the Sagami Sea offshore from Tokyo.

This remarkable story of arc–arc collision is clearly written in the rock record around Fujiyama. Even a cursory glance at the geological map shows an enormous indentation in the rock formations as they bend around Mt. Fuji and the impact zone. As each volcano has accreted, the products of rapid erosion are deposited over its rising flanks—first deep marine, then coastal, and finally river-borne sediments as the volcano is progressively uplifted. Slightly farther away, the striped volcanic rocks that mark the beautiful coastline of the Miura and Boso peninsulas—now a weekend playground for Tokyo residents—can be traced back to their origin in a forearc basin on the deep-sea slope of the Izu–Bonin Arc between 3–12 million years ago. Slices of oceanic crust have also been caught up and emplaced on land in the form of the rock assemblages known as ophiolites.

The boundary between the great Eurasian and North American plates stretches from the Izu Peninsula, south of Tokyo, around the western flank of Fujiyama and right across the breadth of Japan. Though of supreme geological importance, this boundary is mostly featureless at ground level. It is only by diligent mapping of the rocks, and by taking careful note of earthquake occurrences, that we can begin to understand the complex history that lies beneath our feet.

Professor Kazuo Amano, Ibaraki University, Japan

Colliding continents Many of the larger island arcs in the western Pacific have grown in a similar manner to the Andes. However, because continents are carried like rafts on the lithosphere, there comes a time when the ocean between two converging continental landmasses vanishes completely and continent–continent collision ensues. The mountain chains of Europe and the Middle East—the Alps, Apennines, and Carpathians among them—have all formed in this way as Africa continues to plough northward into the Eurasian landmass. The twists and turns of microplates, fragmented at the leading edge of collision, as well as island arcs and back-arc spreading centers, volcanoes, and abandoned trenches, make the jigsaw puzzle of mountain ranges and remnant basins in the Mediterranean region one of the most complex on Earth. In this chaotic situation, slivers of ocean crust are isolated and offscraped onto the continent as ophiolites–the name given

to these typical rock sequences—while whole fragments of continent are detached and thrust high up over the mountains to form nappes ("sheets")—horizontally folded masses of rock.

Part of the geological confusion in the Mediterranean is due to plates simply sliding past one another along transform boundaries, rather than colliding or separating. Although this process conserves plate material, it is by no means passive. The stepped offsets along mid-ocean ridges are examples of transform faults, some of which extend for many hundreds of kilometers on either side of the ridge crest. But it is where such boundaries protrude onto land—as in the San Andreas Fault—that the real turmoil begins. As the two parts of a continent crunch, slip, and grate past one another, they alternately squeeze up mountain ranges and open up deep basins, which in turn become flooded with sea and quickly fill once more with the products of rapid erosion.

The Makran Mountains, stretching inland from the southern coast of Pakistan, have been pushed up by the subduction of the Indian plate beneath the Eurasian continent.

250 million years ago, a single world ocean, Panthalassa, surrounded the supercontinent of Pangea. Since that time new oceans have grown and continents have shifted allegiance. The world as we know it was born.

PLATES ON THE MOVE

Oceans of the Past

We are so familiar with the map of our present world, the shapes of its continents, the breadth of its oceans, that it is easy to believe in its permanence. In fact, appearances are deceptive—continents have grown, collided, fused together, pulled apart again, and moved to other parts of the globe, and oceans have come and gone. The changes, though slow even by geological standards, have often had dramatic effects on the Earth's climate, ocean circulation, and distribution of life. It is therefore an important challenge for oceanography to be able to accurately reconstruct the disposition of past seas and land areas. The study of past oceans, known as paleoceanography, is currently a hot topic for research throughout the world.

Supercontinent, world ocean If we reel back the "assembly line" of ocean formation, removing stripe by magnetic stripe, older and older crust, we progressively close up

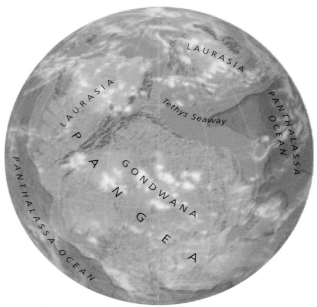

Triassic c.250 mya

most of the major ocean basins—the North and South Atlantic, Indian, Arctic, and Antarctic (or Southern) oceans. A step further back in time takes us to a single supercontinent, known as Pangea. In place of the seas we know, other, unfamiliar oceans appear, and, by unravelling that part of the Pacific subducted over the same time period, we can finally recreate a single super-ocean, known as Panthalassa ("world ocean"). This was the great feature of the global map 250 million years ago, at the end of the Permian period—a time of widespread mountain building, volcanic eruptions, climate change, and mass extinction.

During the ensuing Triassic period, the great, curved, C-shaped landmass of Pangea straddled the ancient equator. The Tethys Seaway was at that time a tongue of Panthalassa that nestled within the cusp of the C. Both poles were under water, the global climate was generally warm, and the continental interior extremely arid. But Pangea probably only remained intact for a few tens of millions of years at most before it began slowly to break up and drift apart. In a remarkable and little understood episode of Earth history, deep fractures or rifts appeared almost simultaneously throughout the supercontinent. Basaltic lavas poured forth in great profusion, the relics of which can be found today as ancient flows and intrusive dikes, including the great Palisades Sill along the Hudson River. These relicts are especially predominant around the margins of the Atlantic.

We have lost all record of the ocean crust at this period, and so will probably never know whether such rifts also appeared in the seafloor. However, those rifts that split the continent, at least the major ones, continued to open until they flooded with seawater and narrow seas invaded the land. Continental lithosphere stretched, thinned, and finally failed completely, allowing the first new oceanic crust to form the floors of deep but narrow ocean basins. Spreading continued, and these new oceans began to take shape during the Jurassic period.

Birth of today's oceans In the first place Pangea split more or less along the equator, with the separation of North and South America and the flooding of the sea across an ancestral Gibraltar gateway to the Tethys. The two large

Stratigraphic timescale

Geological period	Date (million years ago)	Major tectonic events
QUATERNARY	0—1.8	Subduction of Pacific plate, rifting of Africa begins
NEOGENE	1.8—24	Closure of Tethys Sea; India collides with Asia
PALEOGENE	24—65	Atlantic rifting continues; Australia, India drift north
CRETACEOUS	65—144	Pangea drifts apart; Atlantic widens; Tethys closes
JURASSIC	144—205	Rifting of Pangea; birth of Atlantic
TRIASSIC	205—248	Fusion of all continents into supercontinent of Pangea
PERMIAN	248—295	Gondwana and Laurasia fuse; Siberia approaches
PENNSYLVANIAN (late Carboniferous)	295—324	Tethys Ocean shrinks as Gondwana moves north
MISSISSIPPIAN (early Carboniferous)	324—354	Convergence of S. Europe with Baltica and Gondwana
DEVONIAN	354—417	Final closure of Iapetus Ocean
SILURIAN	417—443	Baltica, Avalonia, and Laurentia collide
ORDOVICIAN	443—490	Iapetus Ocean narrows as Baltica moves north
CAMBRIAN	490—545	Rodinia fragments; opening of Iapetus Ocean
PROTEROZOIC	545—2,500	Supercontinent of Rodinia forms
ARCHEAN	2,500—4,550	First microcontinents drift in huge global ocean

The Triassic world (*left*) was dominated by one supercontinent, Pangea, and a single, even larger ocean known as Panthalassa. Over the next 100 million years, Pangea rifted and fragmented, and new oceans and seaways formed (*below*). Of these, the ancestral North Atlantic survived and grew, whereas the great Tethys Ocean slid slowly into oblivion.

Jurassic c.150 mya

Cretaceous c.100 mya

Neogene c.20 mya

continental masses so formed are known as Laurasia and Gondwana, separated by the relatively narrow but geographically important Tethys Ocean. Each of these continents then began to break up along further rifts that had first formed during the Triassic. North America split apart from Eurasia as the North Atlantic began to open, and jagged rift valleys apeared between South America and Africa. India and Australia–Antarctica broke off from Gondwana, heralding the birth of the Indian and Antarctic oceans. The climate began to change, as moisture from this new pattern of oceans penetrated the previously arid continental interiors. Lush plant life flourished in the subtropical conditions and dinosaurs roamed the land. Along the margins of the continents, shallow shelf seas became home to coral reefs and tropical lagoons, sporting myriad new forms of marine life.

Finally, in the early Cretaceous period, South America and Africa began to drift apart and the South Atlantic was born. It is this era that can most truly be described as the Age of the Ocean,

The Cretaceous period (*left*) heralded a warm and watery world. Huge tracts of the continents were flooded with shallow seas, and many of today's oceans were steadily growing. The Tethys Ocean still dominated the equatorial world, and water covered both poles.

The Straits of Gibraltar (*above*) became an important oceanic gateway during the Neogene period. As Africa moved north into Europe, the gateway closed, thus isolating the Tethys Sea, which rapidly evaporated. Further tectonic movements eventually reopened the Straits and the area flooded once more.

By the beginning of the Neogene (*right*), the world had taken on much of its present aspect. India had docked with Asia, Antarctica had drifted over the South Pole, and the Arctic Ocean opened over the North Pole. The Tethys Ocean was fast diminishing in size.

The Lost Tethys Ocean

The Himalayas in the Mustang area of Nepal. The world's greatest mountain chain was born from the closure of the Tethys Ocean, and further uplifted as the Indian plate began to underthrust beneath the Eurasian plate, yielding a double thickness of continental crust.

As the northern edge of India collided with Asia, the vast west–east Tethys Ocean that lay between them was finally obliterated and the Himalaya Mountains grew in its place. Deformed and altered remnants of crust and seafloor from this long-vanished ocean—known as ophiolites—are now found high up in the Himalayas, marking the line of the suture. Thick piles of deep-sea sediment are also found, folded and thrust into place.

Between 60 and 50 million years ago, as the Indian subcontinent drifted northward from the southern supercontinent of Gondwana, the leading edge of the Indian plate began to collide with the northern landmass of Eurasia. Before long, a full continent–continent collision ensued.

As India collided with China, so Africa closed in on central and western Eurasia and South America drifted toward its northern counterpart. The Tethys Ocean slowly disappeared, leaving only slivers of rock from its former floors now caught up in the Himalayas, in China, and in the Middle East. To marine scientists this is the "lost ocean" of Tethys—a great seaway that extended from Mexico across the middle Atlantic and the Mediterranean into central Asia, reaching its zenith around 150 million years ago.

A lost ocean is a great scientific challenge. With careful detective work we have slowly unravelled something of the Tethys's long and intriguing history. The ophiolites of Cyprus,

Oman, and the Himalayas were all once small pieces of ocean crust. The pillow lavas, metal-rich sediments, and fallen stacks of black smokers that make up these rock sequences, together with associated deep-ocean fossils, tell us that they formed as part of the spreading center from which the Tethys was born.

For much of its history the Tethys was rimmed by broad shelf seas, teeming with life. Southern Europe 150 million years ago must have been quite similar to the Bahamas today—a series of carbonate banks and basins, balmy lagoons, and glistening white sands. The lime muds of these shallow Tethyan margins were so fine that even the most delicate structures of feathers were perfectly fossilized. They contain a host of remarkable finds—jellyfish and dragonflies, pterosaurs with their leathery wings intact, sponges, corals, and many small dinosaurs.

The story documented in the fossil record tells of a thriving, colorful ocean, every bit as alive as the oceans of today. Then dramatic change came to the region. Even though the process took millions of years, it was rapid in geological terms, and its repercussions were to be felt throughout the world's oceans. As a result of continental drift, the Tethys Ocean was almost closed. By this stage the Indian subcontinent had already collided with Asia, the Middle East had become sutured together, and Africa had nearly closed in on Europe across the Gibraltar Strait, so all that was left of the once mighty Tethys was an irregularly shaped body of water roughly similar to the present-day Mediterranean and Black seas.

The real trouble started when con-

tinued collision in the west finally cut the Mediterranean off from the Atlantic Ocean, about 6.5 million years ago. The climate at that time was even drier and more arid in the Mediterranean region than it is today, so that the influx of river water was unable to keep pace with the rate of evaporation. Progressively, the level of the water fell. The rivers that still fed into the remnant basins incised deeply into their banks and cut mighty canyons across the former seafloor.

As the water evaporated, the concentration of sea salts increased many times until normal marine life could no longer exist. Salt-tolerant species of red algae survived for longer, but mass death inevitably followed, and still there was no let-up in the burning heat. Natural salts began to precipitate from the concentrated brine pools, gypsum being at first the most abundant; then, as the seas dried up completely, common rock salt (or halite) lined a gaping, blistering white basin. With the exception, perhaps, of halophilic ("salt-loving") bacteria, life was unable to exist where once there had been profusion and diversity. We now know that the last remnants of the Tethys Ocean evaporated to dryness or near dryness many times over, being filled from the Atlantic by a tremendous cascade across the Strait of Gibraltar between each episode. This whole process lasted for about half a million years before a more permanent connection was established through the straits.

And so, the turmoil of successive flooding and drying over, a reign of peace and quiet ensued and the Mediterranean slowly gained its present nature and azure calm. Barely a trace was left, save in the rocks, of this most amazing salinity crisis of all time. The Tethys Ocean had finally disappeared forever.

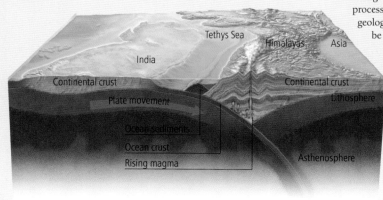

Continental crust
Plate movement
Ocean sediments
Ocean crust
Rising magma
India
Tethys Sea
Himalayas
Asia
Continental crust
Lithosphere
Asthenosphere

for sea levels rose higher than at any time during the past billion years, perhaps as much as 350–650m (1,150–2,150ft) higher than today. Only 18 percent of the globe was dry land, compared with 28 percent today. These extremely high sea levels were a direct result of increased spreading activity—new oceans, new ridges, and faster spreading rates all meant that the mid-ocean ridge systems collectively displaced a greater volume of water than ever before. Global warming accompanied global flooding. Temperatures in the ocean rose to around 30°C (86°F) at the equator and as much as 14°C (57°F) at the poles. The deep ocean waters were far warmer than they are today, so that little convection—the circulation or overturning of water masses as a result of heat transfer—occurred.

The oceans could never quite stagnate, but in these extreme conditions they became dangerously low in oxygen. Between about 120 and 80 million years ago there was a remarkable period (or perhaps several periods) during which oxygen starvation occurred within the oceans, globally or at least semi-globally. These intervals, known as ocean anoxic events, led directly to the death of countless marine organisms and their preservation in the form of undecayed organic matter in seafloor sediments. Wherever oceanographers have drilled into sediments of this age beneath the ocean floor, or, indeed, where rocks of mid-Cretaceous age are now exposed on land, they encounter dark gray rocks rich in organic matter called black shales. Over half the world's oil reserves have been generated from these rich source rocks.

The high sea levels of the Cretaceous flooded most of western Europe and central North America. Vast shelf seas spread over low-lying areas of the continents, glinting with sunlight and teeming with life. Billions of microscopic organisms floated in the warm, shallow water, making up a rich planktonic soup. After death, their delicate skeletons drifted down like snowflakes to cover the seafloor in a creamy white ooze. Such stable and tranquil conditions persisted for many millennia over huge tracts of the globe. Layer upon layer of sediment accumulated, eventually hardening into the white limestones, commonly known as chalk, that we find today along the southeast coast of England, in Denmark, Kansas and surrounding areas, and along the Gulf Coast of the USA.

Toward modern times The changes that were to come at the end of the long Cretaceous period, 65 million years ago, were every bit as dramatic as the age of the chalk seas had been uneventful. The reign of the dinosaurs came to an end, and with them went some 50 percent of known species. Sea levels fell and vast areas of shelf seas vanished to be replaced by dry land, offering a range of different ecological niches for new species to fill. Global cooling set in with a vengeance as temperatures plummeted. Antarctica separated from Australia and drifted across the South Pole, taking up more or less its present position on the globe. This relocation, coupled with the isolation of the southern continent 15 million years ago when the strong Antarctic Circumpolar Current was initiated, caused a fresh bout of global cooling that finally sent Earth's climate spinning into the grip of an ice age.

By the middle of the Paleogene period the principal oceans of the modern world were established and land cover had increased to around 30 percent of the globe. The subtle changes that have occurred since then, as well as the major sea-level variations that accompanied the advance and retreat of the ice-age glaciers, are relatively easy to reconstruct in comparison with the planet-wide dramas that had gone before.

The Red Sea marks the northward extension of the Great African Rift Valley. This 3,000-km (1,850-mi) fissure in the Earth's crust, in places over 2,000m (6,500ft) deep and 50km (30mi) wide, is marked by active volcanism and earthquakes. The separation of the African and Arabian plates has led the rift to submerge at this point, leading to flooding by the sea and the beginning of true ocean spreading. Further north, the Gulf of Aqaba narrows into the Jordan Rift Valley.

The break-up of Pangea that produced the Atlantic Ocean was the result of mantle upwelling, which lifted and fractured the continental surface, producing a rift valley (below). Modern-day examples such as Africa's Great Rift Valley have similar features.

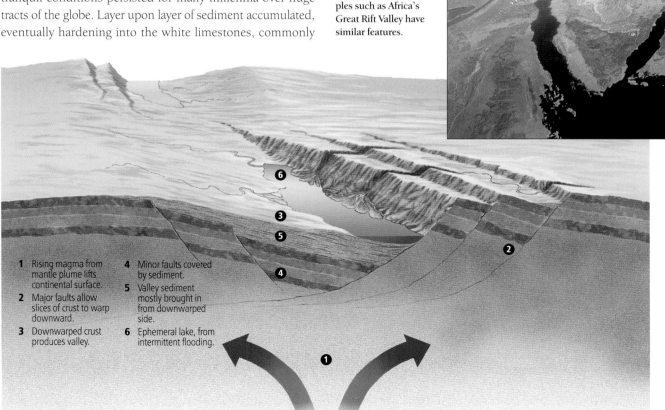

1 Rising magma from mantle plume lifts continental surface.
2 Major faults allow slices of crust to warp downward.
3 Downwarped crust produces valley.
4 Minor faults covered by sediment.
5 Valley sediment mostly brought in from downwarped side.
6 Ephemeral lake, from intermittent flooding.

The Pembrokeshire coast in southwest Wales directly confronts the onslaught of wind and waves from the North Atlantic Ocean. Erosion is intense, but the rocks subjected to this constant pounding are very old and resistant, the core of an ancient continent some 300–500 million years old. The regular bands of sedimentary rock seen here, which have weathered at different rates according to hardness, are the result of repeated cycles of deposition, each reflecting the climate and environment at that time. Such cycles of change continue to be played out in today's world.

PATTERNS & CYCLES

CHAPTER 2

The continuous interaction between sea and land

Amid the amazing variety of shape and form that makes up our world, and through the constant change that is its essence, order and regularity often exist where none at first appear. Patterns of change in the earth–ocean system are forever cyclic, although the timescales may vary and the past is never exactly repeated.

The cycle of the rocks, from the erosion of mountains to deposition in the deep sea, through subduction, collision, and uplift once again, is the longest and most complex of these cycles. It may be played out over tens or even hundreds of millions of years. Along the shoreline, changes are inevitably more rapid. Cliffs are pounded into sand, and sands are swept offshore into submarine canyons that incise the continental shelf. Underwater currents, debris flows, and slides complete the transfer of sediment downslope into the deep oceans, on a scale that has no parallel on land.

Understanding the processes that affect the different types of sediment and their distribution across the ocean basins is still a great challenge for ocean scientists. Where we have drilled into the sediments that have accumulated over millions of years on the seafloor, we find regular cycles of change on seasonal, decadal, and millennial timescales, often intricately linked to climate.

Tens of thousands of islands pierce the ocean's surface, but their location and distinctive trends—linear, arc-shaped, or clustered—reveal just a few distinct types and origins. Each island, however, yields its own intriguing information about the ocean in which it occurs.

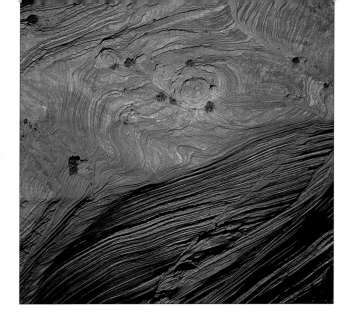

A grain of sand will carry the signature of its origin for thousands of kilometers from the highest mountains down to the deep ocean floor, only to await its return to the summit of some future mountain range.

From Peak to Trough

KEY TERMS

HYDROLOGICAL CYCLE
SEDIMENTARY CYCLE
LITHIFICATION
IGNEOUS ROCK
SEDIMENTARY ROCK
METAMORPHIC ROCK
EROSION
WEATHERING
DENUDATION
TRANSPIRATION
SEAMOUNT
REDBED
CONGLOMERATE

The Earth's surface is constantly changing. Tides ebb and flow, waves pound the shore and cliffs crumble into the sea, rivers flood and break their banks, discharging plumes of muddy water far out to sea. Slowly but surely, mountains become hills and hills in turn become low-lying plains and deltas, as the material we think of as firm, dry land is worn away and washed into the oceans. In the last chapter, we saw change on an even slower timescale—the spreading of oceans and drifting of continents, the uplift of mountains and subduction of oceanic crust. These gradual processes are also punctuated by dramatic events. The immediacy and violence of volcanic eruptions and earth tremors, for example, display just a fraction of the energy involved in mantle convection and plate movement.

Such constant change must be powered by a continuous supply of energy, and that energy comes from a variety of sources. The Earth's tectonic plates are driven by an internal energy source—the remnants of deep cooling, coupled with radioactive decay in mantle and crustal rocks. Tectonic processes cause mountains to rise up and deep basins and trenches to sink down, also providing potential energy to carry off eroded material under the influence of gravity. The tides are also driven by gravity, principally the gravitational pull of the Moon on the Earth's thin envelope of water. But the primary source of energy on Earth is the Sun. Solar radiation bathes its surface in an almost inestimable quantity of perpetual energy, which warms the planet, drives the hydrological cycle, stirs the atmosphere, and allows ocean currents to flow and life to thrive.

Cycles within cycles

The abundance of liquid water on Earth is unique, having no parallel elsewhere in the Solar System. The Earth's water is the single largest sink for incoming solar radiation, and has a profound effect on life, climate, and the shape of the land. The storage of water, its transition between three states of matter—liquid, solid (ice), and gas (water vapor), and its transfer between different reservoirs (the oceans, the atmosphere, and the land) is known as the hydro-

logical cycle. The dominant reservoir is the ocean, which contains over 90 percent of the total supply. Second in importance, but little known and much of it never seen, is subsurface water—some held in soils and the cracks and crevices of rocks on land, but most residing in tiny pore spaces of sediment that still remain open down to several kilometers beneath the seafloor. Icesheets, glaciers, and snow hold most of the fresh water, the remainder cycling through the atmosphere, rivers, and lakes. If the global water reservoir was represented by a garden pond,

5

The sedimentary cycle is part of the much larger rock cycle (*right*). Particles of terrigenous (earth-born) sediment are eroded from the land via wind, water, and glacial and volcanic action. Biogenic particles come from the remains of living organisms (reefs, shells, microplankton). Chemogenic particles are chemical precipitates from seawater. All are then subject to transport, deposition, burial, and lithification (becoming rock).

Sedimentary rock sources
1 Glacier (breccia)
2 Coastal flats (halite)
3 River delta (sandstone)
4 Lake (shale)
5 Coral reef (limestone)
6 Deep sea fan (turbidite)

Metamorphic rock sources
7 Subduction zone (gneiss, schist)
8 Magma body (hornfels, marble)

Rock groups

Rock type	Source	Process	Examples	Typical mineral content
IGNEOUS	Melting of rocks at extreme temperatures in the lower crust or upper mantle	Cooling and crystallization of magma, sub-surface (plutonic) or surface (volcanic)	Granite, rhyolite, obsidian, pumice, andesite, diorite, basalt, gabbro	Quartz, feldspar, mica, pyroxene, amphibole, olivine
SEDIMENTARY	Weathering and erosion of rocks at surface	Deposition by water, ice, or air, compaction, and lithification	Shale, sandstone, conglomerate, breccia, limestone, dolostone, chert	Quartz, clay minerals, feldspar, calcite, dolomite, gypsum, halite
METAMORPHIC	Rocks under high pressure and temperature, in deep crust and upper mantle	Reorientation and re-crystallization to form new minerals	Slate, schist, gneiss, marble, quartzite, hornfels	Quartz, feldspar, mica, garnet, pyroxene, staurolite, hornblende

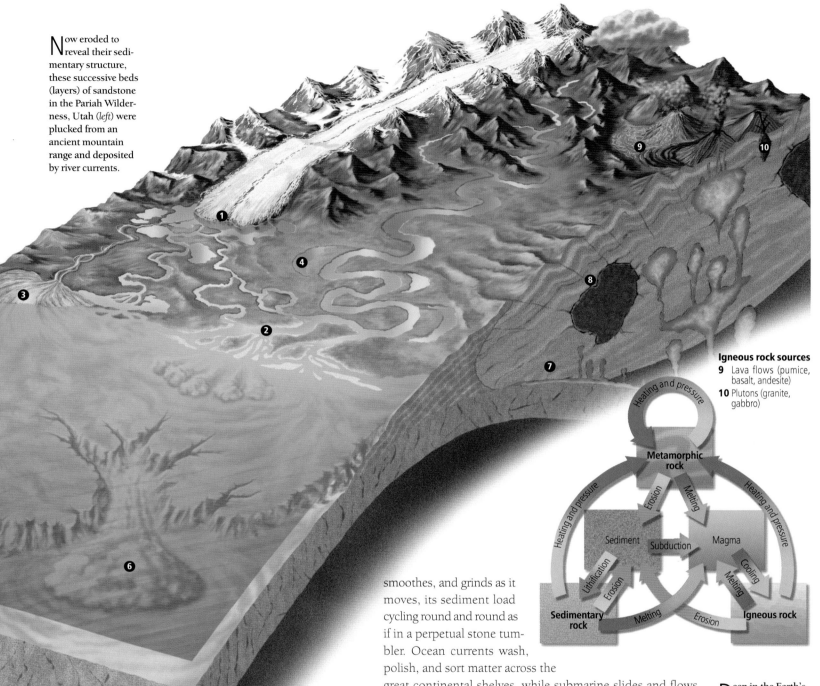

Now eroded to reveal their sedimentary structure, these successive beds (layers) of sandstone in the Pariah Wilderness, Utah (*left*) were plucked from an ancient mountain range and deposited by river currents.

Igneous rock sources
9 Lava flows (pumice, basalt, andesite)
10 Plutons (granite, gabbro)

smoothes, and grinds as it moves, its sediment load cycling round and round as if in a perpetual stone tumbler. Ocean currents wash, polish, and sort matter across the great continental shelves, while submarine slides and flows eventually transport sands and muds deeper and deeper into the ocean basins. Even the burial of sediment beneath the seafloor is closely accompanied by water, and it is the cycling of millions of liters of water through the pore spaces between individual particles that eventually leads to the precipitation of mineral cements and lithification into a hard sedimentary rock.

Such sedimentary rock is rafted across the ocean floor on the great tectonic plates to perch for a moment of geological time at the edge of the abyss before it slides inexorably into the ultimate rock recycling plant—the oceanic trenches. Some will "escape," scraped off onto the wall of the trench, only to be caught later in an episode of mountain-building, subjected to enormous pressures and heated to several hundreds of degrees Celsius. The metamorphic rock that results is then uplifted and incorporated into the body of the mountain, where the processes of weathering (breakdown of the rock due to fracturing and chemical

Deep in the Earth's interior, temperatures of over 1,000°C (1,800°F) melt rock of all kinds to form magma. This molten lava crystallizes into igneous rocks, either plutonic (formed slowly at depth) or volcanic (extruded at the surface). Erosion of these rocks produces sediment, which, when compacted, becomes sedimentary rock. If subjected to great heat and pressure, this becomes the third type of rock—metamorphic.

the amount locked up as ice and snow would equate to little more than a thin sheet over the surface, while the mighty Amazon, Mississippi, Ganges, Brahmaputra, Nile, Chang Jiang, and all the other rivers of the world combined would contain no more than a thimbleful.

Water drives the sedimentary cycle through all its various phases. Rocks crumble under the influence of water and break into their constituent minerals; whole rock fragments and single mineral grains bounce, tumble, and are carried by rivers across the face of the land. The force of flowing water rounds,

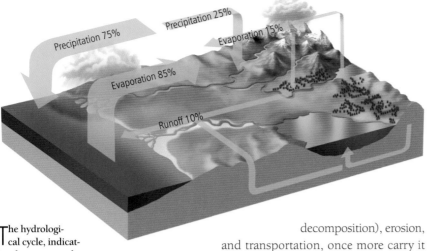

The hydrological cycle, indicating the nature and relative volume of water exchange between different reservoirs. Rainfall into and evaporation from the sea represent the dominant exchange, and involve the single largest reservoir—the oceans.

Spray rises over the Iguaçú Falls, on the Brazil–Argentina border. Water returns to the sea carrying huge volumes of suspended sediment eroded in its course.

decomposition), erosion, and transportation, once more carry it away. Other sediment slides still deeper into the mantle, melting under the intense frictional heat and becoming transformed into igneous rock. This molten material then rises up through the crust to form the granite intrusions at the core of new mountains, or pierces the surface in the form of volcanic eruptions. In this way the sedimentary cycle is linked into the full rock cycle.

The entire cycle of erosion, sedimentation, lithification (rock formation), and mountain-building is only made possible by the fact that water is everywhere constantly on the move. Solar energy causes large-scale evaporation from the oceans, lakes, and rivers, as well as transpiration from plants on land. Atmospheric winds disperse the moisture across the globe until it cools and condenses, falling back to the surface in the form of rain or snow. Some 80 percent of the rain falls directly into the ocean, the rest eventually finding its way back via rivers and groundwater flow.

Tidal inlets snake across the Ganges Delta (*right*). Supplied by the world's muddiest rivers carrying silt eroded from the Himalayas, the delta has built steadily south into the Indian Ocean over tens of millions of years.

Much smaller cycles are played out locally within the grander scheme. Under clear night skies the ground cools rapidly, and nearby water vapor in the air condenses out as tiny droplets of dew. As the sun comes up and temperatures increase, dew evaporates back into the atmosphere. In the tropics, the heat of the morning sun evaporates large quantities of water, leading to high humidity and the build-up of clouds. These rise and cool, resulting in heavy rain and thunder showers by late afternoon.

The average time for which a single molecule of water remains in any one reservoir varies enormously. It may survive for only one night as dew, up to a week in the atmosphere or as part of an organism, two weeks in rivers, and up to a year or more in soils and wetlands. Residence times in the oceans are generally over 3,500 years, and water trapped in ice caps may remain undisturbed for tens of thousands of years.

Mountains into dust The forces of plate tectonics driven by the Earth's internal engine continue to build mountain ranges and to push them high above sea level. But it is the unique properties of water and the relentless turning of natural cycles—rock, sediment, water, chemical, and biological—that reduce even the most majestic peaks to dust. The pebbles people collect along the seashore, the sands that carpet holiday beaches, and the soft mud flats of estuaries uncovered daily by the ebbing tide are all the end result of abrasion and erosion. Each particle of sediment has its own individual tale to tell of a long voyage to the sea.

In mountain ranges there is a constant battle between uplift and erosion. The battle may last for tens of millions of years, but the outcome is never in doubt. Water will win, and the broken mountain will be returned to the sea as dust. The mighty Himalayas have been growing for 50 million years or more, and there is evidence that a proto-Indus River was already draining the western part of the young mountain range over 40 million years ago. Denudation of the Himalayas (the combined effects of weathering, erosion, and transportation of material) has, over millions of years, built up the great fertile plains of southeast Asia and eastern China, the residue being carried out into the East and South China and the Yellow seas.

It is perhaps fitting that the world's greatest mountain range feeds the muddiest rivers, has built up one of the largest deltas—the Ganges Delta, and has then spewed billions of tons of sediment into the deep Indian Ocean. The Bengal Fan, a gigantic submarine delta, is the world's single largest sediment body. It covers an area of over 1 million sq km (almost 400,000 sq mi) and reaches a maximum thickness of at least 15km (9.3mi), thinning to a feather edge 3,000km (1,865mi) out into the Indian Ocean.

In the late 1980s an international scientific expedition drilled into the outermost part of this fan. The deepest hole penetrated 1,000m (3,300ft), reaching back to sediment eroded from the

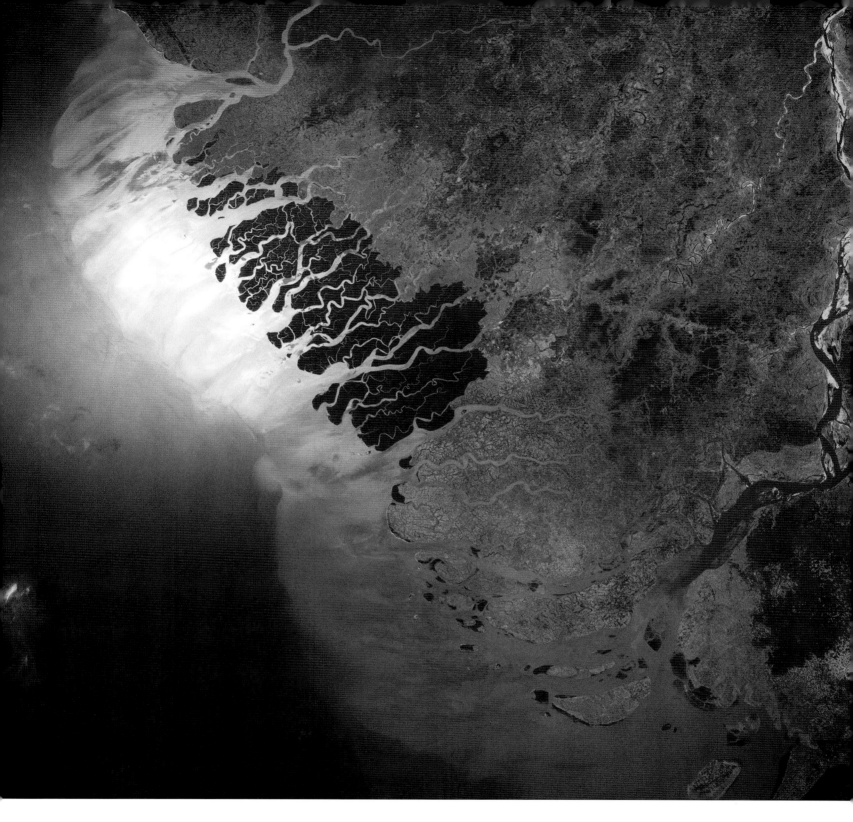

Himalayas some 18 million years ago. Through analysis of the composition of the sediment column, a picture emerged of progressive denudation of the mountains, so that today the deep core of the range is exposed at its summit. Interestingly, some of the mineral grains and clay particles recovered from the drilling sites showed several additional sources of sediment—some supplied from erosion of the volcanic Deccan Traps in central India, some from the warm shelf seas off Sri Lanka, and the smallest amount from the nearby Afanasy Nitikin seamounts.

One of the principal ways we have of reconstructing the history of past oceans, especially those that have long since vanished, and of charting their subduction and recycling, is by seeking out the worn-down remnants of past mountain ranges on land. The Caledonian highlands of Scotland are a good example, being the core of a once-great mountain belt that linked up with the Appalachian Mountains of North America and extended north into Scandinavia. This range was formed just over 400 million years ago by the complete closure and subduction of the former Iapetus Ocean, when the ancient continents of Avalonia, Baltica, and Laurentia collided. The fragments of ocean crust, metamorphosed sediments, igneous intrusions, and extensive redbeds—the sandstones and conglomerates that eroded from the rapidly rising mountains—are all evidence of the seismically active northern margin of the vanished Iapetus.

SEE ALSO

PLATES ON THE MOVE
Plate Tectonics and Time   26–29

A dynamic duel between land and sea is played out along the coastline. Ocean water is the master sculptor, rounding pebbles and grinding sand, molding the rocky shore into magnificent but ephemeral landscapes.

PATTERNS & CYCLES

Where Land Meets Sea

KEY TERMS

CONTINENTAL SHELF
DELTA
SALT PAN
ESTUARY
SEDIMENT LOAD
SUBMARINE FAN
TURBIDITY CURRENT
RIA
FJORD
WAVE-CUT CLIFF
SEA ARCH
SEA STACK
BERM
LONGSHORE BAR

More than a third of the global population lives on or fairly near the coast. This concentration is no coincidence, for throughout history population centers have grown up around maritime trade routes, in areas of abundant fishing stocks, and where fertile coastal plains have yielded rich farming.

Yet the coast is and always has been one of the most changeable zones on the planet. Only 20,000 years ago, at the peak of the last ice age, the coastline lay many tens to hundreds of kilometers out to sea, as great tracts of low-lying continental shelf became dry land. Britain was joined to Europe, Alaska to Siberia. But these changes we know to be cyclic in their expression, so that at the peak of the last interglacial period, some 100,000 years before that, when the mean global temperature was higher than it is today and even less ice was locked up in polar ice caps, coastlines were tens of kilometers inland. Such conditions may provide a foretaste of our future if global warming continues at anything like its present rate.

Deltas Wherever the ocean and continent meet there is change and variety. The restless wash of waves, the ebb and flow of tides, and the violent power of maritime storms gradually mold the land and the sediment it yields into a myriad of different forms—rocky shores and pebbly coves, rippled sands and windblown dunes, mudflats and mangrove swamps. Climate, too, plays its part. Coastlines frozen fast with sea-ice are worlds apart from the palm-fringed shores of coral isles. The lush green Sundabans of the Ganges Delta, lashed by one of the wettest climates on Earth, are quite unique and distinct from the arid shores and salt pans in the Gulf of Kara Bogaz of the Caspian Sea.

A satellite infrared image shows part of the Niger Delta, West Africa (*right*). Sediment plumes dispersed from the mouths of individual distributaries show up as light blue in colour. A combination of wave and tidal activity redistributes the sediment along the shoreline.

Nowhere do we witness so clearly the constant change and incessant conflict as in deltas. A river's task is finally complete, and its energy gainfully spent, when it has carried its full load of water, dissolved salts, and particulate sediment to the coast. If marine conditions are right and the sediment load sufficient, then the river will produce a delta. Variation in delta shapes is due to a delicate interplay of sediment supply by the river and its reworking by marine processes.

The Nile is the classic delta—indeed, its triangular shape first prompted the historian Herodotus to use the term "delta" in 432 BC, on the basis of its similarity to the Greek letter (Δ). We now know that this form results from a well-matched balance between sediment input and the smoothing action of waves and

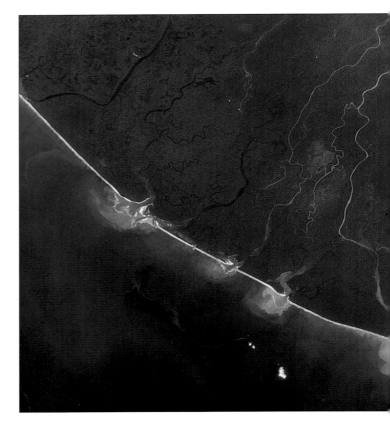

Major river discharge						
River	**Source**	**Outflow**	**Drainage area (km²)**	**Water discharge (m³ per second)**	**Sediment discharge (million tons/yr)**
GANGES–BRAHMAPUTRA (& Meghna)	Himalaya Mts	Bay of Bengal	2,011,000	45,000	1,244
AMAZON	Glacier-fed lakes, Peru	Atlantic Ocean	6,150,000	200,000	1,150
CHANG JIANG (Yangtze)	Tibetan Plateau, China	China Sea	1,940,000	28,500	480
MISSISSIPPI	Lake Itasca, Minnesota, USA	Gulf of Mexico	3,344,000	18,400	400
IRRAWADDY	Confluence of R. Nmai/ R. Mali, Burma	Bay of Bengal	410,000	13,600	260
INDUS	Himalaya Mts	Arabian Sea	960,000	7,600	250
MAGDALENA	Andes Mts, Colombia	Caribbean Sea	260,000	6,890	220
GODAVARI	Western Ghats, India	Bay of Bengal	287,000	2,920	170
MEKONG	Tibetan highlands	South China Sea	810,000	14,900	160
ORINOCO	Sierra Parima Mts, Venezuela	Atlantic Ocean	945,000	34,900	150

currents along the shoreline. Where sediment load is the stronger influence, as in the Mississippi Delta in the Gulf of Mexico, an elongated "birdsfoot" delta is formed. Where waves predominate, as in the Rhône Delta, sediment is swept sideways into spits, bars, and sand ridges. Where the influence of tides is greater, the distributaries are deepened into a series of tidal channels perpendicular to the shoreline, as in the Ganges Delta. All deltas are prone to sudden, rapid shifts in the location of one or more of the main distributary channels, so that the entire locus of deposition along the shore changes periodically. This phenomenon has occurred every thousand years on average for the post-glacial Mississippi Delta.

With shifting sands, migrating channels, and an ever-changing coastline subject to regular flooding and serious problems of subsidence (sinking), deltas are among the most dangerous places to live. Yet delta cities continue to flourish and expand—Calcutta, Alexandria, New Orleans, St. Petersburg, Venice, and Bangkok are only a few of many examples from around the world. Adding to the obvious benefits deltas offer to farming, fishing, and navigation, the latter part of the 20th century revealed their economic importance as hugely rich oil provinces. Nigeria's booming industry is wholly due to prolific oil discoveries beneath the Niger Delta. Although the oil business developed further inland in the Gulf Coast region of the USA, it is now dominated by the Mississippi and is moving further offshore to tap delta sands reworked downslope. Geologically farther back in time, sandstone reservoirs from the 160-million-year-old Brent Delta system triggered the oil boom in the North Sea.

The sea strikes back Surprisingly, some of the world's greatest and muddiest rivers produce no delta where they enter the sea. The Amazon, for example, has the highest water discharge of any river on Earth and is second only to the Ganges–Brahmaputra in terms of sediment load delivered to the ocean, and yet there is no visible delta. In its place, a broad, sediment-laden estuary enters like a gigantic gash into the very heart of the rainforest. Great plumes of muddy water swirl into the equatorial Atlantic, buoyed up by the lower density of warm river water. Offshore, however, a broad submarine "delta" is created across the continental shelf. This is soon reworked off-shelf into the great Amazon submarine fan, which extends 1,000km (620mi) downslope to the abyssal ocean floor.

Across the Atlantic, the mighty Congo River has an even higher ratio of sediment discharge to water—and no delta results. In this case, a deep-sea canyon is incised right into the axis of the river estuary, so that dense, mud-laden flood waters hug the floor and proceed down-canyon as turbidity currents—turbulent submarine flows of mixed sediment and water. Telegraph cables laid across the river gulf that separates Zaire from Angola are frequently broken as these powerful currents carry their sediment load onto the deep Congo submarine fan and beyond.

The diagram below shows the interplay of forces molding the world's major deltas. The Mississippi Delta (1), for example, is shaped mainly by riverborne (fluvial) sediment, the Senegal Delta (8) by ocean waves.

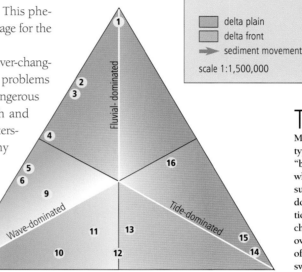

1 Mississippi
2 Po
3 Danube
4 Ebro
5 Nile
6 Rhône
7 São Francisco
8 Senegal
9 Orinoco
10 Burdekin
11 Niger
12 Copper
13 Mekong
14 Gulf of Papua
15 Ganges–Brahmaputra
16 Mahakam

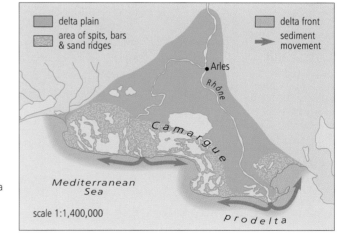

delta plain
delta front
sediment movement
scale 1:1,500,000

The Mississippi Delta in the Gulf of Mexico (*above*) is a typical elongated, "birdsfoot" delta, in which the riverborne supply of sediment is dominant. Its location is known to have changed markedly over time as a result of river-channel switching.

The Rhône Delta (*below*), in southern France, is fed by a strong, sediment-laden river, but much of the load deposited is rapidly reworked into coastal sands by dominant wave action. The effect of tides is generally minimal in the Mediterranean Sea.

delta plain
area of spits, bars & sand ridges
delta front
sediment movement
Arles
Rhône
Camargue
Mediterranean Sea
scale 1:1,400,000
prodelta

In the Gulf of Papua in New Guinea, the river-borne sediment is quickly and efficiently reworked by strong tides. Long creeks and elongated islands are characteristic of such tidal deltas, fringed in this case by extensive mangrove swamps that thrive in the tropical climate.

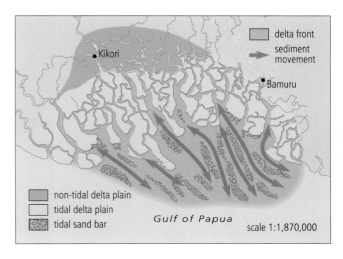

delta front
sediment movement
Kikori
Bamuru
non-tidal delta plain
tidal delta plain
tidal sand bar
Gulf of Papua
scale 1:1,870,000

Fingers of water reach inland from Talbot Bay, Western Australia. These spectacular inlets, fringed with mangroves, once marked the course of river valleys winding their way to the coast. As sea levels rose at the end of the last ice age, the valleys filled with ocean water. Flooded valleys of this type are known as "rias."

A typical high-energy coastline, with rapidly eroding cliffs and a headland pounded on both sides by the waves. The action of continual erosion commonly produces sea caves, arches, and stacks. This type of coast has a negative sediment budget, with more material removed than added. Typically, a neighboring beach will benefit from the deficit, some of which may find its way back to land as a longshore drift of sand.

The relatively warm climate and high sea levels that characterize today's world generally favor the ocean in its battle with the land. The course of former rivers that once meandered their way across broad tracts of exposed continental shelf were flooded when sea levels rose 10,000 years ago, and the valleys drowned by the inflowing ocean waters became estuaries and rias. In colder climes, where glaciers had once carved deep courses through the land, they left steep-sided canyons to be repossessed by the rising seas in the form of fjords. The intricately indented and imposing fjords of Norway and Greenland, Chile and New Zealand offer some of the most beautiful and tranquil coastal scenery in the world today.

At lower latitudes, richly vegetated coastal plains gave way to marshlands and swamps, wetlands and mangrove forests, as the seas marched ever inland. Along arid shorelines, lagoons became trapped behind moving barriers of sand and evaporated

to dryness in the blistering heat. Giant salt pans were created, periodically flooding and spawning life—the saline waters colored red with salt-tolerant algae, the heat haze above shimmering with flocks of pink flamingos.

Shoreline processes Some coastlines can be wild and imposing, just as others are calm and serene. Those with sheer cliffs or jagged, rocky margins are subjected to the impact of the highest energy, as the sea pounds against the rocks, cuts sea stacks and arches, throws sand and rocks against the shore, and sprays salt water high into the winds. The force of storm waves crashing against the land is both furious and unrelenting—whole beaches can be rearranged or even removed in a single storm, concrete promenades smashed, and coastal settlements severely damaged. The force of breaking waves has been measured up to 500kg per sq cm (7,000lb/sq in), but instruments have never successfully recorded the impact of freak waves that can reach over 30m (100ft) in height. Where the Wild Coast of South Africa meets the Skeleton Coast of Namibia off the Cape of Good Hope, one of the world's most dangerous and notorious currents sweeps down from the north directly into the path of strong winds and waves. This region has an infamous record for shipwrecks; its spectacular scenery is testament to the the sea's power to carve the land.

Beach erosion occurs where sediment outputs (sediment transport away from the area, either out to sea, along the shoreline, or blown inland) are greater than the inputs (sediments supplied either from rivers or cliff erosion or carried onshore by the currents) with the result that the "sediment budget" is not in equilibrium, having more "debits" than "credits." More input than output results in beach accretion, while a balance between inputs and outputs produces a "steady state" situation.

Along high-energy coastlines, beaches are absent or greatly restricted. Sea cliffs receive the brunt of storm action—the pounding of waves, the dramatic abrasion by sand and gravel impact—as well as more or less continuous chemical corrosion by seawater. Depending on wave intensity and the resistance of the coastal rocks, cliffs retreat landward at different rates. In the USA, for example, soft glacial drift on the east coast of Cape Cod erodes at up to 30m (100ft) per century, whereas parts of the chemically unstable chalk cliffs of the English south coast are retreating at over 100m (330ft) per century.

Retreat, however, is never uniform, and it is this that provides the charm of endless variety. Some rocks—for instance, granite or basalt—are more resistant than others to erosion, while certain geological features, such as faults and joints, are more easily worn away. The net result is an irregular coastline of quiet bays and exposed headlands. Eroded sediments accumulate in the bay areas to form beaches, whereas headlands are lashed by waves on three sides. Where the cliffs are undermined, notches and sea caves form; these break through to form arches that eventually collapse, leaving sea stacks.

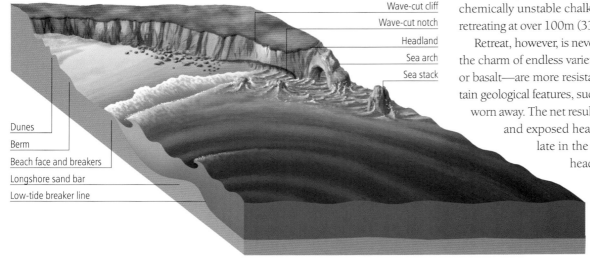

Wave-cut cliff
Wave-cut notch
Headland
Sea arch
Sea stack

Dunes
Berm
Beach face and breakers
Longshore sand bar
Low-tide breaker line

Beach treasure The sandy beaches that attract vacationers to coastal regions around the world have an appeal for geologists and oceanographers as well as for sun-lovers, for these thin strips of pulverized rock are tied closely to their source, each grain having a history that may span many millions of years. The beaches that fringe the Canary Islands are gray-black, the color of the volcanic rocks from which they derive, whereas those of the Maldives are broken fragments of coral that glisten a brilliant white in the equatorial sunshine of the Indian Ocean. Visitors to Sri Lanka would do well to sift carefully through the glorious sands that adorn its coasts, for they hold a myriad of minute gems—rubies, sapphires, emeralds—plucked from the very core of the island's mountainous interior, while the shores of southwest Africa, where the Namibian Desert spills into the Atlantic Ocean, are the scene of a major mining operation for diamonds.

Less valuable perhaps, but with a tale more easily deciphered, are the pretty, water-worn pebbles that accumulate along the shoreline. A pebble's size is a direct function of the energy expended at the seashore: the greater the average energy imparted by the waves, currents, and tides, the larger the pebble. But pick up that pebble, look closely at the different mineral and fossil components that make up its fascinating textures and patterns, and you hold the story of the Earth in your hand. So, for example, a quick examination of the pebbles lining a beach in Cyprus can reveal in microcosm every detail of the Mediterranean island's complex geological history. There will be dark, almost black pebbles of mantle material; green, shining serpentinites from near the Moho; gabbros, dolerites, and basalt lavas from the ocean crust; chalks, cherts, and fossil coral fragments from the lost Tethys Ocean.

One of the world's longest stretches of sandy beach lies where the rippling dunes of the Namib Desert (in southwest Africa) meet the expanse of the South Atlantic Ocean (*above*). A colorful assortment of polished pebbles (*left*) fringe the shores of Namibia's legendary Skeleton Coast.

51

The mosaic of sediments spread over the seabed tells a vibrant tale.
To read the record lying in the sedimentary basins, we must first
understand the surface processes controlling the distribution of material.

Patterns on the Seafloor

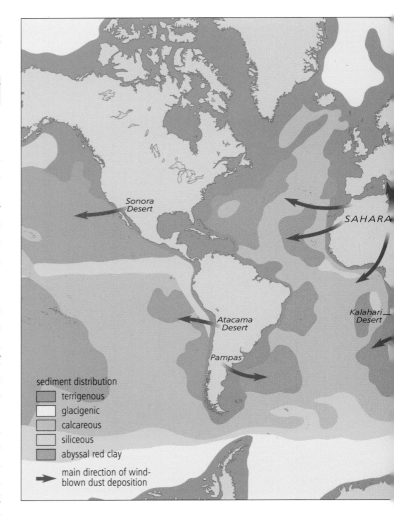

KEY TERMS

TERRIGENOUS SEDIMENT

PRIMARY PRODUCTIVITY

BIOGENIC SEDIMENT

CHEMOGENIC SEDIMENT

TEST

CALCAREOUS

SILICEOUS

COCCOLITHOPHORE

FORAMINIFERAN

DIATOM

RADIOLARIAN

CARBONATE
COMPENSATION DEPTH

CARBON SINK

ABYSSAL CLAY

sediment distribution

☐ terrigenous
☐ glacigenic
☐ calcareous
☐ siliceous
☐ abyssal red clay

➤ main direction of wind-
blown dust deposition

Except for the newly formed volcanic crests of mid-ocean ridges, the whole of the seafloor is covered in sediment. In places the sedimentary layer is no more than a thin veneer. Elsewhere, along stable continental margins or beneath major deltas where deposition has persisted for millions of years, its accumulated thickness can exceed 15km (9.3mi). These are the true sedimentary basins of the world, in which so much hidden wealth resides—both economic and historical.

Sediment types and supplies All sediments are made up of particles of material that accumulate over time in distinct layers on the Earth's surface. Some are deposited on land, but the great majority settle at sea. With the passage of time older sediment is buried to ever greater depths, thus becoming subjected to progressively higher pressures and elevated temperatures. Eventually the grains begin to fuse together, while the fluids (mostly seawater) that were trapped between grains at the time of deposition themselves become highly concentrated. As these pore fluids are squeezed outward and upward to escape from the ever-increasing pressure, certain minerals precipitate out and act as a cementing agent binding together the once-separate particles. This is the process of lithification by which a sediment becomes a sedimentary rock.

But let us return first to the original particles. Many different types of material find their way into the ocean basins. Terrigenous materials—those originating on land—are derived primarily from the weathering of rocks—igneous, metamorphic, and sedimentary. Collectively, the world's rivers transport the bulk of this material across vast distances to the sea. Further debris is broken directly from the rocks and cliffs along the

At low tide, mud flats (*right*) seem sculpted by the waves. Because of the strong tidal influence, estuarine muds are often a mix of river-borne load and fine silt washed in from the continental shelf.

Marine sediments			
Type	**Composition**	**Source**	
TERRIGENOUS (land derived)	Rock fragments, quartz sand and silt, clay, volcanic ash	Rivers, coastal erosion, landslides, glaciers, turbidity currents, wind-blown dust, volcanic eruptions	
BIOGENIC (biological origin)	Calcium carbonate	Coccolithophores, foraminiferans, shelled organisms, coral reefs	
	Silica	Diatoms, radiolarians	
CHEMOGENIC (ocean derived)	Manganese nodules, phosphorite, oolite, metal sulfides, evaporites	Precipitation from sea-water	
COSMOGENIC (extraterrestrial)	Iron-nickel spherules and meteorites, chondrites, tektites	Meteors and space dust	

shoreline, or even from below the waves. These form the gravel, sand, silt, and clay (mud) deposits common to all ocean margins. Dissolved loads add their individual chemistry to that of the global ocean, and may remain in solution for many eons before separating out (precipitating) to become fixed as organic or inorganic additions to the sediment pile.

At high latitudes, melting and calving from glaciers and icebergs add much particulate material to the oceans, but far less solute (dissolved matter). Then again, wind-blown dust from arid and semi-arid land masses, and also from volcanic eruptions, may travel far into a neighboring ocean basin. Saharan

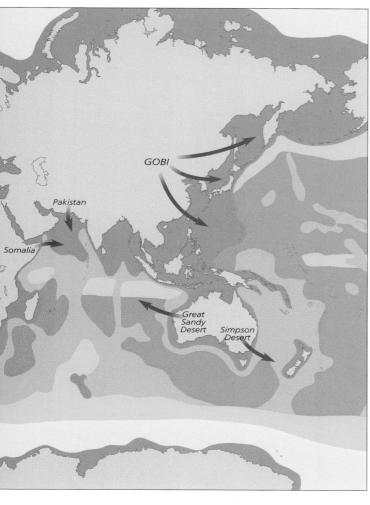

cut off from the global ocean can, for example, rapidly become salt pans if climatic conditions lead to an excess of evaporation over precipitation; natural salts such as halite and gypsum are the result. By a process that is still not fully understood, certain physico-chemical conditions favor the precipitation of iron, manganese, and a host of other metals as nodules and crusts over large areas of the deep sea.

Once material has been supplied to a basin or its margins, a wide range of processes operate to quickly redistribute the sediment, perhaps to erode and deposit it many times over, before ultimately locking it into the permanent sedimentary record.

Bands and belts When we examine a map of sediment distribution across the global ocean, the pattern revealed is an intriguing one. Some of the attributes exhibited can readily be understood in terms of different sediment types and supply routes, but other aspects are much less obvious. The continental margins, including deltas and the submarine fans that form seaward of major rivers, are composed of terrigenous sediments, obviously reflecting the fact that this material is supplied from the continent. Likewise, glacigenic sediments clearly dominate high-latitude seas. But why are calcareous biogenic oozes—formed from the remains of microorganisms rich in calcium carbonate—most common in the central Atlantic and Indian Oceans, and over broad tracts of the Pacific? And why do the siliceous oozes (from silica-containing organisms) form three broad, latitudinal bands? The answers to these questions lie in the complex nature of productivity and preservation in the world's oceans.

A map of the global distribution of seafloor sediments (*left*) reveals that terrigenous (earth-born) sediments (shown in brown and yellow) are closely influenced by their supply routes from the land—glacial, riverine, volcanic, or windblown. Biogenic sediments (the blue and green areas) are controlled by the primary productivity of the oceans' plankton, coupled with chemical dissolution.

Once biogenic ooze on the floor of a past ocean, these white chalk cliffs at Beachy Head in East Sussex, England, are now being eroded by the action of storms and waves. The pebbles on the beach will eventually become fine sediment lining the continental shelves and beyond.

dust is distributed right across the Atlantic, dust from China across the Pacific, while that from the Great Australian Desert is blown across the Indian Ocean.

Biogenic material—that produced from living organisms, including the tests (shells or skeletons) of many billions of microscopic organisms—is supplied to the seafloor from surface waters by the great cycles of primary productivity—the process by which such organisms synthesize organic materials from inorganic substances, notably by photosynthesis. The soft, creamy calcareous or siliceous biogenic oozes that accumulate at depth will eventually become the cherts and chalks so commonly exposed on land after uplift. Biogenic particulates are also derived from the break-up of larger organisms—shells, corals, algae, bryozoans, sea urchins, and starfishes—abundant in the shallow waters of continental margins, where there is an ample nutrient supply. Various types of limestone are the end result.

Chemogenic particles—those precipitated from the ocean's waters—are less important in terms of volume, but they supply useful information about the environmental conditions in which they were laid down. The ocean waters above any sedimentary basin contain an enormous range of dissolved chemicals, but their composition by and large remains constant over long periods of geological time. When such steady-state conditions change, however, then certain chemicals are likely to precipitate out of solution. Coastal lagoons or shallow seas partially

Primary productivity, the first link in the web of ocean life, is ever-present in the surface waters and yields an almost endless supply of calcareous and siliceous skeletal hard parts—the detritus of billions of tiny planktonic organisms—as a steady rain to the seafloor. The calcareous component comes from a variety of sources, chiefly coccolithophores (microscopic algae) and foraminiferans (protozoans). These single-celled organisms

manage, while alive, to recycle calcium ions carried into the oceans from weathered rocks. These ions, chemically combined with bicarbonate ions, also naturally present in seawater, form the organisms' calcium-carbonate tests. When the organisms die, however, these multitudinous shells and skeletons sink. At a certain depth—currently around 4.5km (2.8mi) in the Atlantic—the ocean's waters become more acidic, and this acidity, combined with the increased solubility of calcium carbonate in cold water under pressure, dissolves any tests that reach that far. Calcareous oozes, therefore, can only be preserved where the depth of the seafloor lies above what is known as the "carbonate compensation depth"—where the amount of carbonate produced is equal to the amount being dissolved. Their pattern of distribution closely follows that of the relatively shallow regions above mid-ocean ridges, whose peaks often bear a calcareous-ooze "snow line".

The tiny organisms known as diatoms (algae) and radiolarians (protozoans) bequeath delicate skeletons of opaline silica to the oceans. These tests are even less stable than the calcareous ones in any part of the water column—the vertical profile of the ocean's waters from surface to seabed—once their protective organic coating dies and decays. They can only reach

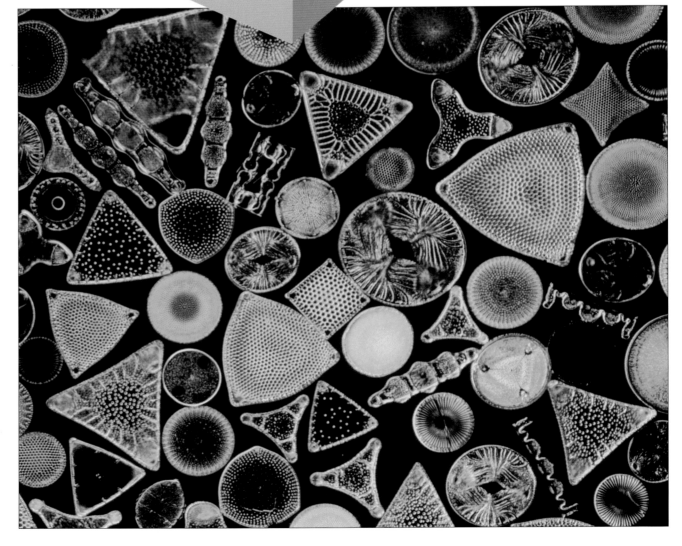

Coral or evaporites
Wind transport
River transport
Shelf sediments
Submarine canyon
Continental shelf and slope
Turbidity current
Volcanic fallout
Windblown dust
Meteorite debris
Ice-rafted sediment
Plankton
Carbonate compensation depth
Turbidite deposit
Abyssal clays
Manganese nodules
Siliceous ooze
Calcareous ooze
Mid-ocean ridge

As shown in the diagram above, different types of sediment reach their home in the ocean in differing ways. Wind, rivers, and floating ice supply terrigenous sediment to the continental margins. Turbidity currents (gravitational flows of sediment-rich water) distribute these materials to the deeper basins. A steady rain of microscopic shells and skeletons from surface waters yields siliceous and calcareous oozes, while chemical precipitation from seawater allows the slow growth of manganese nodules and crusts far out to sea.

Magnified many times, opaline silica reveals a pattern of geometric forms, each no bigger than a grain of sand. They are the lacy skeletons of marine diatoms—microscopic algae that contribute their remains in vast profusion to form siliceous ooze.

The Angola Basin is one of the four main basins in the South Atlantic, having an area of some 1 million sq km (385,000 sq mi) and a maximum depth in excess of 5,500m (18,000ft). In many ways it is typical of any other ocean basin with a flat-lying central abyssal plain, bordered on the west by a mid-ocean-ridge system and on the east by a major continental landmass. To the north and south the topographic barriers are now submerged, but have been at least partially emergent in the past. The South Atlantic has been opening for over 110 million years, with the Angola Basin deepening progressively as the oceanic crust on which it formed cools and subsides. It has also been slowly but steadily filled with sediment shed from its various margins.

This sediment fill is best known from a Deep Sea Drilling Project investigation in the southeast portion of the basin (Site 530), where 1,100m (3,600 ft) of sediment was drilled overlying ocean-floor basalts that are roughly 100 million years old. In that time there have been four different phases of basin history, each of which has left its unique impression in the sedimentary record. Decoding this record has been an exciting challenge involving scientists from some 15 different countries who sailed together aboard the drillship *Glomar Challenger*.

The first phase after formation of the ocean floor comprises fine, organic-rich sediments, known as black shale, deposited at a time of oxygen-depleted bottom waters in an early, narrow but deep Angola Basin. Quite remarkably, the same black-shale event can be recognized clearly throughout the Atlantic–Mediterranean realm and, to a lesser extent, in other ocean basins across the world. This was a significant era of global history—an oceanic anoxic event. It was also economically important, as these black shales form the principal source rocks for many oil reservoirs throughout the world.

The next phase is distinguished by an influx of greenish volcanic sediments (turbidites) derived from the Walvis Ridge to the south, which became active and partly emergent around 85 million years ago. At the same time, the African plate carrying this portion of ocean crust moved slowly northward into the warmer

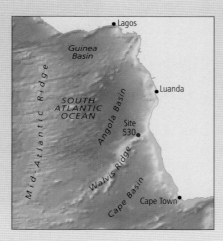

waters of the tropics, so that the emergent volcanic seamounts were rapidly colonized by reef-builders. Soon, however, the wild seas of that ancestral ocean were remorselessly battering the shoreline of this remote island chain. Fragmented remains of reef organisms, mixed with dark volcanic detritus, formed speckled beach sands around the jagged coastline. Periodically the island flanks collapsed, leading to catastrophic slides and submarine flows that carried the beach sands and slope muds deep into the Angola Basin.

During this period of normal deep-sea sedimentation, there occurred one of the most momentous episodes of prehistory—the mass extinction event in which the dinosaurs vanished forever from the face of the Earth. Scientists on board *Glomar Challenger* took great care to examine sediments across the infamous Cretaceous–Tertiary boundary. They found no dramatic changes, but rather a rapid but phased transition from Cretaceous to Tertiary faunas. This finding is not commensurate with a catastrophic end to the dinosaur era, but supports a more gradualist explanation.

Other slow changes were also afoot. Volcanic activity ceased, the seamounts subsided, and the fringing reefs soon drowned. Calcareous skeletal debris that once rained down from the ocean surface dissolved before it reached the seafloor. The continued northward drift of Africa into the desert belt meant that little river-derived material was available for redistribution. These events are reflected in the third phase—one of very slow accumulation of abyssal red clay carpeting the deep-basin floor.

As dramatically as the calcareous sediments had disappeared, so they reappeared some 15 million years ago. This time there was also abundant siliceous material and organic carbon, derived from the Walvis Ridge and African shelf. These clearly mark the onset of the coldwater Benguela Current at the ocean's surface that led to upwelling and enhanced primary biological productivity along the African margin. This fourth phase is still in existence today.

Professor William W. Hay,
Head, Joint Oceanographic Institutions, USA

Angola Basin: 100 million years of ocean history

Part of the sediment core (halved lengthways) recovered at a water depth of over 5,000m (18,000ft) from Site 530 in the southeast Angola Basin. These brownish-green sediments are typical abyssal clays from deep oceanic basins. The thin white layers are calcareous sediments introduced by episodic turbidity currents.

---

the seafloor and become preserved as sediment if they are produced in sufficient quantities to completely swamp the system. This can occur beneath zones of especially high primary productivity—regions of equatorial, coastal, or seasonal upwelling, where the interaction of different water masses and oceanic currents recycle nutrient-rich waters to the surface. This process takes place in a single, broad, equatorial belt, and in two high-latitude belts, around the world, as well as in shelf regions on the western margins of continents. These are exactly the areas where siliceous sediments are found on the seafloor. Where supply is particularly rapid, there is too little time even for the organic coating to fully decay, so that organic carbon is also preserved and buried in the sediment. In just a few million years' time, such carbon sinks will become future sources of oil and gas.

Floor of the abyss In the deepest, most remote parts of the ocean basins lie great swathes of fine-grained clays. These abyssal regions are too deep for calcareous skeletons to survive and are far from areas of high primary productivity, so that only the very finest clays and wind-blown dust particles settle out. At least 70 percent of this material is terrigenous; of the rest, some arrived as space dust, while other particles entered the atmosphere as volcanic material, only to be transported over great distances by the winds. The particles are so sparsely distributed and fall so slowly through the thousands of meters of water column that average rates of accumulation are a matter of a few millimeters every thousand years. All traces of iron minerals are fully oxidized, yielding the characteristic red-brown rust color of these abyssal clays.

Incised into the steep continental slopes that rim the ocean basins are narrow gorges, broad troughs, and tightly meandering channels. These are the Grand Canyons and great hidden rivers of the deep.

PATTERNS & CYCLES

Canyons, Slopes, and Fans

KEY TERMS

CONTINENTAL SLOPE
TURBIDITY CURRENT
OXBOW SEGMENT
CHANNEL-LEVEE SYSTEM
SEDIMENT WAVE
SLIDE
SLUMP
DEBRIS FLOW
SALT DOME
CLATHRATE
TURBIDITE

The transfer of sediment to the deep ocean basins occurs on a scale not witnessed on land, and by a variety of processes, many of which are still poorly understood. The continental slopes down which the transfer occurs are the surface expression of the true transition between continental and oceanic crust. This transition in crustal properties is the fundamental reason behind the relative elevation of the continents and the depression of the ocean floor. On our generalized maps of the ocean floor and in schematic diagrams of plate margins and subduction, the continental slopes are typically represented as extremely steep or even precipitous. In fact nothing could be further from the truth, for the actual gradients mostly lie between 2° and 5°—the sort of slope that, on land, is aggravating to cycle up but that gives little benefit in descent.

A volcanic mudslide cloaks the slopes of Mt. Pinatubo in the Philippines. Similar slides can be triggered below the sea's surface by earthquakes or the eruptions of submarine volcanoes.

Grand Canyons Four main types of continental slope make up the ocean margins. The first is relatively steep and rugged; in some cases gradients can exceed 10°, and locally more than 45°. Such features are typically found off coral reefs and carbonate banks (built up from the rain of calcium-carbonate tests), where seafloor sediment has been stabilized by organic growth and early cementation, and also wherever major faulting has steepened the slope angle. These declivities

are rocky and irregular, patchily covered with submarine rock falls and debris avalanche deposits.

The second type is also irregular in profile but gentler in gradient, although still at the steeper end of average, with slopes of about 5°–10°. They may result from the very rapid build-up of excessively thick, fine sediment piles supplied from glacial margins or river deltas. Often found along active continental margins and subject to volcanism and submarine earthquakes, these areas are characterized by great instability. Slope failure results in the displacement of huge blocks of sediment in the form of slides, together with collapsed areas of more unconsolidated material (slumps), scarring the upper slope and dumped as chaotic masses and debris-flow lobes at the base.

The third and fourth types are the most intriguing, and are currently the subject of intense scrutiny by research teams around the world. These are the slopes cut by channels of every different size, shape, and style. Deep canyons shaped like Vs are incised up to several hundred meters down into the shelf margin, cut into the hard, rocky substrate. In some cases these submarine valleys are clearly linked to an onshore river system, and presumably originated when sea levels were much lower and the rivers flowed out across broad continental shelves.

Off some of the major river deltas, an even deeper, U-shaped trough—the fourth type of slope—has been excavated through the slope muds beyond the delta mouth, going on to carve a meandering pathway far out into the basin plain. The main channel of the Laurentian Fan off eastern Canada is up to 800m (2,600ft) deep, while the Swatch-of-No-Ground gully off the Ganges Delta extends at least 600m (2,000ft) down in parts.

Hidden channels These deep-sea features lie far beyond the penetration of significant sunlight and cannot be reached by scuba-diving. Although some of the large canyons in the upper slope regions are probably well-known to military submarines, most remain completely unexplored. Yet with the advent of deep-towed remote-sensing survey vehicles, great advances have been made in mapping the deep seafloor over the past 25 years. What has become dramatically apparent is that certain slopes are riddled with channels of all sizes, whereas others are apparently featureless, smoothed by the action of contour-hugging bottom currents. Some of the

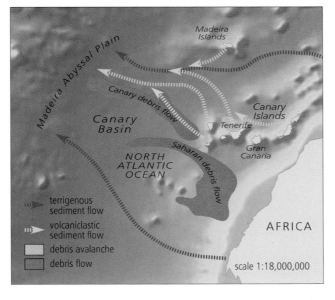

Cloud swirls beneath the rugged peaks of Tenerife (*above*), largest of the Canary Islands. Pico de Teide, the island's still-active volcano, is a source of avalanches, debris flows, and turbidity currents that plaster the continental slope off Tenerife's coasts and flow through sinuous channels that eventually feed into the Madeira Abyssal Plain (map, *right*).

channels are highly sinuous, with tight meanders, cut-off loops, abandoned oxbow segments, and a host of other features very reminiscent of rivers on land, where they meet low-lying flood plains and amble their way slowly toward the sea. Because deep-sea channels are generally thought to be cut and kept open by turbidity currents (fast-flowing currents thick with erosive sediment), it has been surprising to find so many riverlike features in their configuration. Our understanding of deep-sea processes is currently being reassessed in the light of these findings.

Where submersibles have reached the very bottom of these deeply hidden channels, or core samples have been recovered from their narrow floors, coarse gravel and sand have often been observed. In some cases these sediments are shaped into a series of large-scale dunes and waves, or else are cut by deep, horseshoe-shaped scours—all evidence for the passage of very powerful currents using the conduits as major highways for the transport of material into the deep ocean beyond. On either side of the channels are banks, or levees, which can be hundreds of meters high, built up of thick, well-layered deposits of fine silts and clays, material that has spilled out of the channel during the passage of a giant flow—a submarine river flooding its banks. Some remarkable images of these areas show huge fields of regularly spaced giant sediment waves. The wavelength (the distance

The principal types of slope system: (**1**) steep, rugged, coarse-grained slope apron with active faulting, earthquake activity, rock falls, and debris avalanches; (**2**) relatively steep, unstable slope apron characterized by irregular slides, slumps, and debris flows in fine-grained sediments; (**3**) relatively gentle slope apron incised by numerous channels and gullies, along which shelf sands and muds are transported to the foot of the slope; (**4**) shallow, stable slope apron, its mud cover smoothed by alongslope currents, and erratically punctuated by deep channels feeding deep-water fans.

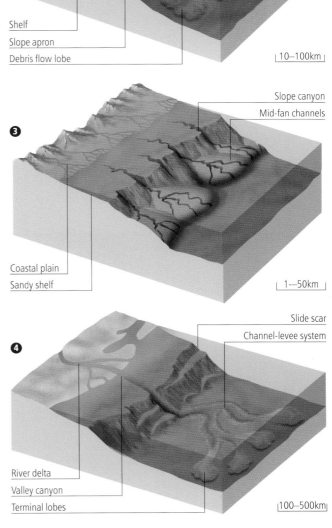

from crest to crest) is typically between 1 and 2km (0.6–1.2mi), while the amplitude (the height from trough to crest) is several tens of meters. A full scientific explanation of their origins has so far eluded us.

Some channels extend beyond the foot of the slope and feed sediment across the great submarine fans. The largest and most impressive of these is the Bengal Fan, whose gently sinuous channels are raised like superhighways well above the level of the rest of the fan surface and extend over 2,500km (1,500mi) into the central Indian Ocean. The Amazon Fan in the central Atlantic is another giant, its surface covered by a series of meandering channels—some still active, others abandoned and slowly filling with sediment.

Other channels simply wander across open ocean floor. The Northwest Atlantic Mid-Ocean Channel is born out of a very extensive network of tributary channels from the Canadian slope of the Labrador Sea, and then flows continuously for a distance of some 2,000km (1,250mi) between Greenland and Newfoundland, around the tip of the Grand Banks and into the Sohm Abyssal Plain.

Phantom currents Much mystery still surrounds the nature of the processes that move material from the basin margins downslope, either through the many channel systems or by some other means. The realm of the deep slope is entirely hidden from view, and the processes that operate there appear to be mainly episodic—short-duration events that are typically dramatic in their nature and effect, but are extremely elusive to view or measure.

Two important large-scale processes that have close analogues on land are submarine slides and slumps. These events, which are very common on most submarine slopes, involve the sudden downslope displacement of the upper layers of sediment along an underlying plane of rupture. In slides, the internal disturbance of these upper layers is minimal, whereas slumps show more disruption. Single-slide masses range from very small, localized displacements to very large and often complex bodies that may be several hundred meters thick and measure over 100cu km (24cu mi) in volume; a dry-land equivalent might involve moving Martha's Vineyard to the Massachusetts coast or Greater London over the Channel to France! Such catastrophic slides have involved the loss of a drilling platform with all hands on board off the Mississippi Delta, the removal of half a new runway at Nice Airport in southern France and its disappearance downslope toward the Alboran Basin, and the overnight vanishing of a Ukrainian village from the Crimean Peninsula, as it slid to a watery grave in the Black Sea.

Sediment instability on basin slopes is affected by a combination of interacting variables—high slope gradients, high rates of sedimentation, seismic shock, high primary productivity, the intrusion of salt domes or mud volcanoes, and sudden degassing in the sediment from clathrates (frozen water–gas mixtures). There is recent data to suggest that gas-triggered slides on the Bermudan slope remove sufficient of the upper sediment layers

The Amazon submarine fan, stretching from South America's continental shelf to the Demerara Abyssal Plain in the equatorial Atlantic Ocean (*right*), is one of the world's largest. These gigantic sinks of sediment in the oceans have been intensively studied in recent years. Deep-tow sidescan sonar imagery (*below*) reveals one of many tightly meandering channels crossing the mid-fan region. Such features bear a close resemblance to the channels carved out by rivers on land.

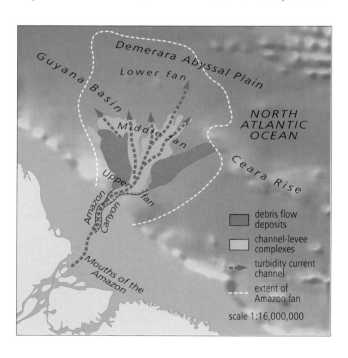

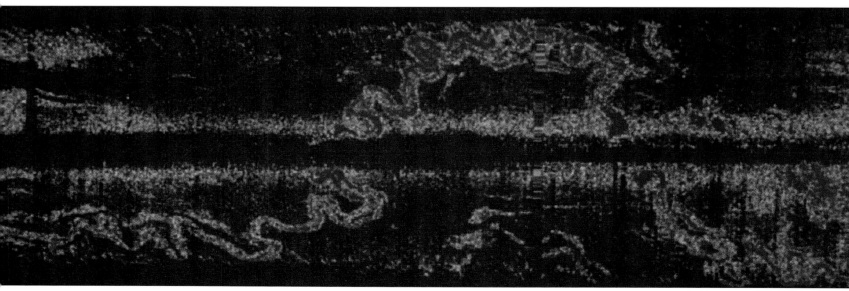

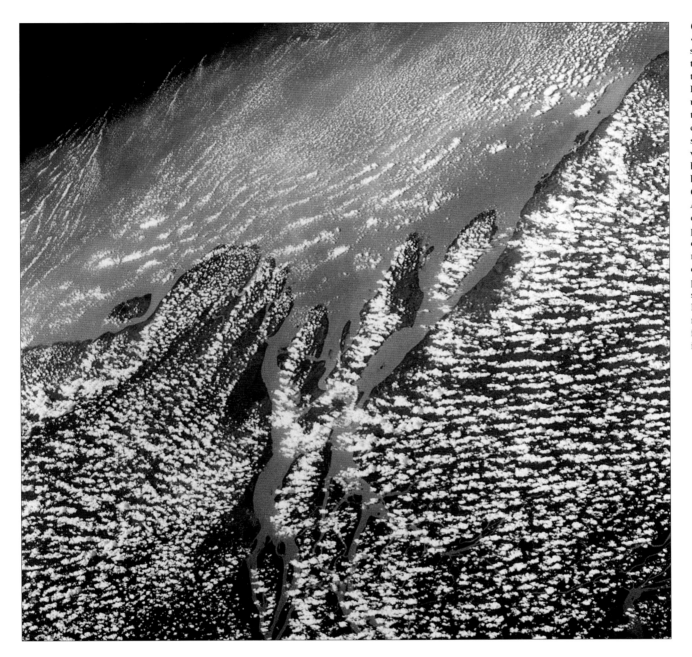

Striated cloud patterns dapple the sky high above one of the many mouths of the Amazon River in Brazil (the river runs up from bottom center). Large quantities of brownish-colored sediment are clearly visible in the estuary, being swept far out by the prevailing Guyana current. The Amazon River discharges more water per second to the ocean than any other river in the world, carrying with it suspended particles of soil eroded from the High Andes and carried through the rainforest swamps far out into the Atlantic.

to decapitate large areas of buried clathrates and hence release giant bubbles of methane gas. Speculation links these natural events with the unexplained disappearances of ships and even planes within the "Bermuda Triangle".

During particularly chaotic slide events, the sediment mass absorbs a lot of water, with progressive disintegration and transformation into debris flows and turbidity currents. Turbidity currents are believed to be one of the most common processes of the deep-sea environment, although, paradoxically, a full-sized prototype has never yet been observed in nature. They remain a phantom of the deep, known only from their dramatic effects, such as the breaking of submarine telegraph cables, and from their characteristic, graded sediment deposits—turbidites.

One of the scientifically intriguing aspects of turbidity currents is their ability to achieve a state of self-maintained flow (known as autosuspension) in which they can travel for several thousands of kilometers down the gentlest of slopes and across almost flat abyssal plains, even moving upslope for short distances. Truly large turbidity currents, generated perhaps from gigantic basin-margin slides, may reach over 500m (1,650ft) in thickness and 10km (6mi) in length, and may travel at speeds of over 70 km/h (40mph), eventually depositing a layer of sand and mud 2m (6.5ft) thick over an area the size of England! If the receiving basin is small and confined, or if the flow does not reach autosuspension, then megaturbidites up to 25m (80ft) thick, grading from coarse gravel and even small boulders at the base to fine silt and clay at the top, may be deposited by single flows. Such large events are relatively rare, perhaps occurring once every 1,000–3,000 years, interspersed with more frequent small-scale flows. They have been found to be more common at times of lowered sea levels and in areas of high tectonic activity. All, however, form geologically instantaneous deposits; gravel and sand are deposited within a matter of minutes to hours, and even the very finest material settles over a period of days.

Islands are oases of life in a marine desert, providing fragments of information about the world beneath the waves. They are also footprints in time, yielding clues to past oceans.

Islands in Time

KEY TERMS

PYROCLASTIC

LAVA

CORAL REEF

TALUS SLOPE

ATOLL

HOTSPOT

ISLAND ARC

MANTLE PLUME

To the ocean scientist, the study of islands can yield a detailed record of a long and complex past. Some islands, like the young volcanic island of Surtsey, provide only a tiny (but vital) snapshot of that history; others yield a far greater richness and diversity, charting whole chapters in the development of the oceans. The islands that dot the globe today are the fragments of an amazing story played out over many millions of years as continents collided, joined, and split apart again, and oceans formed and disappeared.

Island origins It is rare in the scale of human history to witness the birth of an island. Indeed, geological events, which follow their own slow beat, are so infrequent in human terms that even the most well-known may never occur within the lifetime of an individual. Major events in

Surtsey, off the southwest coast of Iceland, is one of the youngest islands in the world. It was "born" in a series of spectacular volcanic eruptions between 1963 and 1967.

the oceans are equally elusive, usually occurring far from view beneath the surface.

The dramatic appearance of the small volcanic island of Surtsey off the south coast of Iceland on November 14, 1963 was, therefore, a remarkable occurrence, and one that has led to a greater understanding about how oceanic islands develop. The beginning of the four-year-long eruption that actually gave birth to Surtsey went unnoticed because it occurred over 100m (330ft) below the sea surface. But slowly the submarine volcano was growing, until one day in early November when fishermen from the coastal village of Holt noticed, as they cast their nets, that the waters were uncommonly warm, while an eerie orange glow was emanating from far beneath the pre-dawn sea. Sensing danger, they returned to harbor. The following day the island of Surtsey appeared, in a column of ash, steam, and volcanic gases that rose some 4km (2.5mi) above the ocean surface and showered pyroclastic (volcanic) materials over the fishermen and the rest of southern Iceland as well. The eruption continued for the next five months, the column of ash reaching as high as 10km (6mi), so that all air traffic had to be diverted.

During these first months, the eruption was especially explosive because seawater had come into direct contact with the molten lava rising up the vent. Basaltic lavas, typical of mid-ocean ridges, flow at temperatures of around 1,200°C (2,200°F), whereas the waters off southern Iceland during the winter months are only a few degrees above zero. Two other, smaller island cones were established briefly above adjacent eruptive centers, but both of these were washed away in the severe winter storms of 1963–64. The volume of material produced by the Surtsey volcano was marginally greater than the amount removed and so the island continued to grow.

A critical turning point came in April 1964 when, for the first time, the sea could not reach the eruptive vent, and the explosive combination of ice-cold water with red-hot lava no longer determined the style of eruption. Instead, molten lavas flowed quietly across the loose, rugged land, mantling the surface and protecting the coasts from battering waves. Surtsey had sur-

Principal types of islands

Generic type	Origin	Type of location	Examples
VOLCANIC HOTSPOTS	Formed directly over a hotspot by persistent volcanic activity	Oceanic ridge and mid-ocean	Hawaiian, Marquesas, and Tuamotu chains (Central Pacific); Galápagos (E. Pacific); Canaries (Atlantic); Surtsey (N. Atlantic)
ISLAND ARCS	Formed by volcanic activity over a subducting tectonic plate	Convergent plate margin, typically ocean margin	Aleutians and Japanese islands (N. Pacific); Marianas (W. Pacific); Antilles (Atlantic)
CONTINENTAL FRAGMENT	Formed by the separation of fragments of crust from the mass of a continent	Ocean margin, typically close to a landmass	Orkneys, Shetlands, Outer Hebrides (N.E. Atlantic); Seychelles (Indian Ocean); Corsica (Mediterranean)
CORAL REEFS	Formed by corals and other reef-building organisms on any substrate	Ocean margin and at mid-ocean volcanic hotspots	Maldives (Indian Ocean); Great Barrier Reef (S.W. Pacific)
COMPLEX ISLANDS	Long-lived and larger islands formed by a combination of processes	Mainly ocean margins	Sicily, Cyprus (Mediterranean)

vived, to grow steadily in size over the next three-and-a-half years while eruptions continued intermittently.

Not all islands, however, are formed in the same way as Surtsey, nor in the same ocean setting. They come in a wide variety of shapes, sizes, and complexity, many with a volcanic core, others as tiny fragments of continent, still others as atolls formed from coral reefs.

Coral islands Lying just above sea level and built on any suitable rocky substrate in warm, shallow, clear water, coral islands are created over hundreds of thousands of years by the collective industry of millions of tiny marine organisms. Reef corals, aided by zooxanthellae—the even tinier symbiotic algae living in their tissues—take up the readily available calcium and carbonate ions in the seawater and secrete limestone. Each coral animal constructs its own tiny dwelling or protective coating and, together with bryozoans, sponges, mollusks, and the cementing action of coralline algae, builds a composite skeletal mass or platform just below the surface of the sea. These reef communities eventually form massive limestone deposits, covered with a thin layer of living organisms. In time, the pounding of waves and the ebb and flow of tidal currents break and erode the more exposed parts of this reef. A protective fringe of coarse debris—a talus slope—forms on the ocean side, while much of the finer sandy and silty material collects on the sheltered side to form a low-lying island, soon colonized and stabilized by seagrasses and other plants.

There are several main types of coral reef, including fringing reefs, barrier reefs, and atolls. Although reef corals can only survive in sunlit, shallow waters, coral atolls are found in the open ocean. They originate where reef organisms have created a fringing reef around the slopes of newly formed volcanic islands.

The Maldive Islands in the southwestern Indian Ocean make up an archipelago of some 1,200 coral islands. Twenty-six are natural atolls, in which a circular reef protects an inner lagoon; these contain numerous patch reefs, as above. Perched on volcanic seamounts and stretched out along an ocean ridge, these picturebook tropical paradises never rise more than 2m (6ft) above sea level.

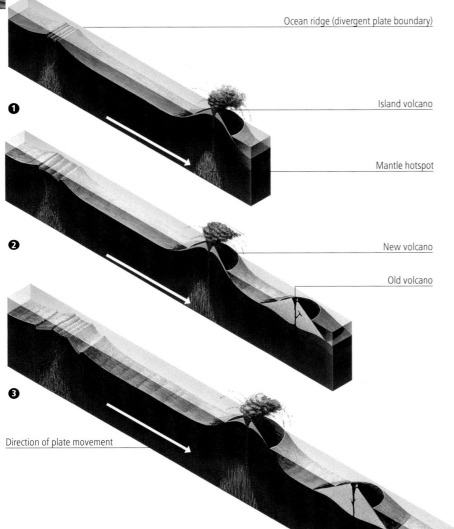

Ocean ridge (divergent plate boundary)

❶

Island volcano

Mantle hotspot

❷

New volcano

Old volcano

❸

Direction of plate movement

Islands form chains when an ocean plate moves over a hotspot in the underlying mantle, producing a volcano (**1**). As the newly formed volcano is carried away by tectonic plate motion, volcanic activity ceases and a new active volcano appears in the fresh crust above the hotspot (**2**). The extinct dome gradually subsides, together with the cooling ocean crust on which it is riding (**3**). A fringing coral reef may form and grow upward to form an atoll as the volcanic island sinks. Completely submerged volcanic islands are called seamounts or guyots.

When volcanic activity ceases and the island subsides, the fringing reef on its flanks continues to grow upward, becoming first a barrier reef—separated from the land by a lagoon, and then an atoll—a ring of coral islands with no central landmass. This is the case for many of the islands that form the Maldive archipelago in the Indian Ocean, as well as for numerous isolated atolls in the Pacific. Reefs will also develop around continental landmasses, provided that the seas are not swamped by mud-charged river runoff. Almost 3,000 individual reefs make up the Great Barrier Reef off northwestern Australia—the world's largest living structure.

Volcanic islands Many islands in mid-ocean locations are the result of volcanic activity linked to "hotspots." These are thermal plumes in the upper mantle of the Earth's interior that lead to continuous upwelling of molten lava over long periods of geological history. They may be located at the crest of a mid-ocean ridge, as in the case of Surtsey and Iceland, or they may be isolated in the open ocean (or even found beneath continents, as at Yellowstone National Park, Wyoming). Because the hotspot remains stationary while the ocean plate is moving, linear chains and clusters of volcanic islands are formed, exhibiting regular changes in age and size

along the chain. An active volcano lies over the hotspot, closest to the ridge axis, and shows the highest elevation above sea level. As the moving plate takes the island away from the center of volcanic activity, it naturally subsides as the ocean crust itself cools, becoming less buoyant, and sinks. Still further from the hotspot, the island drops below sea level; it is then known as a seamount. Those seamounts furthest away from the hotspot are the oldest and deepest.

Another type of island, also associated with active volcanism as well as with abundant earthquakes, is found along convergent plate boundaries at some distance behind the subduction zone (trench). Water released from the subducting crust flows into the upper mantle, leading to partial melting, magma formation, and extrusion through the overlying crust. The lava here is mostly andesite (fine-grained, gray igneous rock), which tends to produce more episodic and violent eruptions than those of mid-ocean hotspots, rapidly building up steep-flanked cones. Strings of such islands commonly assume great curvilinear patterns across the ocean surface and so have become known as island arcs. This form is the natural trace of a subducting straight-edged plate on part of the Earth's spherical surface.

Santorini, quietly nestling in the central Mediterranean, is part of the Aegean island arc, which is believed to be the result of continued subduction of the African plate beneath Europe. Its dark volcanic sands tell a dramatic story. The sheltered sea lagoon around which sheer, reddened cliffs of oxidized volcanic debris now lie was once the crater of a mighty volcano. It was part of an island arc rather than a mid-ocean ridge, as is evident from the chemical composition (mainly andesitic) of the sand grains. In approximately 1500 BC the island, then a huge volcanic dome, began a 50-year period of repeated catastrophic eruptions. Ash spread over the entire eastern Mediterranean, and dust filtered through the air across the world.

It is believed that this prolonged period of intense volcanic activity, coupled with the

Pico in the Azores is one of a string of volcanic islands along the Mid-Atlantic Ridge. There is still debate regarding their origin: over a large hotspot, or along an oceanic fracture zone.

Today's Mediterranean Sea was once a salt desert; and before that it was another ocean. The transition from sea to desert and back to sea may be seen in the Troödos Mountains of Cyprus. Some 100 million years ago, the slopes of the Troödos range were part of the deep-rooted core of a mid-ocean volcano more than 2,000m (6,500ft) beneath the ocean surface. This in turn was a small piece of the spreading center from which the ancient, now-extinct Tethys Ocean was born. Through the gradual upwelling of millions upon millions of tons of lava, the ocean floor grew wider, forcing apart the continents on either side.

When the Tethys had ceased growing, the great continental masses on either side began drifting together, and much of the newly formed ocean crust was subducted. Some fragments, however, were squeezed upward, and so it was that the Troödos crust eventually produced a new island south of the landmass that now includes Turkey.

Slivers of ocean crust preserved in mountain terrain are known as ophiolite, and the Troödos ophiolite was one of the first to be recognized for what it is. These mountains are still among the few places in the world where it is possible to walk along a path and view layers of rocks that sequentially link the outer crust of the ocean to the deep interior of the Earth's mantle.

On other parts of the island there are hectares of rugged basalt lava flows that were once part of the ocean floor. One can see the rounded, pillowlike form of the basalt that was extruded molten onto the seafloor, the glassy pillow rims from sudden quenching of the hot lava, and millions of tiny holes formed by trapped volcanic gases. In large depressions on the lava surface, there are distinctive, brown-colored, metal-rich sediments. These formed where hot water escaped through cracks in the ocean crust, pumping thousands of tons of dissolved iron, manganese, and other metals into the cold

Cyprus: a small slice of the past

Skorniotissa copper mine in Cyprus. Copper deposits are often found in the upper layers of basaltic ocean crust and in lower layers of pelagic (open-ocean) sediment. They have been mined on Cyprus (named for the element) since Roman times, but are currently uneconomic.

waters, and great plumes of metal precipitates settled to the seafloor. Present-day examples of these hydrothermal vents, known as "black smokers" for their dark clouds of minerals, have only recently been discovered on modern mid-ocean ridges, together with a bizarre assemblage of deep-ocean lifeforms previously completely unknown to marine biologists.

Soon after the ocean crust was formed, the whole area that is now Cyprus was very slowly covered in a veneer of sediment, a thin blanket that became thicker and thicker as the eons wore on. Spiralling through the sunlit surface waters were billions of tiny creatures that made up the plankton, microscopic organisms different but not dissimilar to those that fill the same ecological niche today. It is their remains, often the most delicate and intricate skeletons made of pure opaline silica or pristine white spar, that accumulate with painstaking slowness over the

deep-sea floor and, in time, harden to form chalk, chert, and marl.

However, the ocean was closing, and already that portion of ocean crust that was to become Cyprus was beginning to push upward. The rocks demonstrate progressive shallowing of the sea as shelly limestones and reef-building organisms colonized the substrate. Some of these fossilized reefs, fringed by broken talus slopes and encircling lime muds originally laid down in quiet lagoons, remain almost intact in the form of knolls to this day, although they grew and flourished between 10 and 15 million years ago and have now risen high above the sea.

Between 5 and 10 million years ago the Tethys was cut off from the world ocean and rapidly evaporated, not once but many times over, so that great thicknesses of salt deposits formed a mantle over much of the seafloor. Some of these deposits are still preserved in Cyprus.

associated earthquakes and tsunami, finally wiped out the great Minoan civilization on Crete which had dominated the area for almost 500 years. In the process Santorini brought about its own destruction, as it completely blew off its top, leaving a gigantic crater that was soon filled again by the sea. Many scholars think that this remarkable event might explain the legend of the lost continent of Atlantis, involving a great civilization that suddenly disappeared, leaving behind no tangible evidence.

Bits and pieces The many islands, large and small, that decorate the coastlines of continents are small fragments of continental crust separated from the mainland. The Scottish isles of Orkney, Shetland, and the Outer Hebrides were once part of the 350-million-year-old Caledonian continent, formed by the collision and fusion of northern Europe with North America. The same is true of the barren but beautiful islands that fringe the Norwegian fjords. Even the granitic rocks of the Seychelles, far out in the Indian Ocean,

were formerly part of a continent that contained what are now India and Madagascar. And that remote patchwork of icebound islands stretching from Hudson Bay to the northern tip of Greenland can be readily reassembled as part of the great North American continental plate.

An island's separation from the mainland most typically occurs as a rising sea level floods low-lying land, leaving isolated ridges, hills, and mountains as newly formed islands. Greater separation occurs where the movement of tectonic plates has ripped continents apart, in some cases followed by limited seafloor spreading, which creates a steadily widening seabed between the split pieces of land.

Apart from a small number that have a purely volcanic origin, most of the Mediterranean islands are continental fragments, exhibiting different ages and different rock types perhaps, but all broken off the adjacent mainland at some time in the history of the region. Some, such as Sicily and Cyprus, have long and complex histories that involve a combination of processes.

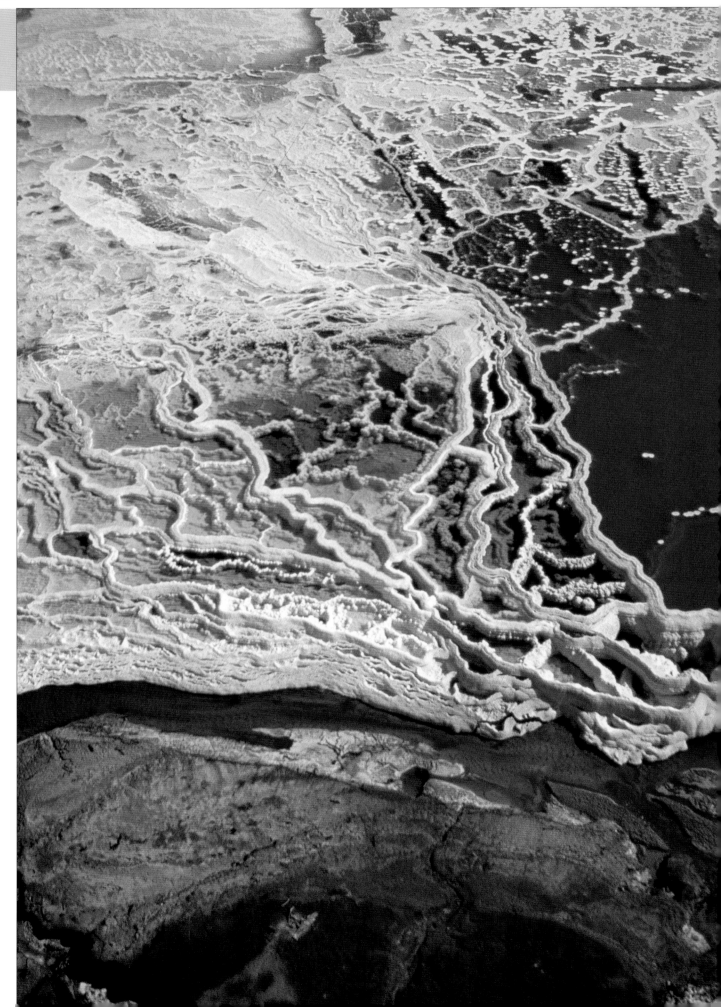

Known in Biblical times as the Salt Sea, the Dead Sea is actually a salt lake, 76km (47mi) long, 18km (11mi) wide, and up to 400m (1,300ft) deep. It is located in the Jordan rift valley—the deepest depression on Earth—with its shoreline lying about 400m (1,300ft) below sea level. In this region of blistering heat and almost endless sunshine on the northern edge of the Negev Desert, evaporation far exceeds rainfall, with the result that a body of water that began life as a freshwater lake is now ten times saltier than seawater. These salts create strange and beautiful crystalline forms as they precipitate from such a hypersaline solution.

SALT, SUN, & SEA LEVEL

CHAPTER 3

Maintaining a global balance

From beyond our own world, the Earth hangs like a blue sapphire in the velvet-black canopy of space. Streaked by swirling white clouds and dotted with the green-brown of continents and islands, it is a single entity linked by water. The remarkable properties of the seemingly simple water molecule have allowed the Earth to retain the integrity of its hydrosphere, its seas to develop their saltiness and complex hierarchy, and life on the blue planet to flourish. Water is a supersolvent, absorbing gases from the atmosphere and leaching salts from the continents. About 3 billion metric tons of dissolved chemicals are delivered by rivers to the oceans each year, yet the salinity of seawater remains much the same. Some elements remain in seawater for millions of years, others for only a few hundred, but all are eventually cycled through the rocks and the chemical equilibrium is maintained.

Just as the oceans act as a chemical buffer for planet Earth, so they control its temperature and moderate its climate. Colossal quantities of heat energy are transferred every second from the equator to the poles via ocean currents, helping to maintain the Earth in its present fragile interglacial period with relatively high sea levels. But the ocean also insulates Antarctica, allowing the build-up of over 4,000m (13,100ft) of ice and snow above the South Pole. Although tectonic and volcanic activity have affected sea levels throughout geological time, independently of ice ages, cyclic variations in the Earth's orbit will inevitably lead us into the next glacial period, with ice caps across one-third of the land and sea levels over 100m (330ft) lower than today.

Seawater is a chemical cocktail of nearly 100 different elements. Originally "fresh" water, the oceans now contain over 5 trillion metric tons of dissolved salts, and have reached a remarkable equilibrium.

SALT, SUN, & SEA LEVEL

Why the Sea Is Salt

The ubiquitous substance that makes Earth so special has a very simple and well-known chemistry. Water (H_2O) is made up of two atoms of hydrogen, the most abundant element in the universe, and one atom of oxygen, an element that accounts for nearly half the weight of the Earth's crust. The way in which these elements combine yields a substance whose unique and remarkable properties are crucial in determining physical and chemical conditions across the surface of the world.

The oceans formed early in the planet's history as gases escaping from the crust and mantle (a process called outgassing) condensed into water. The rains that fell to earth and filled the seas were freshwater in the first instance—if rather acidic —together with some dissolved gases. Today, however, the world ocean contains over 5 trillion metric tons of dissolved salts, and nearly 100 different naturally occurring elements, including some 5 billion kg (over 11 billion lb) of gold—the medieval alchemists' dream of obtaining gold from seawater was, perhaps, not so ill-founded! In fact, if the oceans' water evaporated completely, the dried residue of salts would be equivalent to a layer 45m (150ft) thick over the entire planet.

Small islands of salt near the shores of Lake Eyre, South Australia. Although normally a dry salt-pan, this vast inland lake floods two or three times a century. The waters quickly evaporate under a scorching desert sun, leaving a crust of salts, originally leached from rocks in the drainage basin.

Major constituents of seawater

Solutes (3.5% salinity)	Concentration (parts per 1000)	Percent of total salts	Residence time (years)
Chloride (Cl^-)	19.3‰	55.0%	100,000,000
Sodium (Na^+)	10.7‰	30.6%	68,000,000
Sulfate (SO_4^{2-})	2.7‰	7.7%	11,000,000
Magnesium (Mg^{2+})	1.3‰	3.7%	13,000,000
Calcium (Ca^{2+})	0.41‰	1.2%	1,000,000
Potassium (K^+)	0.38‰	1.0%	12,000,000
Bicarbonate (HCO_3)	0.14‰	0.4%	unknown

Marine chemists today are challenged to understand the nature of ocean salinity and the flux of elements into and out of solution. The chemical elements that form the building blocks of life—carbon, nitrogen, phosphorus, hydrogen, and oxygen, as well as several biogenic trace elements—all occur in water. The global cycling of these elements is both vital and extremely complex, and research scientists studying this process face a task almost as difficult as that of the alchemists who once tried to extract gold from seawater.

Supersolvent The effectiveness of water as a solvent is due to an asymmetrical bonding between its hydrogen and oxygen atoms. The resultant water molecule has an angular or kinked shape with weakly charged positive and negative ends, rather like magnetic poles. This polar structure is especially significant when water comes into contact with substances whose elements are held together by ionic bonds—the attraction of opposite electrical charges in atoms. Such bonding is typical of many salts, such as sodium chloride (common salt), in which positive sodium ions are attracted to negative chloride ions, forming a regular array or crystal. Water molecules quickly infiltrate the solid compound, the positive hydrogen end being attracted to the chloride and the negative oxygen end to the sodium, surrounding and then isolating the individual ions, thereby dissociating the solid—an apparently simple process, but extremely efficient.

changed little for much longer than that—perhaps for the past 200–300 million years. But this poses a conundrum—why does the ocean not get progressively saltier with age when outgassing along the mid-ocean ridges continues unabated and when the world's rivers annually supply millions of tons of dissolved chemicals? There is also a related puzzle—why is the chemistry of river water so different from that of seawater?

Cycles and sinks Although seawater appears to be in a steady state, in which both the relative proportion and amounts of dissolved elements per unit volume are nearly constant, this fact belies an underlying chemical turmoil. The vast supplies of elements from mantle outgassing and continental runoff must be exactly balanced by their removal from the oceans into temporary or more permanent sinks. The principal sink is the sediment, and the principal agent removing elements and ions from solution is biological.

Although only a small proportion of soft tissue and skeletal (hard) parts from dead organisms finds its way into the

As the first rains fell onto a primordial Earth, they began dissolving the more abundant and most easily dissociated materials that had started to differentiate out as primitive crust. Certain gases in the atmosphere, notably carbon dioxide and sulphur dioxide, dissolved in the water droplets as they fell, thus forming weakly acidic rain, an even more effective solvent than pure water. Over the eons, through untold cycles of evaporation, rainfall, and runoff, the oceans slowly accumulated dissolved chemical ions (solutes) of almost every element present in the crust and atmosphere. At the same time, outgassing from the deeper mantle has continued to this day to add a variety of other substances to the mix, including new (juvenile) water, from volcanoes and vents along the mid-ocean ridges. Most elements are present in extremely small amounts, generally measured in parts per thousand (‰) for the seven most abundant solutes—chloride, sodium, sulfate, magnesium, calcium, potassium, and bicarbonate—and parts per million (ppm) for the next five. All the rest are present only in trace amounts.

The total concentration of dissolved (inorganic) substances in water is referred to as its salinity. This varies typically from about 3.3 percent to 3.7 percent by weight (or 33–37‰), depending on local factors such as evaporation, precipitation, and river runoff from land. The average for open water is 3.5 percent (35‰), a figure that has remained more or less constant for as long as measured records are available. In fact, the best evidence we have of past ocean chemistry suggests that it has

Blackman River sweeps through a broad estuary into Blackman Bay, Tasmania (*left*). In addition to the load of sedimentary material, enormous volumes of dissolved salts are added daily to the global ocean by tens of thousands of rivers, large and small, across the world. It was in this way that seawater gained its current level of salinity long ago; since then its composition appears to have remained in a steady state.

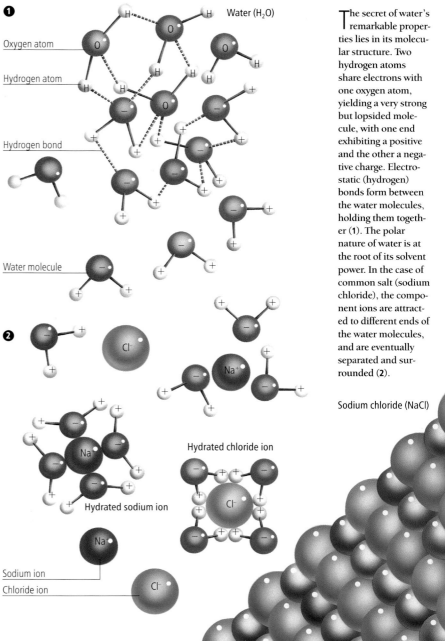

❶
Water (H₂O)
Oxygen atom
Hydrogen atom
Hydrogen bond
Water molecule

❷

Cl⁻

Na⁺

Hydrated chloride ion

Na

Hydrated sodium ion

Na

Sodium ion
Chloride ion

Cl⁻

Sodium chloride (NaCl)

The secret of water's remarkable properties lies in its molecular structure. Two hydrogen atoms share electrons with one oxygen atom, yielding a very strong but lopsided molecule, with one end exhibiting a positive and the other a negative charge. Electrostatic (hydrogen) bonds form between the water molecules, holding them together (1). The polar nature of water is at the root of its solvent power. In the case of common salt (sodium chloride), the component ions are attracted to different ends of the water molecules, and are eventually separated and surrounded (2).

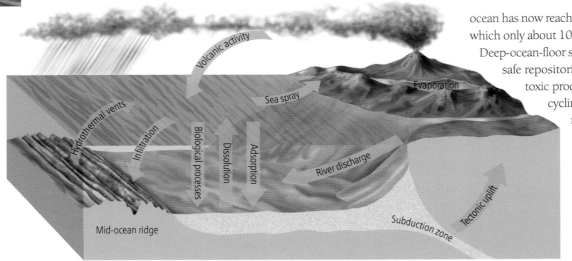

ocean has now reached around 6 million metric tons per year, of which only about 10 percent is from natural seeps.

Deep-ocean-floor sediments are currently being considered as safe repositories for radioactive wastes and other highly toxic products. But research shows clearly the active cycling of water and chemicals through the sediments and so the potential for rapid leaching from buried waste. In fact, the sudden influx of radioactive materials into the oceans from the testing of nuclear bombs during the 1950s and 1960s provided oceanographers with a series of chemical tracers that could be used to study the rates of natural processes. The swift dissemination of such materials led to two important findings: first, that there is a very active interchange between the oceans and atmosphere through a sea-surface microlayer; and, secondly, that vigorous current activity ensures relatively rapid mixing throughout the oceans. The average mixing time for complete cycling through the ocean is estimated at about 1,000 years.

The carbon dioxide sink One of the fundamental exchanges between ocean and atmosphere is that of carbon dioxide. Ocean life is very sensitive to the amount of dissolved carbon dioxide in the water, where there is 60 times the quantity of the gas found in the air. Like land plants, all marine plants and also many bacteria take in carbon dioxide during the process of photosynthesis and release oxygen.

Earth, oceans, and atmosphere act as a giant carousel, recycling 100 different elements, at differing rates, through air, water, and sediment (*above*). The sediments beneath the oceans are the single largest sink for many salts, while biological processes are the most effective at locking dissolved chemicals into sedimentary particles.

Notwithstanding the variation in chemical inputs and outputs, the cycles and sinks, the global ocean is a well-mixed entity. Surface salinity shows a narrow range between 33 and 37 parts per thousand (3.3–3.7 percent), but is distinctly affected by latitude, with maximum salinity in the subtropics and minimum values near the equator and poles (*right*). These salinity variations are caused by seawater dilution, due to precipitation, river runoff, and ice melting, and concentration through evaporation or freezing.

sediment, the sheer volume of organic activity in the oceans nonetheless ensures that the accumulation represents a huge quantity of material. Other elements in seawater are removed into the sediment through inorganic processes: by adsorption ("sticking" of ions) onto the surface of clay particles as they fall through the water column; by infiltration and entrapment at mid-ocean ridges; by incorporation into manganese nodules; and by deposition as evaporites along arid shorelines. Very slowly, soft, waterlogged sediment lithifies into hard sedimentary rock. Some is drawn down into the mantle at subduction zones; most is ultimately uplifted into mountain ranges on land, so that the cycle of weathering, dissolution, and runoff begins again.

As more and more data are collected by ocean scientists, we can begin to build up quantitative measurements of the rates at which different elements are added to or removed from the ocean reservoir. It then becomes possible to calculate the average length of time an element spends in the ocean—its residence time—by dividing the total amount of the element present by the rate at which it enters or leaves. Interestingly, the results for different elements vary enormously, from tens of millions of years for chloride and sodium to only a few hundred years for manganese, aluminum, and iron. Ocean water itself has a residence time, which we can estimate from measured rates of evaporation, precipitation, and runoff, of around 3,500 years. The oceans are more than 4 billion years old, so that, on average, individual water molecules have cycled through the atmosphere and returned to the seas more than a million times since the world ocean formed.

Disposal or exchange Such facts are not just intriguing, they are also of great practical significance. As pollution of the ocean accelerates, with an ever-increasing input from human activity—lead from gasoline, mercury and other heavy metals, radioactive elements, pesticides, petroleum, plastics, sewage—we desperately need to know the effects this will have and the timescales involved. Natural oil seepage was, for millions of years, the main source of immiscible (unmixable) petroleum in seawater, most finding its way back into the sediment or being washed up along the shoreline. Oil entering the

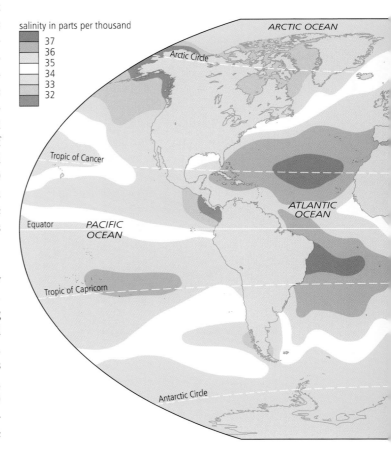

Eyes in space: satellite oceanography

Since their pioneering development in the 1960s, the use of satellites has permitted spectacular advances in oceanography. Satellites allow us to observe large parts of the ocean simultaneously, as well as to monitor specific areas repeatedly over an extended time period. An ocean research ship typically covers 450km (280mi) a day, whereas satellites can travel at 30,000km/hr (18,650mph). Depending on the use for which they are required, they can be placed in very high stationary orbits in order to gather 24-hour information on changes at any one Earth location, or in lower orbits with various inclinations to the equatorial plane, so that they circle the Earth but cover a different area with each orbit. Complete coverage of the globe then becomes possible in about 12 hours.

Satellites are now routinely used for accurate navigation at sea, and in weather observation, prediction, and warning. Ships can be advised to avoid storms or iceberg tracks, the progress of tropical cyclones can be monitored, and fishing vessels can even be directed to large shoals of fish in their vicinity. Dedicated satellites for scientific observation of the Earth, oceans, and atmosphere have made enormous contributions to the study of many aspects of chemical and physical oceanography, as well as to understanding the marine biological environment.

Remote observation from space utilizes electromagnetic radiation conveyed from the sea to a sensor on the satellite. Whereas most wavelengths are absorbed or scattered by the atmosphere, there are three main spectral windows that can penetrate it more readily—the visible waveband, parts of the infrared spectrum, and microwaves, including radar. Sensors may be either passive receivers or active emittors of electromagnetic energy, which then measure the radiation reflected back from the sea surface.

Sea-surface temperature maps can be constructed in enough detail to show the development of the notorious El Niño climatic event over the Pacific, the rapid temperature gradients that exist across oceanic fronts between different water masses, and zones of cold-water upwelling that lead to nutrient factories in the oceans. Remote sensing of ocean color, as well as of the brightness of back-scattered light, can be used to infer the concentration of phytoplankton in the upper layers, and to monitor the distribution of suspended sediment—data sets that can act as tracers for patterns of oceanic circulation. Very sensitive instruments producing Synthetic Aperture Radar images (SARs) are used to gather data on sea-surface roughness or topography. These figures can yield information on sea state, wave height, internal wave characteristics, mid-ocean tides, and currents.

The data gathered by satellite observation and experimentation are set to revolutionize the way we look at ocean space. The challenges ahead are both demanding and very exciting.

Professor Ian S. Robinson,
Southampton Oceanography Centre, UK

Satellite image of Hurricane Andrew, fully formed and menacing over the Gulf of Mexico, west of Florida. Such images, revealing a wide range of different properties of land, sea, and air, have completely revolutionized our understanding of the ocean–atmosphere couple. For the first time, they permit an almost instantaneous view of the global ocean—something that cannot be achieved through conventional sea-going research.

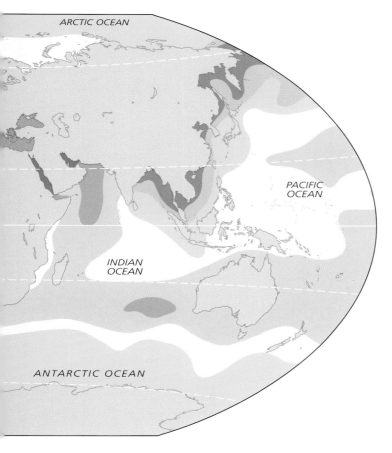

During respiration, the reverse is true; they take in oxygen and release carbon dioxide. This gas is particularly soluble in water, forming a weak acid (carbonic acid), so that, together with other natural atmospheric gases, great quantities are delivered to the oceans daily in rainfall. Human pollution, principally from burning fossil fuels, adds around 3 billion extra tons of carbon dioxide to the oceans each year. Other dissolved airborne pollutants, such as sulfur dioxide, cause great damage to freshwater rivers and lakes, increasing their acidity until they become barren of life. By contrast, dissolved carbon dioxide in the ocean has almost no effect on its acidity, due to a series of chemical reactions (a buffering system) that maintain seawater as a mildly alkaline solution. We use the pH scale (from 0 to 14) to reflect the acidity/alkalinity of any solution. Pure water is an exactly neutral solution, with a pH of 7; lower values are more acidic and higher values more alkaline. Seawater has an average value of 7.8, with a range generally between 7.5 and 8.4.

Seawater's weakly alkaline nature is beneficial for animals that construct carbonate shells and skeletons, as more acidic solutions would dissolve them. However, the greater solubility of carbon dioxide at low temperatures and high pressures makes deep-ocean waters slightly more acidic, so that they do dissolve any skeletal carbonates that drift to the seafloor. As a result, carbonate sediments accumulate in warm, shallow water but not in the deep ocean.

An almost infinite supply of solar energy reaches the Earth.
Its heat is stored in the oceans, regulating global temperature, its light
initiates the ocean food chain, and its blue wavelengths color the water.

SALT, SUN, & SEA LEVEL

Heat, Light, and Sound

KEY TERMS

PHYTOPLANKTON
SPECIFIC HEAT
WAVELENGTH (LIGHT)
VISIBLE SPECTRUM
PHOTOSYNTHESIS
PHOTIC ZONE
TWILIGHT ZONE
SOFAR CHANNEL
SONAR
DOPPLER EFFECT
INTERNAL WAVE

The remarkable chemical and physical properties of water are vital to life on Earth. Water has an important thermostatic balancing effect, preventing extreme temperature ranges over much of the planet, ensuring that the equatorial oceans don't vaporize, or the polar oceans freeze solid. It is fundamental to all life processes, and is the primary component of living organisms. The penetration of light into the upper reaches of the water column energizes the great phytoplankton factory of the Earth, and the rapid transmission of sound waves through seawater allows long-distance communication across the world.

Controlling temperature Because of its polar structure and the hydrogen bonding between individual molecules, water has a high capacity for storing large amounts of heat: one of the highest specific heat values of all known substances. As a consequence, water can absorb (or release) large amounts of heat energy while changing relatively little in temperature. Just try waiting by a stove for a pan of water to boil! Beach sand, by contrast, has a specific heat five times lower than water, which explains why, on sunny days, beaches soon become too hot for bare feet while the sea remains cool.

Solar radiation is by far the dominant source of heat energy for the ocean and for the Earth as a whole. The marked differential in solar input with latitude is the main driver for atmospheric

winds and ocean currents, and these in turn are the prime means of equalizing the polar–tropical heat imbalance. In this way, the polar oceans do not freeze solid, nor the equatorial oceans gently simmer—a bizarre concept, it may seem, but the actual heat transfer is quite astonishing. For example, the Gulf Stream transports some 550 trillion calories of energy from the offshore eastern USA northward across the North Atlantic *each second*. Such heat transfer has a dramatic warming effect on the climate of northwestern Europe. Any small change in the global atmosphere–ocean regime induced by human or natural causes can have a potentially significant impact on heat transfer and therefore on regional climate.

Ocean color For anyone who has spent time at sea, or indeed simply walking along the coast, the sea is a chameleon of color change. From space, however, the global ocean is mostly blue, save for the swirling white clouds that hover above and the plumes of brown-yellow river-borne sediment near the coasts. Why should this be so?

Sunlight shines on the sea with all its wavelengths, including the rainbow of colors that make up the visible spectrum. The longer (red) and very short (ultraviolet) wavelengths are preferentially absorbed by water, leaving near-monochromatic blue light to penetrate furthest before it, too, is absorbed. The dominant hue that is scattered back to the surface, therefore, is blue. In coastal waters, particulate matter from sediment and dissolved organic debris absorbs additional short blue wavelengths, resulting in a greener hue. The crystal lattice structure of ice behaves in a similar way to shallow, inshore waters, yielding the mystically green cliff faces of Antarctic iceshelves and icebergs.

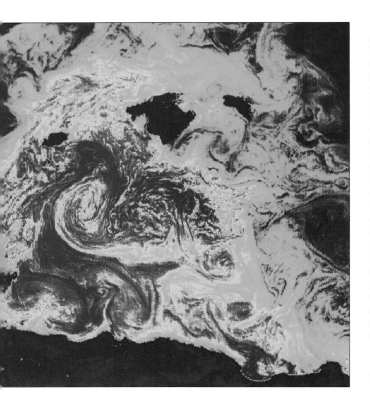

A night-time image taken by the ESA ERS-1 European Satellite of sea-surface temperatures near the Balearic Islands in the western Mediterranean. The coldest areas are shown in purple and blue—including all the land—while yellow and orange represent successively warmer temperatures. However, the most striking feature of this image is the intricate pattern of eddies, created as cooler and warmer waters spiral around each other. Eddy structures ranging in size from less than 10km (6mi) to nearly 100km (60mi) cover the entire sea area.

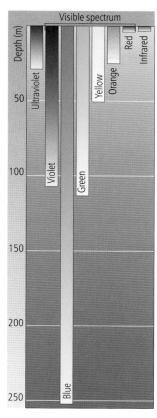

very efficient because the sound is trapped within the layer. Even moderate amounts of energy can be transmitted over great distances, while larger field tests have resulted in sound waves traveling for 25,000km (15,500mi) from Australia to Bermuda.

Experiments have been proposed that would use measurement of sound velocity in the SOFAR channel over a long time-period to infer trends of change in ocean temperature linked to climate warming. As sound travels more rapidly through warmer water, fine differences in temperature can be calculated by recording the time taken for sound transmissions to reach receivers around the world. However, sound propagation at sea also carries the potential for disturbance to marine life, for it now seems certain that some species of whales use this channel as a means of long-distance communication.

Oceanographers have developed an impressive array of scientific instruments for generating sound impulses of different frequencies, and of receivers for recording the signals that return. These underwater sonar devices first produced a simple record of bathymetry (ocean depth), usually as a single line beneath a ship's track. Now, however, we are in the process of mapping great swathes of seafloor in much greater detail, pushing resolution down to the scale of sediment waves and dunes, shipwrecks, and vanished cities. Exploration geophysicists have become equally sophisticated in using sound to look through the water and into the sedimentary record beneath the seafloor.

We can also use sound to investigate ocean currents. The Doppler principle—the apparent change in the pitch of sound waves due to the relative motion between the source and the receiver—is now used in sophisticated Acoustic Doppler Current Profilers (ADCP). These devices yield measurements of current velocities accurate to within 1cm (0.4in) per second for as many as 128 different depth slices through an ocean current up to 1km (0.6mi) thick, gradually building up a profile. Such studies have greatly improved our understanding of otherwise "invisible" subsurface waves, which travel along the boundaries between water masses of different densities; they have also been used through Arctic ice floes, and to study the distribution of zooplankton in the surface layers of the open ocean.

Water transmits sound and absorbs light. While these Short-finned pilot whales (*far left*) "chat" to each other as they swim along in the sunlit surface waters, the eerily beautiful song of the lone Humpback whale can be heard over thousands of kilometers through the deep ocean. Of the visible light spectrum (*left*), only blue light can penetrate much more than 100m (330ft) below the surface before it, too, is absorbed; the result is a mainly blue ocean.

More significant than the color of light is the energy it provides to feed the planet. At the base of the ocean food chain are the microscopic phytoplankton which utilize sunlight for photosynthesis. But sunlight of all wavelengths is rapidly absorbed, so that even in the clearest open-ocean waters photosynthesizing organisms cannot live below about 100m (330ft). This upper layer of the ocean, the photic zone (that penetrated by light), may be as little as 20m (65ft) deep in nearshore waters. There is a twilight zone beneath it, very dark and gloomy but with the barest hint of energy from the Sun, penetrating 250–750m (800–2,500ft) below the surface. Beyond that all is pitch black, except for the bioluminescence generated by some rare and remarkable creatures of the deep.

Underwater sound Marine mammals communicate with sound, while some crustaceans such as pistol shrimps are known to click their claws, and cod can grunt. Most fishes have lines of sensors along their bodies that pick up vibrations from sound and movement, although they cannot generally create sound themselves.

The speed of sound in seawater is about 1,500m per second (3,355 mph), almost five times faster than in air. It is still faster where the water is warmer or more salty, and shows a slow but steady increase with depth (related to increasing water pressure). But at around 1,000m (3,300ft) depth there is a zone of minimum sound velocity, known as the deep-sound channel (or SOFAR—for *sound fixing and ranging*). Sound generated in this channel is focused by refraction from above and below, with correspondingly little loss of energy due to dispersion. Although the speed is relatively slow within this zone, the transmission is

The speed of sound in water varies with changing temperature and pressure (*below*). At a depth of around 1000m (3,300ft) there is a zone of minimum velocity (the SOFAR channel). Sound waves bend inward, becoming focused and able to travel very long distances. Sound refraction also creates a "shadow" zone in shallower waters.

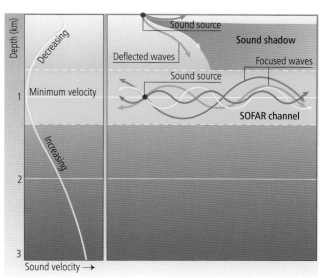

Far from being a single well-mixed bathtub, the oceans are arranged in a hierarchy of ever-changing layers, each with its own physical or chemical properties that are slightly different from those above or below.

SALT, SUN, & SEA LEVEL

Ocean Layers

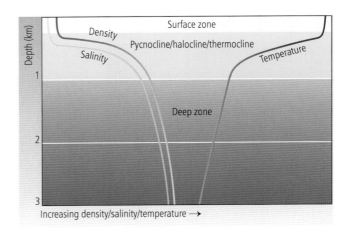

Increasing density/salinity/temperature →

From top to bottom, the ocean is organized into layers, but these layers, like the ocean itself, are dynamic and changing. As well as showing marked variation across the surface of the globe, the physical and chemical properties of the ocean—salinity, temperature, density, light penetration—show strong vertical segregation. Part of this layer-cake structure is old and immutable like the ocean itself, part is constantly stirred and shaken, whereas other parts are subject to complete breakdown and cyclic change.

Ocean hierarchy Almost all properties of seawater vary with depth. Light penetration is attenuated by absorption and scattering, giving an upper photic and a lower aphotic (lightless) zone, with a more or less well defined twilight region in between the two. Absorption of incoming solar energy also heats the surface waters, although with marked variations between latitudes and seasons. This circumstance results in a warm surface layer, a transition layer—the thermocline— through which the temperature decreases rapidly with depth, and a cold, deep homogeneous zone reaching to the ocean floor. The same broad, three-tiered structure is also true for salinity,

except that at high latitudes salinity increases with depth, while in the tropics the high-salinity water remains at the surface: the transition zone is termed the halocline.

Temperature, together with pressure and salinity, controls the density of seawater, with the effect that colder, saltier, and deeper waters are all more dense. A rapid density change, known as the pycnocline, is therefore found at approximately the same depth as the thermocline and halocline. The exact location varies from about 10 to 500m (30–1,650ft), although it is often completely absent at the highest latitudes. Winds and waves thoroughly stir and mix the upper layers, even destroying the layered structure during major storms, but barely touch the stable waters of the deep zone which make up some 80 percent of the global ocean.

This deep realm is made up of different water masses—bodies of water with distinct physical characteristics—generated at

The ocean is many-layered, with temperature, salinity, and, thus, density, varying with depth (*above*). The most rapid changes take place just below the mixed surface waters in a transition zone known as the thermocline (temperature), halocline (salinity), or pycnocline (density).

Freshwater icebergs float on the denser Arctic Ocean (*left*). The polar waters are the coldest and densest produced, and sink to the very bottom of the ocean pile.

Density stratification in the ocean creates a series of cascades, as billions of tiny particles, carried from the land or due to biogenic productivity, drift to the deep seafloor (*right*).

high latitudes by the cooling and sinking of polar waters, which then spread out at depth towards the equator. Their properties differ depending on just where they were formed (in the Arctic, Antarctic, Labrador, or Greenland Seas, for example) and how much subsequent mixing has taken place. The coldest, most saline waters are densest, and find their density equilibrium at the very bottom of the ocean pile.

But the layer that is perhaps of greatest importance is also the most difficult to study, for it is a microlayer at the sea surface only a few millimeters thick. It is through this thinnest of horizons that 70 percent of the Earth's solar energy is absorbed, most of the water vapor, carbon dioxide, and oxygen are exchanged, and enormous volumes of particulate matter and pollutants are passed—with consequent influence on the ocean–climate link and on the whole of marine life.

Suspension cascading One of the intriguing results of such a multi-layered ocean is its effect on the transport and settling of fine particulate matter through the water column. The microscopic skeletons of plankton often have such a delicate porous or open-network structure that they settle only very slowly indeed. Wind-blown dust and riverborne clays are equally fine, and can linger for long periods at the surface before beginning their slow descent.

As this constant but dispersed deluge of material rains downward, it progressively meets water layers of greater density and becomes held up temporarily at the interface between layers. Dispersion along these horizons may then take place in the slow drift of water masses.

The result is a series of mist-like cascades of suspended material between density layers within the ocean, in the course of which individual particles can be displaced many tens of kilometers laterally from their surface starting point before they finally come to rest on the seafloor. The ocean margins are blanketed with thick deposits from these suspension cascades. In the deeper parts of the ocean basins, much less material is delivered in this way, while many of the skeletal particles are dissolved by the aggressive chemical action of deep waters before they reach the floor. Deposits of the open oceans, therefore, are typically thin

and accumulate with painstaking slowness—sometimes only a few millimeters every thousand years.

Stagnant pools Another effect of layering can be to isolate the deep sea from the well-mixed, airy surface layer. Thriving blooms of plankton at the surface contribute tons of dead organic matter to the deeper waters. The decomposition of this material as it sinks, by the action of oxygen-breathing bacteria, can seriously deplete dissolved oxygen, leading first to an oxygen-minimum layer—a mid-water zone of depleted oxygen—and eventually to a completely anoxic basin. Such stagnant basins, or even pools within basins, often occur in marginal seas (semi-enclosed basins adjacent to continents), such as the Black Sea, the eastern Mediterranean, and the Gulf of California, but are less common in the open ocean. When they do occur, organic matter settles on the seafloor and accumulates as carbon-rich sediment—a critical sink within the global carbon cycle, as well as a future source of oil.

Over the past million years there have been at least ten episodes of stagnation in the eastern Mediterranean, each yielding a thin layer of black, carbon-rich sediment (black shale, or sapropel) across the whole basin. In the more distant past, as for example at the end of the Silurian and Permian periods, much larger parts of the global ocean appear to have reached such low levels of oxygen that widespread black shales accumulated during intervals known as ocean anoxic events. But the ocean is inherently unstable, and such stagnation cannot persist for long before its waters are overturned and oxygen is once more introduced into the deep. Layers of black shale are therefore interbedded with layers of pale-colored, carbon-poor sediment.

Satellite view (looking southwest) of the Black Sea, the Bosporus, and the Sea of Marmara; the eastern Mediterranean lies toward the top of the picture, the Black Sea to the bottom right. Such semi-enclosed, marginal seas, dominated by local climates, are subject to pronounced layering. In the Black Sea, rainfall and river discharge create a buoyant surface layer of low-salinity, oxygen-rich water, while summer heating produces a distinct thermocline. A sharp density gradient (pycnocline) results, isolating the surface water from the deeper water, which is completely anoxic and almost devoid of life.

River sediments

Windblown dust

River plume

Plankton bloom

Shelf-edge sediment

Detachments from current

Suspension cascade

Turbidity current

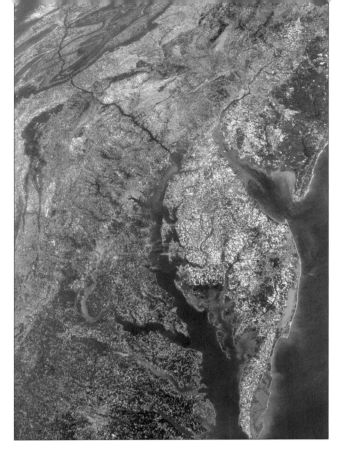

As far back in time as we can trace, the height of the sea relative to land has changed. Such sea-level oscillations are driven by a complex interplay of volcanic activity, plate movements, and climate.

SALT, SUN, & SEA LEVEL

Sea Level Rise and Fall

KEY TERMS

TECTONIC UPLIFT
SUBSIDENCE
TRANSGRESSION
REGRESSION
GREENHOUSE EFFECT
ICEHOUSE EFFECT
ICE AGE
INTERGLACIAL PERIOD
ISOSTATIC REBOUND
MILANKOVITCH CYCLE

One of the great unsolved puzzles of a few hundred years ago was how to explain the remains of what were clearly marine creatures found in rocks many meters above sea level. "Sports of the devil," they were called at a time when Biblical creationism held sway. The weight of evidence demonstrating that onetime seafloor can be found on land is now, of course, overwhelming. Marine rocks and their fossils are commonly exposed in cliffs along the coastline—for example in the chalk cliffs along the English south coast—or even on the tops of mountains; the floor of the ancient Tethys Ocean can be seen some 2,000m (6,500ft) up in the Troodos Mountains of Cyprus. Even so, such dramatic changes to the Earth as we know it are not easily understood, nor do they seem intuitive to most people.

Sea level variation and climate change

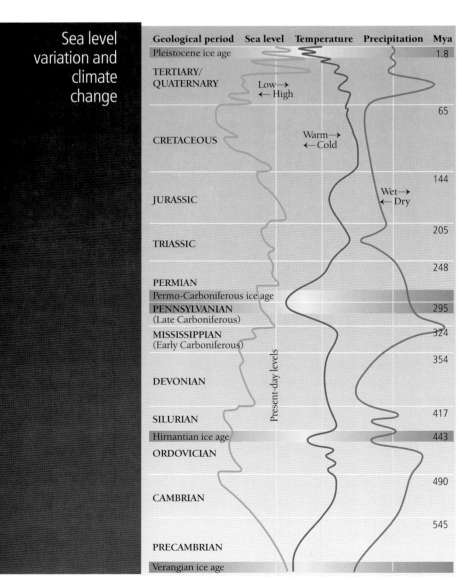

Geological period	Sea level	Temperature	Precipitation	Mya
Pleistocene ice age				1.8
TERTIARY/ QUATERNARY	Low→ ←High			
				65
CRETACEOUS		Warm→ ←Cold		
				144
JURASSIC			Wet→ ←Dry	
				205
TRIASSIC				
				248
PERMIAN				
Permo-Carboniferous ice age				
PENNSYLVANIAN (Late Carboniferous)				295
MISSISSIPPIAN (Early Carboniferous)				324
				354
DEVONIAN				
SILURIAN				417
Hirnantian ice age				443
ORDOVICIAN				
				490
CAMBRIAN				
				545
PRECAMBRIAN				
Verangian ice age				

Signs of past upheaval Raised beaches with sea-worn pebbles and broken shells, wave-cut marine terraces and lithified sand dunes lying at heights of a few to many tens of meters above present sea level have been identified by geologists in many parts of the world. In contrast, some of our most impressive coastlines are the sea-drowned remnants of former rivers or glacially carved valleys. Archaeologists who venture out to sea have discovered the remains of ancient cities almost completely buried by sediment several kilometers offshore. Evidence of human and animal migration between Alaska and Siberia, or between Britain and continental Europe, points clearly to the former existence of land bridges across what are now the Bering Sea and the English Channel.

Such signs of past upheavals clearly reflect more than just the daily rise and fall of tides or the occasional incursion inland of storm floods and tsunami. Some other explanation is warranted. Mean sea level itself must have risen and fallen, or else mountains have been uplifted and land has subsided, or land masses have moved. But which of these explanations is valid? What has caused the changes to occur, and to what extent?

The full answer to these questions, like much of ocean and earth science, is extremely complicated, with many interacting variables, and oceanographers are still working to unravel the full story. It is not the intrigue of scientific enquiry alone that drives research forward, but also the fact that the rise and fall of sea level is known to be very closely linked with climate change. Understanding past sea-level changes may, therefore, help explain the causes and effects of past climate change, and hence cast light on the looming problems of global warming and sea-level rise at the present time.

Although the causes of global changes in sea level (known as eustatic changes) are many and complex, it can help if we simplify our visualization of the ocean as a whole. For present purposes we can assume that it is essentially like a bathtub that has already been filled, in this case around 4 billion years ago, and with the taps now firmly closed. The water cycle operates too rapidly and continuously to effect any change in sea level. Even the addition of juvenile (new) water along mid-ocean ridges can be assumed to balance that being consumed as wet sediment is drawn back down to the mantle in subduction zones. Only if water is transferred to the land in the form of ice and snow will the ocean volume change significantly.

Sea floor spreading Ocean basins grow along the Earth's great mid-ocean ridge system. Hot molten rock rises from deep within the mantle and pours out onto the seafloor. But hot rock expands and is less dense than the cold mantle material it is forcing apart, so that the mid-ocean ridges have domed upwards to form the world's greatest mountain range. Tens of thousands of kilometers of this broad chain girdle the Earth below the oceans, displacing ocean water up and over the edge of the continents to cover the broad continental shelves we know today. However, the rate of seafloor spreading through geological time is not exactly constant. Periods of more rapid spreading result in larger-volume ridges, greater displacement of ocean water, and a rising sea level. Slower spreading, in contrast, leads to a fall in sea level.

Volcanic activity of a similar sort, releasing deep-seated mantle materials and heat, also occurs at isolated hotspots fed by mantle plumes, such as those that lie below Iceland and Hawaii today. Superplumes may tap even deeper sources close to the core–mantle boundary. Gigantic volcanic provinces are produced in short periods of geological time, piercing through ocean and continent alike. The Ontong–Java Plateau in the west central Pacific was active some 120 million years ago, the Deccan Traps poured lava over 500,000 sq km (190,000 sq mi) in southern India around 65 million years ago, and a North Atlantic Volcanic Province reached from Greenland through Iceland to Scotland about 10 million years later, forming extensive basalt plateaus 1.5km (1mi) thick.

Not only do these hotspots result in large-scale uplift or doming of the Earth's crust and a consequent rise in sea level, but they also lead to excessive emission, into the oceans and atmosphere, of common greenhouse gases. It has been estimated that the outpouring of 1,000 cu km (240 cu mi) of lava is associated with the emission of 16 trillion kg (35 trillion lbs) of carbon dioxide, 3 trillion kg (6.6 trillion lbs) of sulfur, and 30 billion kg (66 billion lbs) of halogen gases (fluorine, chlorine, and bromine). This sort of event over a relatively short period of time must have a profound forcing effect on global climate.

A good example of high rates of seafloor spreading, combined with a peak of volcanic activity centered in the Pacific Ocean, occurred during the Cretaceous period about 120 million years ago. Independent evidence shows that mean global temperatures were at least 10°C (18°F) warmer, and sea levels may have been as much as 200m (650ft) higher, than today.

It is perhaps significant that the timing of the formation of large volcanic provinces in the past was often closely associated with mass extinctions in Earth history. The biggest such event

Satellite picture (*far left*) showing drowned coastline along the eastern United States (Washington DC is at center left). Delaware Bay (right) and Chesapeake Bay (center) were both formed as sea levels rose at the end of the last ice age. The lower reaches of river valleys and coastal lowland areas were rapidly flooded as seas transgressed inland. Sediment supply is effectively trapped within the newly formed estuaries and bay regions.

This broad marine terrace on the California coast was cut and smoothed by the past action of waves and currents at a time when sea level was considerably higher. Tectonic uplift (mountain building) along this coast has caused a relative fall in sea level, despite the global rise in sea level that followed the great ice melt 10,000 years ago.

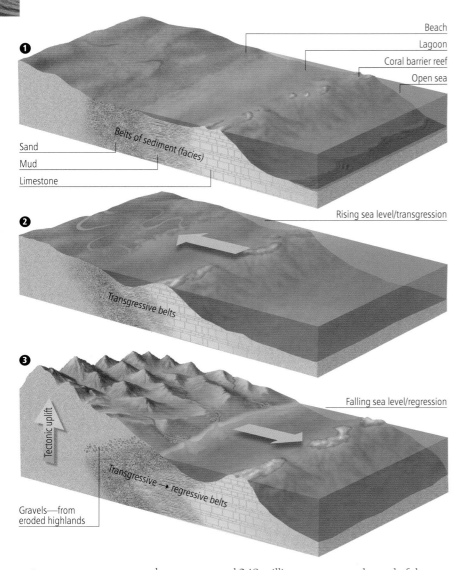

①

Beach
Lagoon
Coral barrier reef
Open sea

Sand
Mud
Limestone

Belts of sediment (facies)

②

Rising sea level/transgression

Transgressive belts

③

Falling sea level/regression

Tectonic uplift

Transgressive → regressive belts

Gravels—from
eroded highlands

Clues to past sea-levels can sometimes be found in the pattern of shoreline sediments. A static coastline (1) deposits belts of sediments (facies) according to the marine environment. Transgression by the sea moves the boundaries of these belts inland (2). Subsequent tectonic uplift causes a withdrawal of the sea, with the facies boundaries moving in the same direction as the shoreline (3).

Low-lying islands, such as Male (Maldives) in the Indian Ocean (*right*), are in danger of complete submergence should sea level rise by only 1–2m (3–6ft).

ever known occurred 248 million years ago at the end of the Permian period—a cataclysm that coincided with the eruption of the voluminous lava flows of the Siberian Traps. The eruption that created the Deccan Traps 65 million years ago was coincident with another major extinction event, the one that witnessed the demise of the dinosaurs, which in that case was possibly also linked to asteroid impact. There is little doubt, however, that sea level, global climate, and life on Earth are inextricably connected.

Tectonic effects Yet even sea levels up to 200m (650ft) higher than at present, during one of the warmest periods of Earth history, are not enough to explain the ocean-crust and seafloor sediments that have been found high in the Himalayas. In such cases, it is mountain-building and tectonic uplift that best explain our observations.

Indeed, the movement of the Earth's great tectonic plates causes both uplift and subsidence (sinking). Continental rupture and rifting, followed by subsidence, allow seas to spread over what was formerly land—today's Red Sea is one example, just as the Labrador Sea and the North Sea were before it. As mountain ranges are uplifted where plates collide, so their erosion dumps billions of tons of sediment into low-lying and

coastal areas. At first, the land builds upward and outward, and the world's great delta plains—the Ganges, Nile, Chang Jiang (Yangtze), and others—march ever seaward. There comes a time, however, when sediment overloading occurs and regional subsidence sets in. The once-mighty delta begins to sink under its own weight, and great slices at the seaward end may even detach and slide into the ocean. The Ganges delta, which makes up most of southern Bangladesh, is already very close to this point. Heavy monsoon rains, coupled with tropical storms at sea, conspired to flood more than half the country in 1955.

Many tectonic effects result in local or regional change in relative sea level; in other words, as plate movements push up the land, there is an apparent fall in sea level, and vice versa, but without influencing sea level worldwide. But mountain-building can have a much more widespread effect on mean global sea level, by influencing climate. There is considerable evidence now that the uplift of the Tibetan Plateau and the high Himalayas altered atmospheric circulation, which in turn helped induce global cooling. As more water became locked up in both mountain and polar ice caps, so global sea level fell.

Even more significant is the movement of continents with respect to one another ("continental drift") and hence the gradually changing positions and proportions of land and sea. The lack of any polar landmass during the entire Cretaceous period prevented the build-up of a permanent ice cap, and so would have kept sea levels relatively high. This circumstance augmented the "greenhouse" (warm-climate) conditions caused by intense volcanic activity and rapid seafloor spreading at the time.

It was only later, during the Paleogene and Neogene, as the supercontinent of Gondwana finally broke up, that Antarctica drifted over the South Pole and the planet very slowly began to cool. The Drake Passage opened between South America and Antarctica and severed the last connection between the southern continent and the rest of the world. The subsequent northward movement of Australia left Antarctica completely isolated in the Southern Ocean, causing the formation of a circumpolar

current that shielded the continent from any equatorial heat transfer; the big freeze set in and ice sheets began to grow, locking up huge amounts of water and lowering global sea levels. This tectonic reshuffle of continent and ocean was probably the biggest single factor in developing the "icehouse" (cool-climate) Earth conditions that then ensued—the climate mode that, despite global warming, is still in action today.

Ice age oscillations The story of sea-level change is closely bound up with climatic oscillations, at least during ice ages. However, as far as the scarce and imperfect rock record allows us to judge, the major spread of continental ice sheets across the planet has been a relatively rare event in Earth history. Certain features of the Gowganda rock series in Canada, deposited more than 2 billion years ago, are thought to indicate the very first ice age. Since that time, we can be more confident in identifying six principal episodes—very approximately at 950, 750, 650, 450, and 250 million years ago, with

the last beginning around 2 million years ago. We are still in the grip of this last ice age, although at present we are in one of its relatively brief, warmer, interglacial phases.

Geologists may decide that the current icehouse climate began with the major cooling of the southern landmass some 10 to 15 million years ago. They may also, in some future time, find a more definitive explanation for the root causes of all these episodes. But for now, the details and variation through the current ice age, during which parts of the Earth have been almost constantly icebound, have been studied extensively and are quite well understood.

The current ice age arrived quite suddenly at the end of the Neogene and with dramatic effect. The global climate began to fluctuate wildly, but with an overall cooling trend. Winds grew stronger, rain and snow increased, and the inevitable cloud build-up served to reflect more of the Sun's heat away from the Earth's surface. World temperatures dropped by up to 8°C (14.4°F), and sea surface temperatures by as much as 6°C

Fierce blizzards are part of everyday life along the Antarctic margin. At the height of past glacial episodes these seas were frozen solid as the Antarctic ice shelves expanded and polar ice caps grew to an amazing 7,000m (23,000ft) thick. Global sea level dropped by over 100m (330ft) as water was locked up in the ice. With the return to an interglacial climate, the ice melted and sea levels rose once more.

The fjordland of Norway (*right*) includes some of the world's most beautiful coastal scenery. U-shaped valleys, sculpted by the action of glaciers, were flooded as sea levels rose at the end of the last ice age, creating deep, steep-sided fjords. But the post-glacial story of Scandinavia and the Baltic Sea (*below*) is a complex see-saw between the alternating dominance of sea-level rise—caused by the melting of the Scandinavian ice cap—and isostatic rebound of the land, caused by the reduction in weight of the ice. Isostatic rebound still continues today, most markedly in the Gulf of Bothnia.

(10.8°F). Polar ice caps spread from Antarctica and Greenland in the form of huge floating ice shelves over the Antarctic and Arctic oceans, the Norwegian, Labrador, Barents, and Bering seas, and Hudson Bay. Ice sheets formed in mountain regions and then spread into temperate lowland areas of the northern hemisphere, covering half of the land area of North America and a quarter of Eurasia beneath ice 2–3km (1.2–1.9mi) thick. At times almost one third of the continental land mass was covered by ice, with thicknesses at the heart of polar ice caps reaching an amazing 6–7km (3.7–4.3mi). As the ice spread and thickened, sea levels dropped by up to 120m (400ft), and coastlines moved right out across many of the world's continental shelves, often by as much as 100–200km (60–125mi). Britain was joined to continental Europe, and Alaska to Siberia.

Fortunately for us today, as well as for the early humans that had to adapt to the severe climate, the most extreme conditions were not continuous. Within the last 2 million years there have been at least 10 major glacial periods and around 40 shorter, less severe periods. In the major periods, the ice stayed and sea levels remained low for 80,000–100,000 years, giving way to inter-glacial episodes that lasted between 10,000 and 20,000 years. These latter periods were warm, often a little warmer than today's global mean of 15°C (59°F), and sea levels were higher.

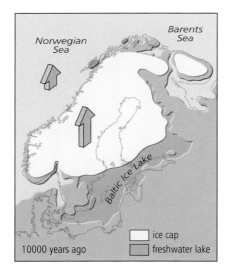

10000 years ago

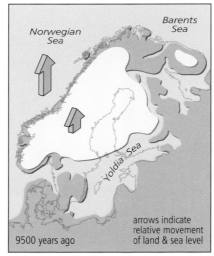

9500 years ago

arrows indicate relative movement of land & sea level

ice cap
freshwater lake

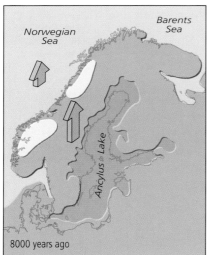

8000 years ago

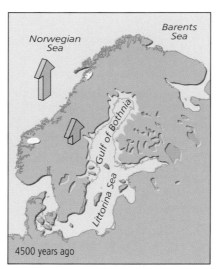

4500 years ago

Another effect of ice build-up over the continents is very significant locally in terms of relative sea-level change. It is due quite simply to the additional loading caused by several kilometers of ice. The land on which the ice forms slowly subsides, and relative sea level rises. Even more evident is the uplift that occurs when the ice melts and the depressed continent bobs up again, almost like an air mattress in water when a sunbather rolls off for a refreshing dip in the pool. This phenomenon, known as isostatic rebound, is currently taking place in Scandinavia, which was covered by a thick ice sheet until about 10,000 years ago. It is now rebounding at a rate of nearly 1m (3.3ft) per century, such that the docks of "coastal" cities built only a few centuries ago are now far inland. Isostatic rebound can be still more rapid, being close to 2m (6.6ft) per century in parts of eastern Canada.

Charting change In the 1970s, oil company geophysicists, using seismic reflection techniques to survey the continent–ocean margin, developed an important technique for recognizing past sea-level changes. Seismic profiles of the sediment layers below the surface showed high rates of sediment supply from land when sea levels were low, resulting in gradual movement of the shoreline toward the sea (progradation). A thick wedge of sediment accumulated at the foot of the continental slope, with the coarser-grained material towards the top. When sea levels were high, less material was supplied from land and a thin series of reworked sediments covered the broad continental shelves as seas pushed inland. The stratigraphic sequence was reversed, with the finer offshore sediments now on top of the coarser landward sediments.

Following analysis and correlation of data from many passive (tectonically stable) continental margins, the cyclic and global

Two very important and completely different scientific concepts have provided invaluable insight into the nature and timing of past changes in climate and sea level. Bringing together these diverse approaches has allowed oceanographers to unravel a unique record locked up in deep-sea sediment cores.

The first breakthrough was provided by a Serbian geophysicist, Milutin Milankovitch, who in 1924 proposed that the cyclic alternation of glacial and interglacial periods was the result of slow changes in the Earth's orbit about the Sun. The shape of the Earth's orbit changes from nearly circular to distinctly elliptical (or eccentric) over a 97,000-year period. This cycle of eccentricity results in long-term variation in the amount of heat received from the Sun. Furthermore, every 41,000 years the tilt of the Earth's axis moves between 24.5° and 22.1°. The greater the tilt, the more sunlight is received at the poles, and the more marked the seasons become. Finally, the Earth actually wobbles in space during its orbit about the Sun, rather like a slowly spinning top, so that the time of year when the Earth is closest to the Sun varies. The period of this change is 21,000 years. These "Milankovitch cycles" are now a well-established cause of climatic change, not only for the recent ice age but throughout Earth history.

The second important discovery was that the two different forms of oxygen (known as oxygen-18 and oxygen-16 isotopes) occur in very slightly different proportions in the water molecules of ice and of normal seawater. When large amounts of water are locked up in the polar ice caps, then the isotopic composition of seawater is preferentially enriched in the heavier oxygen-18 isotope. Quite remarkably, we can read the changes in oxygen isotope ratios that have occurred through past sea level and climate cycles by measuring subtle changes in the chemical make-up of fossil shell material secreted by

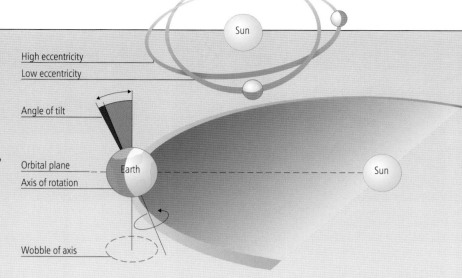

High eccentricity
Low eccentricity
Angle of tilt
Orbital plane
Axis of rotation
Wobble of axis
Sun
Earth
Sun

marine organisms. From the tiny skeletons of microplankton to the thick-walled shells of oysters or giant clams, the record is there. These organisms have used oxygen from seawater to build their protective hard parts of calcium carbonate, and so have recorded the isotopic proportions of the oceans in which they lived. By carefully picking microfossils from sediment cores and analyzing their chemistry, we have slowly built up a complete record of seawater

variation through the past 140 million years. This marine oxygen isotope stratigraphy is now a worldwide standard that tells of past climate change and sea-level variation. It compares very closely with the climate variations predicted by the theory of Milankovitch cycles.

Professor Sir Nick Shackleton,
Cambridge University, UK

Deep-sea sediments: a record of change

The Earth's orbit varies in its eccentricity, tilt, and precession (wobble) (*left*). The fluctuating amount of solar radiation received triggers changes from glacial to interglacial climates. Vital clues to past climatic conditions can be discovered in the ratio of oxygen isotopes found in the fossil skeletons of microscopic marine animals called foraminiferans (*below*).

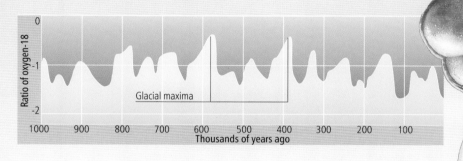

Ratio of oxygen-18

Glacial maxima

1000 900 800 700 600 500 400 300 200 100
Thousands of years ago

nature of these characteristic successions became strikingly apparent. They represented the repeated advance (transgression) of the sea across the shelves, followed by its withdrawal (regression). Over the years, from seismic reflection profiles, we have been able to construct a detailed chart of global sea-level change reaching back some 200 million years. Beyond that time, geologists have to employ a range of different techniques, so that the record becomes more generalized and speculative.

The effects of change

The effects of sea level change are many and profound. The altered shape and position of the coastline is, of course, one of the most evident, together with changing sediment input to the deep basins. Conditions within the oceans also vary—ocean currents are initiated or switched off, shelf seas are subject to more or less warming and evaporation, and the position of the carbonate compensation depth (where the amount of calcium carbonate produced is equal to that dissolved) moves up or down.

Most dramatic of all is the impact on marine life. The broad, shallow shelf seas fringing the continents provide the environ-

ment for some of the most rich, diverse, and spectacular life on the planet, and it is these areas that see the most significant changes as sea levels rise and fall. At times of high sea level, when the shelves are covered with water, they can support a wide range of marine organisms. Shelf seas also create the best production factories for marine phytoplankton and so spawn a profusion of life in the oceans. These sorts of conditions, created by long-term high sea levels, were most likely responsible for the exceptional diversity and fecundity of marine life during the Ordovician and Cretaceous periods of Earth history.

But just as high sea levels can bring renewed life to the oceans, so a marked fall can be catastrophic. The marine extinctions at the end of the Permian period, in which about 55 percent of families and 95 percent of species died out in the course of a prolonged demise that lasted some 8 million years, constituted the largest mass extinction ever to have occurred. One favored explanation is that, coupled with climate change and ocean anoxia, falling sea levels led to a major reduction in shelf space. The fossil record suggests that significant falls in sea level present a major threat to the diversity of life on Earth.

Stirred by winds that raged over some distant part of the ocean, these waves have tracked hundreds of kilometers across the sea. As they foam furiously against the shore, they exert instantaneous pressures measured in many tons per meter of coast. Small wonder, then, that continents erode as eons pass, concrete piers are destroyed by single storms, and seafarers remain ever wary of the changing face and force of waves.

SILENT, SWIFT, & STRONG

CHAPTER 4

The powerful rhythms of waves, tides, and weather

The charm of the oceans lies partly in their ever-changing moods. Not only do they present a kaleidoscope of color and motion to the delighted spectator, but they seem forever restless, embracing a spectrum that ranges from the tranquil calm of a coral lagoon to the roaring winds and monstrous waves that make the Antarctic Ocean so savage.

The pounding of waves on a rock-strewn shore, and the rhythm of tides sweeping back and forth across golden sands, are the very heartbeat of the ocean. Constant and regular since the seas first formed 4 billion years ago, they can be seen and heard by us all. But the mighty currents that stir the seas, although thousands of times more powerful than any river on land, are silent and formless, driven by hidden forces of nature. Deep below the ocean's surface, in the eerie silence of the abyss, there are waterfalls without sound, rivers without banks, and storms that rage unnoticed for weeks at a time.

The oceans and the atmosphere are intricately coupled. Their joint power creates the weather and regulates Earth's climate. Winds drive the currents that redistribute heat across the globe—a transfer of energy essential to the maintenance of a habitable world. But the climate is no more constant than its ocean regulator is simple. Over time, we swing from greenhouse to icehouse conditions with an unpredictability we do not yet understand. And now the pace of global warming, brought about by the carelessness with which we are treating our environment, is affecting even these great natural cycles of change.

Waves are as familiar and changeable as the color of the sea.
They are at once curiously hypnotic and a source of endless delight,
dangerously unpredictable and hugely powerful.

SILENT, SWIFT,
& STRONG

Riding the Waves

KEY TERMS

WAVELENGTH

WAVE PERIOD

FETCH

ORBITAL MOTION

WAVE TRAIN

WAVE INTERFERENCE

LONGSHORE CURRENT

RIP CURRENT

TSUNAMI

ROGUE WAVE

The awesome power of waves stirs a profound response in the deepest recesses of the human soul. Better than anyone else perhaps, surfers who catch a breaking wave at its crest share a sense of the rhythm and mystery of the ocean and respond instinctively to its thrill.

For those who know the sea as mariners, fishermen, leisure-time sailors, or sometime passengers, however, the endless motion of its surface is a force to be reckoned with. Whether in a disquieting calm or a reassuring swell, the potential fury of the sea is always only just over the horizon. There are parts of the world ocean that any sailor would still prefer to avoid, most

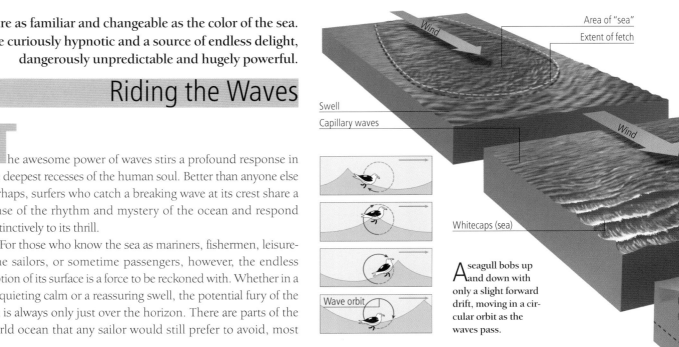

Area of "sea"

Extent of fetch

Wind

Swell

Capillary waves

Whitecaps (sea)

A seagull bobs up and down with only a slight forward drift, moving in a circular orbit as the waves pass.

Wave orbit

Americas Cup challenger, racing off Fremantle, Western Australia. The prevailing westerly wind belt in the southern hemisphere produces the highest average wind speeds and the largest average wave heights on Earth—a serious test of both yacht design and sailing skill!

notably the great swathe of Southern Ocean that encircles Antarctica, stirred by the full force of the westerlies and swept by the Antarctic Circumpolar Current. Wave heights in this region are consistently the highest of any ocean. Yet freak waves of exceptional height may strike anywhere at any time.

For ocean scientists, the challenge of understanding the physical state of the sea, and even attempting to predict the occurrence of freak waves, is as exciting as it is daunting. A stormy ocean is the ideal breeding ground for apparently random monster waves, yet others generated from this same chaos have such distinctive signatures that they can be tracked from one side of the ocean to the other. Within this realm of fuzzy logic, we are now making some scientific advance.

The Beaufort Wind Scale

Force	Knots	Km/hr	Wave height (m)	Description	Effects at sea
0	0–1	0–2	—	Calm	Sea like a mirror
1	1–3	3–6	0–0.1	Light air	Ripples with appearance of scales; no foam crests
2	4–6	7–11	0.1–0.5	Light breeze	Small wavelets with glassy appearance; crests do not break
3	7–10	12–19	0.5–1.0	Gentle breeze	Large wavelets; crests begin to break; scattered whitecaps
4	11–16	20–28	1.0–1.5	Moderate breeze	Small waves, becoming longer; fairly frequent whitecaps
5	17–21	29–38	1.5–2.5	Fresh breeze	Moderate waves, taking longer to form; many whitecaps; some spray
6	22–27	39–49	2.5–4.0	Strong breeze	Larger waves; whitecaps everywhere; more spray
7	28–33	50–61	4.0–5.5	Near gale	Sea heaps up; white foam from breaking waves blown in streaks
8	34–40	62–74	5.5–7.0	Gale	Moderately high waves of greater length; edges of crests begin to break into spindrift; foam blown into well-marked streaks in wind direction
9	41–47	75–88	7–9	Strong gale	High waves; crests of waves begin to topple and roll over; dense foam streaks; spray may affect visibility
10	48–55	89–102	9–12	Storm	Very high waves with overhanging crests; foam streaks give sea white appearance; heavy rolling; reduced visibility
11	56–63	103–117	12–15	Violent storm	Exceptionally high waves; sea covered with foam patches; edges of wave crests blown into froth; visibility further reduced
12	64+	118+	15+	Hurricane	Air filled with foam; sea completely white with driving spray; visibility very seriously reduced

When two separate wave trains overlap (*right*), they create an interference pattern. Wave height is reinforced where they are in phase and diminished where they are out of phase.

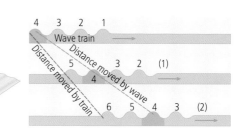

More complex wave patterns (*above center*) result from the interference of wave trains traveling from different directions. From the apparently random zone of mixing, there emerge new regular wave trains, with a modified pattern and different direction.

As the leading edge of a wave train advances (*above*), half its energy is lost in forward motion and the front wave (1, 2) disappears. The remaining energy is transferred to the rear of the train to create a new wave (5, 6) and the whole train moves forward—at half the speed of any individual wave (4).

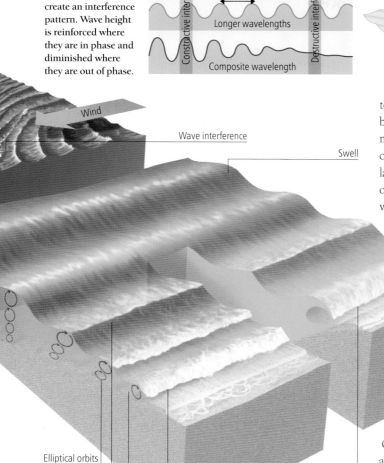

Waves develop (*far left*) as winds blow across an area of sea surface. Tiny capillary waves are the first to form, gradually building to chaotic seas in the area of fetch where wave interference is dominant. Beyond the fetch, a smooth ocean swell develops as waves separate and sort themselves into similar groups. When they reach shallow water the waves "feel" the bottom, as the circular motion of the water particles is interrupted. The waves are slowed down, building in height until, unstable, they spill over in the surf zone as breakers.

to pebbles, and pebbles to sand. Individual storm waves have been measured exerting instantaneous pressures of up to 30 metric tons per square meter (6,000lb per sq ft). Concrete piers can be removed by a single, massive wave; whole houses and villages can be washed away in the course of a major storm. In open waters, even large ships that are caught broadside by such waves will be tipped over, and those that struggle to face the wave head on can be tossed head over heels like matchsticks, their mighty, steel-clad backs snapped like kindling for a fire.

The rate at which energy is transferred across the ocean is the same as the velocity of the wave. The exact figure can be determined by a simple equation in which the wave velocity of deep-water waves —the speed at which they travel across the ocean—equals wavelength (measured as the distance between crests) divided by the wave period (the time interval between waves). On average, successive waves are 60–150m (200–500ft) apart, and the interval between their crests is 10–15 seconds. It follows that waves typically travel at speeds of 20–35km/hr (12–22mph), and that waves with a greater wavelength will travel faster than those with a shorter wavelength.

Energy in motion

Waves are among the most familiar features of the ocean. We have almost all watched a large wave advancing on the shore, waiting to see it crash and spill against a barrier of rock or sand. Yet waves are not what they appear. As a wave travels across the ocean surface, it is transferring energy, rather than actual water, until the moment when it reaches land. Mechanical energy, created by the original disturbance that caused the wave, travels through the ocean at the speed of the wave, whereas water does not. Individual molecules of water do move, but quite slowly and in a generally circular motion—back and forth, up and down. This phenomenon can be observed quite clearly by watching a floating bottle or a resting seagull, which will bob up and down as the waves pass, moving only very slightly forward with each one.

Most waves are created by the wind, forming far out at sea. The greater the force of the wind, and the further it travels, the bigger the wave will be, and in consequence the more energy that will be stored within its bulk and released when it eventually breaks. The sheer amount of energy involved can be quite awesome. Over long periods of time, whole coastlines retreat before the pounding of the sea—cliffs crumble, rocks are worn

Sea and swell

Wind-generated waves form on the sea surface by the transfer of energy from the wind to the water. The smallest waves, formed by a gentle puff of wind that is barely sufficient to break the surface tension, are tiny, diamond-shaped ripples sometimes known as "cats' paws." With increases in wind speed and the length of time for which the wind blows, the wave height, period, and length all also increase. The distance over which the wind blows, known as the fetch, is also a crucial factor in influencing the growth of waves—the greater the area of ocean over which a storm blows, the larger and more powerful are the waves generated. The Beaufort Wind Scale, which defines wind force on a scale of 0 (for glassy calm) to 12 for hurricanes, gives a rudimentary guide to the likely condition of the sea, but a more rigorous scale involving speed, duration, and fetch is required to define the resulting wave state more accurately.

Only some waves survive the hurly-burly of early formation to build into maturity. These waves play out their lives in three stages: sea, swell, and surf. The state of random choppiness, building to larger but still irregular waves with no systematic pattern, that develops in the region of fetch is known simply as

"sea." As the waves leave the region where they were generated, the longer ones outpace the shorter because their velocity is greater. Gradually, they fall in with other waves traveling at similar speed—where different waves are in phase they reinforce each other, and where out of phase they are reduced. Eventually, a regular pattern of high and low waves (or swell) is developed that remains constant as it travels out across the ocean.

One of many remarkable properties of the ocean is its efficiency at transmitting energy. Ocean waters offer so little resistance to the small orbital motion of water particles in waves that individual wave trains may continue for hundreds or thousands of kilometers. They decrease in height a little as they lose some energy, but the same regular pattern remains. Major storm waves generated in the Antarctic Ocean can take nearly a week crossing the Pacific before they break as foaming surf along the shores of Hawaii. Those that miss the islands travel on for another three or four days to wash ashore, subdued, on the cold, remote beaches of Alaska. Quite uncannily, the original pattern of swell that first left the Antarctic Ocean remains intact, providing each wave train with its own unique signature.

Tsunami: swift and stealthy

Japan has a long history of being struck by exceptionally large and devastating waves, with reliable accounts dating back nearly 1,500 years. Understandably, the name for these waves —tsunami— has a Japanese derivation, from the words *tsu* and *nami* for "harbor wave." Legends and children's stories, poetry and great art have all been inspired by their awesome beauty. Somewhere in the world, a tsunami with destructive power will strike on average once a year— but just where and when is still terrifyingly unpredictable.

Tsunami (incorrectly called "tidal waves") are unusual waves of such extreme wavelength—up to 200km (120 mi)—that they effectively behave as shallow-water waves even in the deepest parts of ocean basins. Their origin is seismic, caused by subsea earthquakes creating vertical movements of the seabed along faults, or by giant submarine slides, or else by a sudden displacement of the ocean surface. This can result from a major landslide or debris avalanche into the sea, or from massive icebergs falling from the edge of a glacier or ice-cliff, or from major explosive volcanic flows into water.

Whether the principal movement is on the seabed or at the surface, it is the enormous volume of water displaced that gives tsunami such devastating power and velocity. They travel radially away from their point of origin at a speed of around 750km/hr (480 mph)— similar to that of a long-haul passenger jet. An earthquake off Colombia in 1979 generated a tsunami that traveled clear across the Pacific Ocean, reaching Hawaii in 11 hours and the north coast of Japan just 8 hours later. From the Aleutian Trench in the North Pacific, another hotspot for submarine earthquakes, both Japan and Hawaii are only about 5 hours away by tsunami.

Remarkably, the passage of such high-velocity, high-power waves may go almost completely unnoticed by most of the ships, large and small, in their path. For, in the open ocean, seismic waves have a very low ratio of height to wavelength, combined with a very long period (5–20 minutes). A ship in open waters that encountered a tsunami with a period of 16 minutes would rise imperceptibly for 8 minutes to a crest only about 0.6m (2 ft) in height, and then sink equally slowly to a trough lagging 8 minutes behind. Unlike rogue waves, tsunami are neither feared nor noticed on the open ocean.

It is only when tsunami approach land that the wave parameters change rapidly. The velocity slows and the wave height increases dramatically, reaching anything up to 30m (100ft) in a very short period of time. By this point the tsunami is both dangerous and deafening. First the frontal trough approaches, drawing water away from the shore, often to well below the level of the lowest ebb tide, exposing long-forgotten wrecks and other mysteries. Then the hissing and clattering begins as the full force of the mighty wave sweeps forward, rushing onshore as a surging flood rather than the huge, plunging breaker of popular folklore. Everything is caught up in this amazing display of the sea's unforgiving power. Mighty steel ships are tossed hundreds of meters onshore, trees and houses snap like matchsticks. And the nightmare continues, for tsunami mostly arrive in groups of between 5 and 10 great waves over a period of one or more hours.

There is little chance of escape for those caught unawares, and stories abound of suffering and loss of life. Tsunami have even changed history, for it was one generated by the eruption of the Aegean island of Santorini in 1500 BC that helped destroy Minoan Crete. Similarly, the biblical story of the Flood has been linked to geological evidence of a tsunami that inundated the fertile plains of the Tigris–Euphrates basin some 3,000 years BCE.

Professor Kazuo Amano, Ibaraki University, Japan

Bewildered citizens survey the destruction of their homes and livelihoods at Kentappu village, Japan, in the aftermath of a horrifying tsunami that struck in 1952.

Giant plunging breakers are among the most sought after by surfers across the world. Just as the surfers have traveled the globe to experience the challenge, so the waves have crossed an ocean to find land. The enormous store of energy carried in the wave is suddenly dissipated as it meets the steeply sloping beach, the water particles curling over to form a crest. Spilling breakers, which result from a more gradual beach slope, last longer, offering a more extended, if less exciting, ride.

has some basis in reality, although the exact periodicity and number of larger waves is, in fact, random. Refraction (or bending) of waves around and toward coastal headlands leads to a convergence of energy and hence extreme battering of the shoreline. Divergence and calmer waters occur over deep, hidden canyons in the seafloor and in sheltered bay areas.

Very often, waves approach obliquely to the coast and set up a significant transfer of water and sediment along the shoreline. The longshore currents so developed can be very powerful, removing beach sand and building spits and bars out across open bays or the mouths of estuaries. The build-up of water and energy in longshore currents can overcome the power of incoming waves at an irregular spacing along the coast, leading to the development of strong, seaward-moving rip-currents, which are especially dangerous to swimmers and particularly important for the offshore transport of sediment.

Catastrophic waves Day after day, the energy stored and transmitted by normal, wind-generated waves for thousands of kilometers across the ocean, and the power released as they break along the coast, represent a phenomenal redistribution of solar energy across the surface of the Earth. Even more awesome and terrifying to encounter, however, are the truly gigantic individual waves known as rogue (or freak) waves and tsunami—also called seismic ocean waves.

Rogue waves cannot be predicted. They strike erratically, often in mid-ocean, and then disappear without trace. They occur when two or more storm-wave crests merge, or when a series of storm waves encounters an opposing current. When the speed of the current exceeds 4 knots (8km/hr, or 5mph), opposing storm waves can suddenly burst upward to four times their original height, hurtle forward, and break at sea under their own instability. Examples are well-known from many of the world's wilder ocean waters. Freak waves have frequently been reported from the Agulhas Current off South Africa's Wild Coast, where shipwrecks litter the seabed, as well as from the Kuroshio Current off Japan and the Gulf Stream in the North Atlantic.

Particular climatic conditions that lead to higher than normal storm waves include hurricanes, typhoons, tornadoes, thunderstorms, downbursts, and the sudden pressure drops known as "meteorological bombs." All of these systems can whip up a frenzied wind field within an almost circular melting-pot at sea. As the wave trains form they are pushed toward the center, merging chaotically. In so doing, they reinforce one anothers' individual force to produce towering super-waves that may far exceed normal expectation in their proportions.

The surf zone As the wave train finally encounters shallow water, it moves inexorably toward the end of its life—as foaming surf upon the shore. Shallow water for waves means about half their wavelength—say 40m (130ft) for a wavelength of 80m (260ft)—for this is the maximum depth below the surface at which water molecules still orbit in response to the passage of the wave. At this depth the wave first feels the bottom and begins to slow down in response to frictional resistance—everything except the wave period begins to change. Wavelength decreases as the waves slow. The crests tend to bunch closer together and wave height increases dramatically, until the wave becomes unstable and topples forwards as surf. Violent, plunging waves, with that characteristic tube of air that daredevil surfers love to ride, form when large waves approach a steeply sloping bottom. A more gently sloping bottom generates a milder, spilling wave.

The surf zone, as well as the whole region of waves in shallow water, is complex and enthralling. This is where all the energy stored for so long is finally unleashed—where fun and fear run hand in hand. While most energy is dissipated in the power of the surf, some is reflected directly back toward the incoming train. This interference can serve either to build or to nullify individual waves. Together with the inherent pattern of the incoming train, wave interference lends the incoming surf its characteristic beat—calmer seas followed by a group of higher waves. The common myth that "one in seven waves is a big one" therefore

Satellite imagery can now measure irregularities in the sea surface from areas of high and persistent waves. The image (*left*), taken by the *TOPEX/Poseidon* satellite, shows the sea height in the eastern Pacific during the 1997 El Niño event. A tongue of warm water (white) has built up off the coast of Central America and northwestern South America where thermal expansion of the water has increased sea level by as much as 20cm (8in) above normal.

The pull of the Moon and Sun controls the tides—the daily pulse of the ocean. This same beat is felt by life at sea and, strangely, the Moon also affects us; we are lightest beneath a full or new moon.

SILENT, SWIFT, & STRONG

Rhythm of the Tides

Different peoples and places in the world experience quite different tides. Julius Caesar was born and brought up by the almost tideless Mediterranean Sea, and this fact almost cost him the conquest of Britain. When he arrived off the Kentish coast to invade England in 55 BCE, being quite unaware of the existence of tides, he moored his ships at the peak of a 6-meter (20-foot) spring tide with a strong following wind. After a short reconnaissance inland, he returned to find the entire fleet left stranded high and dry on the tidal flats, and his army was forced to face a ferocious onslaught from the Britons.

Although the tidal beat may differ from place to place, it is always present. The daily rise and fall of sea level has been a feature of planet Earth since the oceans first formed 4 billion years ago. The tides' influence is most keenly felt by organisms living in those shoreline habitats that are alternately submerged and exposed at least once each day. But life offshore and in the open oceans also uses the tides to synchronize certain behavior patterns, especially reproductive cycles.

We may be far less conscious of these influences on land, but there is little doubting the Moon's effect, even on humans. Two-thirds of our body weight being water, we are lightest when the Moon is either very new or full, and heaviest at the first and third quarters, when it is either waxing or waning.

Low tide on the Washington coast, east Pacific Ocean (*above right*). The rise and fall of tides across the world is a constant rhythm that has not missed a beat for 4 billion years.

Sun and Moon duet **F**rom the earliest times people have struggled to explain the cause and rhythm of tides, and many fanciful ideas were sported over the years. However, it was not until Isaac Newton's brilliant exposition of the fundamental laws of physics and of gravity in the mid-17th century that the riddle of the tides was finally solved.

Tides are created by the gravitational pull of the Moon and the Sun and by the rotation of the Earth and the Moon. Gravity acts as a force of attraction between everything in the universe, the force exerted by an object being greater the larger its mass and the closer its proximity. Gravitational collapse of the Solar System is prevented, however, by a centrifugal (or inertial) force that counterbalances gravity and tends to push outward as a result of planetary rotation and orbit.

These two forces are the primary key to daily tides. Because the ocean waters are mobile with respect to the solid Earth, lunar gravity is more easily able to pull the water toward it, creating a very slight bulge on the side of the Earth nearest the Moon. On the opposite side, centrifugal force is the greater and so pulls water away from the Moon, creating a second bulge. These opposing bulges are the high tides. In between, where the water has been pulled away, are the areas of low tide.

Since the Earth spins on its axis every 24 hours, each point on it should pass through each of these opposing bulges, thus experiencing two high tides and two intervening low tides in a daily cycle. Indeed, this is very nearly true in most of the world. But friction slows the movement of water a little, while the Moon's orbit around the Earth is not quite in phase with the Earth's spin, gaining by 50 minutes each day. The consequence is that any spot on Earth will be 50 minutes into the next day before it is in line with the Moon again and so experiencing high

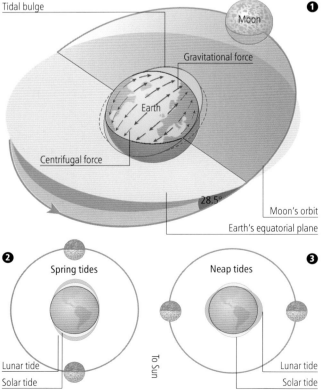

Tidal bulge

Moon

Gravitational force

Earth

Centrifugal force

28.5°

Moon's orbit

Earth's equatorial plane

❷ Spring tides

Lunar tide

Solar tide

To Sun

❸ Neap tides

Lunar tide

Solar tide

❶ **O**cean water is drawn toward the Moon by gravitational attraction. At the same time an opposing force due to planetary rotation creates a second bulge on the side of the Earth away from the Moon. As the Earth spins on its axis, so progression of these bulges causes the daily rise and fall of tides. Being so much further away, the Sun's gravitational attraction has a less pronounced but still important influence on the height of tides. When the Sun and the Moon are in line, then tides are at their highest (spring), and when at right angles at their lowest (neap).

tide. So tides slip by 50 minutes each day, or each of the semidiurnal (twice daily) tides by 25 minutes. This fact explains why, if a vacationer starts a seaside holiday with high tide at noon, then by the end of the first week low tide will be around noon and the high tides at dawn and dusk.

But what about the Sun? It may lie some 400 times further away from Earth than the Moon, yet its mass is about 27 million times that of the Moon. In effect, the mathematics translate to the fact that the Sun exerts a gravitational pull on the ocean waters that is about 40 percent as strong as the lunar influence. The tidal bulges caused by the Sun's gravity, and the opposing centrifugal force, move around the Earth with a quite different periodicity than those caused by the Moon—one measured in solar years rather than in lunar months. And so it is that when the Earth, Moon, and Sun are aligned, yielding either a full moon or new moon in the night sky, then the tidal bulges are in phase, leading to extra high and extra low tides. These twice-monthly extremes of tidal range are known as "spring tides" because they seem to well up like a spring, and not through any connection with spring as a season. When the Moon and Sun are at right angles to each other with respect to the Earth, then the gravitational attractions are opposed and the tidal range is least, producing neap tides.

Tidal phenomena The pulse of the ocean is both compelling and constant. There is no possibility that the ocean has missed, or will ever miss, a single beat since the tidal symphony first began. Its music will play on forever, or at least until some cosmic catastrophe that we cannot foresee. Yet the reality of tides is beset with complexities and variability. The main complications arise from both frictional and topographic restrictions to the flow of water over the Earth's surface. Continents, mid-ocean ridges, oceanic gateways, and other seafloor irregularities create obstacles to tidal flow. When these complications are coupled with variation in the line of the Moon around the Earth and the effects of the Earth's rotation, it is little wonder that the calculation of precise times and ranges for tides at every point in the global ocean is fraught with difficulty.

Most of the Atlantic Ocean experiences semidiurnal tides that are approximately equal in height. Much of the Pacific also has semidiurnal tides, but one is significantly higher than the other, while the shores of the Gulf of Mexico, Antarctica, and some western Pacific lands experience only one high and one low tide per day. Tidal ranges across most of the world vary from 1–3m (3.3–10ft), although in semi-enclosed bodies of water such as the Mediterranean, Black, and Red Seas, they are almost imperceptible. Semi-restricted bays or funnel-shaped estuaries, on the other hand, serve to greatly exaggerate tidal range, the maximum recorded being 16m (53ft) in the Bay of Fundy off eastern Canada, and over 12m (40ft) along the Bristol Channel and Severn Estuary in southwestern England.

A spectacular event in some river estuaries is the tidal bore—a wall of water that surges upstream with the flood tide, increasing in speed and height as the estuary tapers. Most bores are quite small, reaching less than 20cm (8in) in height, and so are barely noticed. However, the Severn bore and the St John's River bore at the head of the Bay of Fundy are powerful wave fronts up to 2m (6.6ft) in height. Even more impressive and dangerous are the 5-meter (16.5ft) bores in the Amazon and Fuchun rivers. Historical accounts of this latter phenomenon in eastern China suggest that, prior to recent dredging, the wave front was sometimes 7m (23ft) high and traveled at over 25km/hr (15mph).

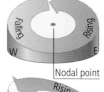
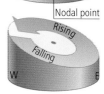

Under the influence of the Earth's rotation, a tidal bulge within an ocean basin becomes a rotary wave, with a crest (high tide) at one side and a trough (low tide) at the other. The pile of water circulates counterclockwise in the northern hemisphere and clockwise in the southern hemisphere, causing a progression of high tides along the coast. The level does not change at the center, or nodal point, of the basin.

Most coastlines experience a semidiurnal tide pattern (two high and two low tides per day). Successive high tides or low tides can be of more or less equal height or markedly different (mixed tides). Some areas have only one high and one low tide per day (diurnal). Tidal ranges are even more varied—from negligible to over 3m (10ft).

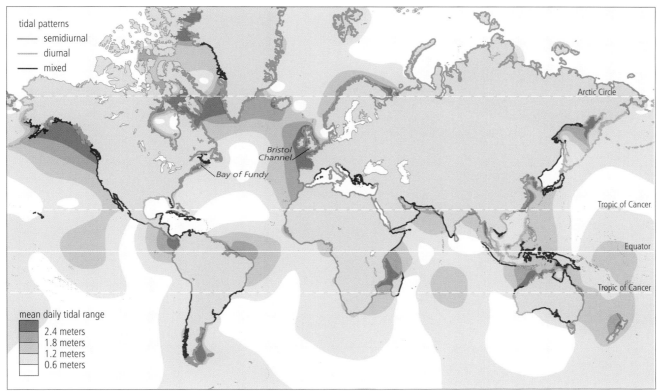

tidal patterns
— semidiurnal
— diurnal
— mixed

Arctic Circle

Bristol Channel

Bay of Fundy

Tropic of Cancer

Equator

Tropic of Cancer

mean daily tidal range
2.4 meters
1.8 meters
1.2 meters
0.6 meters

The mighty currents that stir the oceans are over 2,000 times more powerful than any rivers on land. Their flow is continuous, transporting a phenomenal quantity of water in a massive circulatory system.

SILENT, SWIFT, & STRONG

Great Surface Currents

The constant turmoil of the atmosphere, and winds that are powered by a never-ending supply of solar energy, keep the ocean waters in perpetual motion. Just as wind drag whips the sea surface into a frenzy of waves, large and small, so too does it drive great currents across the face of the ocean. Even compared to the Earth's mightiest rivers these flows of water are immense. The Gulf Stream in the western North Atlantic, for example, transports over 55 million cu m (72 million cu yd) of water every second—which is around 1,000 times the total discharge rate of the world's top 20 rivers added together. This powerful current is but one of many that stir the global ocean. It forms the narrow western arm of the great circuit of water that is the North Atlantic Gyre, with the broader, weaker Canary Current providing the return flow along the eastern boundary of the ocean.

Very early in human history, those who ventured to sea quickly learned the nature of ocean currents, their relative strengths and direction, and how they closely mirror the winds that carry them. The ways of reading the sea were mastered *par excellence* by the Polynesians over 2,500 years ago. They spread across more than 50 million sq km (20 million sq mi) of the Pacific Ocean, discovering and inhabiting even the most remote islands. There followed a great tradition of seafaring by the Egyptians, Minoans, Phoenicians, Vikings, and Chinese—all making skillful use of the winds and currents—well before the famous European expeditions of the 15th and 16th centuries.

Only much more recently, however, have we begun to understand the intimate link between currents and climate: how the great gyres transfer heat from equator to pole; how the warm Gulf Stream ensures that northwest Europe enjoys a mild climate for its latitude, keeping Norwegian and Icelandic harbors ice-free in winter months; how the icy Humboldt and Benguela currents, flowing from the Antarctic regions, cool the shores of western South America and West Africa, bringing sea fog, rich nutrients, and shoals of cold-water fish.

Mounds of water Although wind is the principal force that drives surface currents in the ocean, the actual pattern of circulation results from a more complex interaction of wind drag, pressure gradients, and Coriolis effect. Wind drag is a very inefficient process by which the momentum associated with moving air molecules is transmitted to water molecules at the ocean surface, setting them in motion. A large amount of wind energy is used in the generation of waves, and a similar amount is used to drive currents. The actual speed of a current, initially in the direction of the wind, is only about 3–4 percent of the total wind speed. The mathematics mean that a wind blowing constantly over a period of time at 50 km/hr (30 mph) would produce a water current of about 1.85 km/hr, equivalent to 1 knot, or 1.15 mph.

The second principal force influencing how the pattern of surface currents develops is that caused by the phenomenon of seawater being piled up into mounds. Such an occurrence may seem counter-intuitive but, far from being flat once the local ups and downs of waves are removed, the ocean surface is actually warped into broad mounds and depressions. Converging currents and persistent onshore winds tend to pile water up faster than it can flow away, while diverging currents result in a drawdown of water and the creation of lows.

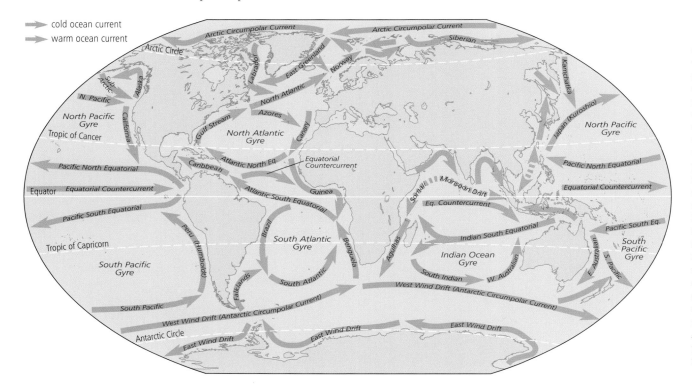

Global ocean surface circulation consists of a series of wind-driven loops, or gyres, north and south of the equator, in each of the major ocean basins. Equatorial currents are driven west by the trade winds, deflected poleward when they reach land, blown east by the westerlies that prevail at mid-latitudes, and then turned once more toward the equator. In the Indian Ocean the seasonal wind reversal of the monsoon affects current direction, while, to the south, the Antarctic Circumpolar Current flows uninterrupted.

The waters of the Gulf Stream as viewed from the Space Shuttle, looking southwest. The sun glinting on the water surface in this optical image reveals the boundary between the warm, fast-flowing waters of the Gulf Stream (lower half) with its complex eddies, and the calmer coastal waters off the eastern USA.

Traveling across the North Atlantic from New York to Africa involves a steady uphill climb over the first 1,000km (620mi) or so, and then a gentle descent the rest of the way. Not that the sea hill off eastern North America is very high—in fact, it rises only about 1m (3.3ft) above the coastal low—but the sheer volume of water involved is enough to set up a significant pressure gradient. The water pressure exerted beneath the crest of the mound is higher than in the adjacent depressions, forcing water molecules to move down the pressure gradient under the influence of gravity—just as a bucket of marbles tipped out on a gentle incline will role downslope.

The spinning Earth The third factor involved, Coriolis deviation, is more difficult to understand, but is of paramount importance to ocean circulation. It is significant in the atmosphere for determining wind patterns and direction, and also in the deep sea for the little-known currents that hug the ocean floor. The Coriolis effect is the result of planetary rotation and is felt by all objects (water, wind, birds, aircraft) that move freely above the ground. If its effect were not taken into account by pilots, then the tens of thousands of planes that criss-cross our skies each day would all land in a place different from that intended, and chaos would result.

The reason for the apparent deviation is the difference in velocity of rotation at different points on the Earth's surface. As the Earth spins on its axis every 24 hours, a point on the equator describes a circle with a larger circumference, and therefore rotates at a faster velocity, than one at middle or high latitudes. So a jet leaving from Stockholm, Sweden (latitude 60°N), and flying more or less due south to Lagos, Nigeria, is already at take-off moving eastward with the planet at a speed of 800km/hr (500mph), whereas its destination near the equator is moving in the same direction at 1,600km/hr (1,000mph). As the plane flies due south, apparently on a direct route for Lagos, it retains its original rotational velocity of 800km/hr, while the ground beneath it moves progressively faster. In a six-hour flight, Lagos will have moved 9,600km (6,000mi) to the east while the plane moves only half that distance. The net result is that, although the plane is flying a perfectly straight course, if allowance were not made for Coriolis deflection, it would miss its destination by 4,800km (3,000mi) and land somewhere near the mouth of the Amazon in Brazil! Moving currents of water and of air are affected in exactly the same way by the Coriolis effect, bending to the right of their intended path (whether north, south, west, or eastward) in the northern hemisphere and to the left in the southern.

Although the Earth spins eastward at a constant speed, the velocity of any individual point on its surface varies with latitude, from zero at the poles to about 1,600kmh (1,000 mph) at the equator. Viewed with reference to the surface of the Earth, a moving, suspended object, such as an airplane, would appear, to a stationary observer, to be deflected from its original course—clockwise in the northern hemisphere and counterclockwise in the southern hemisphere (*left*). This deflection is known as the Coriolis effect in honor of the French scientist who first calculated it.

A "hill" of water up to 2m (6.6ft) high builds up in the center of each ocean gyre. Although water tends to flow downhill, due to the force of gravity, this is balanced by Coriolis deflection in the opposite direction, creating a geostrophic current that flows around the hill.

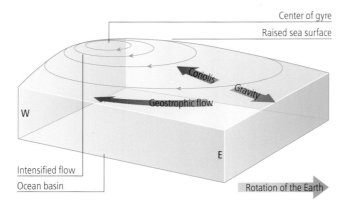

Feeding frenzy as shark packs (swirling central area) scatter a large shoal of anchovy (dark mass to either side) in this aerial photograph taken at Cape Cuvier, Western Australia (*right*). The anchovy are drawn to the Cape to feed on an abundance of plankton, the result of coastal upwelling. Cold, deep, nutrient-rich water is brought to the surface to replace the surface waters carried offshore by Ekman transport (*far right*). This new water contains nutrients recycled from the decomposing remains of dead plankton slowly sinking through the water column—a remarkable cyclic food chain, ultimately driven by the winds and currents that move water across the surface of the seas.

Gyres, loops, and rings Working in close consort, these various forces control the surface movement of ocean water. Although the movement of wind may seem random, apparently changing from one day to the next, surface winds actually blow in a very regular pattern on a planetary scale. The tropics are known for the trade winds, with their strong easterly component, and the mid latitudes for persistent westerlies. Wind drag set in motion by such large-scale wind systems sets the ocean waters in motion. The trade winds produce a pair of equatorial currents moving to the west in each ocean, while the westerlies drive a belt of currents that flow to the east at mid latitudes in both hemispheres. Confusingly, sailors and oceanographers call a current flowing in an eastward direction an "easterly" current, while landlubbers and meteorologists refer to a wind blowing eastward from the west as a "westerly"! Deflection by Coriolis force and ultimately by the position of the continents creates very large, oval circulation patterns or gyres in each ocean.

There is a mirror image of flow about the equator yielding opposing gyres in the northern and southern parts of the oceans. However, without any continental barrier between Antarctica and the rest of the world, the Antarctic Circumpolar Current continues its wild and lonely journey around and around the planet, driven by the relentless action of winds across latitudes in the 40°–60° range in the Antarctic Ocean—nicknamed the "Roaring Forties".

As the Coriolis force continues to take effect, and in so doing deflects these major, wind-generated currents, so it evokes a net transfer of water at right angles to the direction of the wind—to the right in the northern hemisphere and to the left in the southern hemisphere. This in turn creates a gigantic mound of water—only one or two meters (3–6 feet) high but thousands of kilometers long—within the central region of each ocean gyre. Water inevitably tries to flow back down the pressure gradient so produced, and as it does so is further deflected by Coriolis force. In a truly remarkable balancing act between the pressure gradient and Coriolis deflection, a stable clockwise current develops along the slope of the mound, known as a geostrophic ("earth-turning") current. The water at the center of these geostrophic gyres is eerily quiet, as the main current flow swirls around the outer ring. Within the North Atlantic gyre, for example, this central area is known as the Sargasso Sea, named after the unique brown seaweed *Sargassum* that floats in its calm water.

Because of the Earth's rotation, water in the ocean basins tends to pile up against the continental landmasses along their western margins. Here the geostrophic flows become confined and thereby intensified with respect to the broader, weaker return flows on the eastern margins. Such western intensification produces further flow instabilities, so that western boundary currents, in particular, are seen to weave and meander along their course in snakelike fashion. Tight meander loops form, sometimes pinching off completely to form large, swirling eddies or rings that spin away from the main flow, mixing into the adjacent seas.

Spiraling waters Wind drag affects the water molecules at the sea surface. The motion of these in turn creates a fluid drag on those just below the surface, and so on downward until the wind energy is fully dissipated. All the time, Coriolis deflection is driving water motion to the right of the drag effect caused by the layer above (in the northern hemisphere), so that each successive layer deviates very slightly

The oceans are vast and inhospitable, unwelcoming enough for ancient mariners to have called all uncharted waters the Sea of Darkness. There are no road markings, and the only street lamps are the stars at night. Yet all successful seafaring nations have learned to read the signs of the sea—the wind and waves, the storms and calm, and even the silent, unmarked currents that form the highways and byways of the deep.

An amazing discovery on the Indonesian island of Flores in 1998 has led archaeologists to push the first seafaring exploits of humankind far back in time, and so has revolutionized thinking on early human intelligence. The primitive stone tools found on the island could only have been brought there by one of our distant ancestors, *Homo erectus*, between 800,000 and 900,000 years ago. These early hominids had clearly mastered the art of building and sailing seagoing rafts much earlier than had previously been thought possible.

So began a seafaring tradition that continued through the ages, gradually charting the ways of the ocean. Until well into the 20th century, the principal means by which surface currents were measured was by studying the drift of ships and of debris, the flotsam and jetsam of the ocean. As long ago as 310 BCE, the Greek philosopher Theophrastus threw bottles into the sea to prove that the Mediterranean had been formed by inflow from the Atlantic. He was correct with regard to the direction of the flow, although his theory as a whole contained, along with an element of truth, rather more of fantasy.

In 16th-century Britain there was an "Official Opener of Ocean Bottles" in the royal court, whose job was to find messages sent by officers of the crown on secret missions abroad. By the 1770s, having compared the contrasting out-and-return journey times of ships crossing the North Atlantic, the American statesman and scientist Benjamin Franklin was able to publish the first-ever chart of the Gulf Stream.

Even within the last decade, we have continued to learn about current patterns from unlikely and unplanned events. Over 1,000 containers are lost at sea every year as high winds and tempestuous seas toss the ships that carry them. In 1992, a cargo of 40,000 pairs of sports shoes was lost in the Pacific from the *Hansa Carrier*.

Six to twelve months later, they began to wash ashore along the coast of North America, barnacle-encrusted and tarnished but still quite useable. Also in the Pacific, the *Hyundai Seattle* caught fire in 1994 and lost about 50 containers overboard, including some 38,000 distinctive pieces of hockey equipment. These eventually washed ashore on Vancouver Island. Plugging such data into a computer model allows scientists to continually refine our understanding of ocean circulation. The evidence from these two examples, for instance, showed that the North Pacific Gyre completes a full clockwise circulation every seven years—but irregularly, with stops, starts, loops, and eddies.

Current measuring techniques have, of course, reached a much higher level of sophistication. The two most direct methods presently employed are current meters and current drifters. An array of current meters can be attached at different depths to a fixed cable. Direct measurement of current speed and direction can then be made over a period of time at different levels within the water column. Current drifters with remote signaling capability are simply set loose in a current system and tracked acoustically, using different buoyancies to follow currents at different depths. The tracks observed in this way are highly sinuous and irregular; when compiled graphically, they are known as "spaghetti plots." Satellite imagery of sea-surface properties shows clearly that this meso-scale variation, in the form of meanders, loops, and eddies, is the norm.

Flotsam and jetsam: tracking the ocean currents

One of a load of tennis shoes lost in the Pacific Ocean, which the University of Washington tracked to study the Pacific's currents, comes to land at Prince William Sound, Alaska.

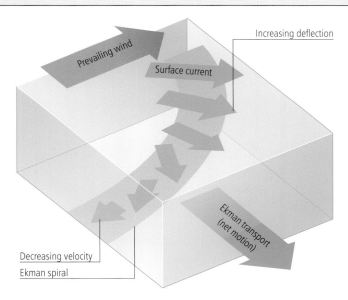

Decreasing velocity
Ekman spiral

from that above. The result is quite remarkable—a downward-spiraling flow pattern of decreasing velocity, like a whirlpool with an almost imperceptibly slow spin, in some cases extending to a depth of 200m (660ft) below the surface. This phenomenon is known as the Ekman spiral, after the Scandinavian physicist who first explained it, while the net transfer of water at right angles to the wind direction is called Ekman transport.

Ekman transport is responsible for another extremely important effect—that of coastal upwelling and downwelling. In the former case, the net transfer of surface water is directed offshore and is therefore replaced by an upwelling of water from below. These deep-sourced waters are high in nutrients, derived from the decomposition of slowly sinking organisms, and so can potentially support a great profusion of marine life. It is these rich wellsprings of productivity that support some of the richest fishing grounds of the oceans—off the coasts of Peru and Namibia, for example.

Similar upwelling occurs also in mid-ocean regions where the pattern of current flow and Ekman transport leads to divergence in the water column. High productivity belts are found along the equator and between latitudes 50° and 60° north and south. Their effects are noted in the seafloor sediment through the preservation of billions upon billions of tiny opaline skeletons, the remains of creatures that once thrived in the surface plankton. Zones of convergence between different water masses or currents in the oceans can have a markedly different effect. Where the the opposing currents differ in temperature and salinity, the convergence zone is readily visible at the surface as a distinct color change, often marked by thick fogs and great turbulence. Water is constantly sinking and rising, violent eddies swirl and mix, foam lines heave at the surface, and the sea can appear strangely menacing. It is little wonder that mariners in ancient times would return from voyages to such seas with tales of marine dragons fighting below the waves.

Wind blowing across the surface of the ocean sets in motion the layers of water beneath in a flow pattern known as an Ekman spiral (*left*). As the depth increases, so the energy given by the wind is absorbed, and the current speed decreases. Each successive layer is also increasingly deflected by Coriolis force, with the average flow at a 90° angle to the wind.

In the deep silence of the abyss there are waterfalls without sound, rivers without banks, currents without end, and storms that rage unnoticed for weeks at a time.

SILENT, SWIFT, & STRONG

Silent Cycles of the Deep

KEY TERMS

THERMOHALINE CIRCULATION

ANTARCTIC BOTTOM WATER

NORTH ATLANTIC DEEP WATER

COMMON WATER

SUBMARINE WATERFALL

MEDITERRANEAN OUTFLOW WATER

OCEANIC GATEWAY

WESTERN BOUNDARY UNDERCURRENT

SEDIMENT WAVE

ABYSSAL STORM

Hidden from view, flowing through even the deepest ocean basins, are powerful and omnipresent, but very slow-moving, currents. Linked with the great surface gyres, these currents form a vast network of circulating water that transfers energy, nutrients, and sediments around the world. They are among the least-known wonders of the ocean world, and are still poorly understood by physical oceanographers.

Driven not by the winds but by density differences linked to thermohaline factors (the joint effects of water temperature and salinity), these bottom currents are part of the global thermohaline circulation system. Subsurface water masses, with identifiable characteristics, flow above and below each other, with the denser masses of water underriding the less dense. They travel across the most spectacular seafloor landscapes—towering

The Atlantic Ocean comprises several layers of slowly moving water. Each layer or water mass has distinct salinity/temperature characteristics inherited from its source. Deep-sea transport is generated largely by formation of cold, dense water in polar regions.

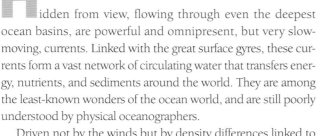

Global water mass properties

Types (depth range)	Water mass	Temp. (°C)	Salinity (‰)
CENTRAL (0–1km)	Pacific Central Water	7–20	34.1–36.2
	Atlantic Central Water	4–20	34.3–36.8
	South Indian Central Water	6–16	34.5–35.6
INTERMEDIATE (1–2km)	North Pacific Intermediate Water	4–10	34.0–34.5
	Red Sea Intermediate Water	23	40.0
	Mediterranean Intermediate Water	6–11.9	35.3–36.5
	Arctic Intermediate Water	0–2	34.9
	Antarctic Intermediate Water	2.2–5	33.8–34.6
DEEP/BOTTOM (over 2km)	Common Water	0.6–9	33.5–34.7
	North Atlantic Deep/Bottom Water	2.5–4	34.9–35.0
	Antarctic Deep/Bottom Water	-0.4–4	34.6–35.0

mountains and vast, endless plains, undulating hills and precipitous ravines. It can take hundreds of years, but gradually, along this journey, the water masses blend, and the deep water is mixed upward, returning to the surface once more.

The deep currents of the ocean realm, although slow and secret, affect 90 percent of the total volume of water in the ocean and have a tremendous effect on marine life. The dissolved oxygen they carry from the surface to the deep sea allows life to exist at all depths, while the nutrients they recycle to the surface enable the phytoplankton production that underpins the whole of ocean life. Sediment distribution is also affected; the currents carry dissolved carbon dioxide to the deeper layers of the ocean, creating the mildly acidic conditions that dissolve carbonate debris. Calcareous oozes are prevented from building up and, instead, pelagic red clays spread across the vast abyssal plains.

Global conveyor belt The pattern of deep thermohaline circulation begins in the surface waters at very high latitudes. During the long polar winters when the seas as well as the land areas are shrouded in a cloak of darkness for 24 hours a day, seawater becomes extremely cold. With never a glimpse of sunlight for weeks on end, the waters at the surface become colder and colder, making them denser.

At the same time, sea ice expands out from the continents and further increases the density of the water beneath. This happens because the freezing process incorporates only freshwater and leaves behind all the salts, making the underlying water more saline. The cold, saline, dense water sinks rapidly and spreads out across the ocean floors, moving very slowly toward lower latitudes. At the surface, currents flowing from the equator replace the waters that have sunk with warmer seas, which in turn will cool and sink. The cycle is apparently without end.

plunge beneath the warmer subpolar waters at the Antarctic Convergence zone around 60°S.

Cold bottom waters that form beneath sea ice in the Arctic Ocean are mainly trapped by the high sills (submerged ridges) that surround this isolated region. The principal deep-water source in the northern hemisphere is the winter cooling of surface waters in the Norwegian and Greenland Seas. As they spill across the Denmark Strait and through the Faeroe–Shetland Channel, they mix with a small amount generated in the Labrador Sea to form North Atlantic Deep Water. This flows southward until, somewhere in the Antarctic Ocean, it blends with Antarctic Bottom Water to form a water mass known simply as "Common Water." It is this mixture that dominates the deep Indian Ocean and much of the Pacific as well.

In order to better understand this complex pattern of thermohaline circulation, physical oceanographers track the flow of nutrients and other important substances through the global ocean. The whole dramatic overturn and cycling of water from its deep-water origin, its slow rise through ocean layers of different densities, and its eventual return, swept along by the great surface currents, to its inhospitable birthplace, is modelled as an immense conveyor belt. The engine that drives the belt is located in the cold-water factory of the North Atlantic, where cooling water sinks and flows southward. Further mixed and driven in the Antarctic Ocean, the deep waters find their way north into the Indian and Pacific Oceans. The shallow flow of warm waters from these regions forms the return loop of the belt.

One complete cycle takes about 1,000 years. Along the way a single water molecule experiences the near-freezing surface conditions of polar regions, the unimaginable dark and tremendous pressures of ocean depths, and the sunlit warmth of balmy tropical seas. But the conveyor belt never stops; the mixing of ocean waters continues unabated.

Surprisingly, there are very few locations where cold bottom waters are generated. The principal source off Antarctica is the Weddell Sea, where the coldest and densest of all water masses is produced under some of the most extreme conditions on the planet. Antarctic Bottom Water, as it is called, cascades down the steep continental slope off Antarctica and spreads out across the floor of the ocean. It streams north into the Atlantic, flows across the equator, and finally mixes upward, losing its identity somewhere east of Newfoundland. Slightly less dense waters are formed in a broader region of the Antarctic Ocean, before they

The Weddell Sea, in summer, lit by the low gleam of the midnight sun (*left*). The cold, dense, deep water masses that spread out slowly northward across the floor of the world's oceans are generated beneath the sea ice around Antarctica. A large proportion derives from the Weddell Sea, especially during the long winter months of perpetual dark.

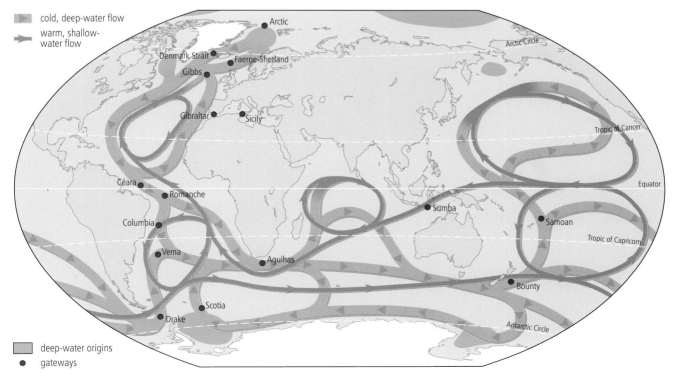

The complex pattern of deep-water circulation in the global ocean can be portrayed as a giant conveyor belt. Generation of cold, deep waters at the poles, particularly in the North Atlantic, is the main engine driving the system. As the dense polar waters sink, so they must be replaced by warm, shallow waters flowing from the equator poleward. Although these deep water flows are, on the whole, very slow-moving, localized stronger currents can occur where topographic restrictions or "gateways" channel the flow of water, speeding it up.

cold, deep-water flow

warm, shallow-water flow

deep-water origins

gateways

Hot and salty: Mediterranean and Red Sea outflows

Deep-water circulation is currently dominated by cold, dense waters formed at high latitudes. Ocean modelers simplify what is clearly an enormously complex physical process by depicting a single global ocean conveyor belt, driven from the North Atlantic and fed also from the Antarctic margin. But this is not the full story, nor has it always been the case in the geological past. At times when the world's climate was more equable, when warm seas lapped the poles and palm trees fringed the northern shores of Siberia, the deep oceans were swept by different currents.

During the mid-Cretaceous period around 100 million years ago, for example, bottom currents were warm and salty, and probably considerably weaker than at present. They formed along the margins of the Tethys Ocean, which girdled the equatorial world at that time. Intense evaporation along arid shorelines created waters that were dense because of their excess salinity. These sank and spread out over the global ocean floor, much as the cold polar waters do today.

There are still two regions of the world today where we can observe this phenomenon occurring—in the semi-enclosed basins of the Mediterranean and Red Seas. The Red Sea is the most salty of the world's marginal seas, having minimum input of fresh water, maximum evaporation by a hot, tropical sun, and an elevated topographic barrier in the south preventing easy access and flushing by the global ocean. Nevertheless, a thin, high-velocity flow of dense, saline water does escape through the Bab el Mandeb gateway and spreads out as a tongue of warm water (with a salinity of 40 parts per thousand) in the Indian Ocean. This is the Red Sea Intermediate Water, which finds its equilibrium density level at around 3,000m (10,000ft) within the Common Water mass that makes up the bulk of deep Indian Ocean water.

An excess of evaporation over rainfall and river runoff also characterizes the Mediterranean Sea, especially the eastern basin. Here, warm, salty Levantine Sea Water forms and sinks, spreading out through the entire Mediterranean and piling up in the west against the Gibraltar gateway. As the spill point is reached, so the dense, warm water is forced through the narrow straits, resulting in some of the highest-velocity bottom currents measured anywhere in the deep ocean. Beneath a core velocity of 2 and even up to 3 meters per second (5 knots), the floor of the gateway is scraped clean of sediment, leaving a bare rock and gravel pavement. Cooler, less saline water from the Atlantic pours into the Mediterranean in return as a surface inflow. As the bottom current emerges into the Atlantic, it tumbles and cascades downslope in a gigantic submarine waterfall, eventually reaching equilibrium at a depth of between 500 and 1,500m (1,600–5,000ft).

Mediterranean Outflow Water spreads across the Atlantic Ocean, and also veers northward under the influence of Coriolis force, sweeping along the continental slope off Portugal and Spain and even reaching as far north as the west of Ireland. A remarkable network of channels and drifts has been eroded and deposited under the influence of this flow in the Gulf of Cádiz. Several teams of European and American scientists are currently working to unravel this complex architecture, and to decode the sedimentary record in terms of variation in bottom current flow linked to past climatic change.

In a remarkable new high-resolution seismic profile from across the Faro Drift off southern Portugal, we can identify 10 distinct cycles of coarser and finer-grained sediment. The coarser, sandy sediments were deposited during warmer interglacial periods (like the present) when rainfall was low and evaporation intense, leading to greater outflow of warm, salty water through the Mediterranean gateway. If the sediment pile is approximately 1 million years old, then each cycle averages 100,000 years in duration, corresponding with other measures of glacial—interglacial cyclicity.

Professor Jean-Claude Faugères, University of Bordeaux, France

View from space of the Strait of Gibraltar. Concentric waves moving into the Mediterranean are formed by tidal pulses of water from the Atlantic. Beneath this surface movement, high-velocity bottom currents scour the ocean floor.

Atlantic

Cool, low-salinity water

Gibraltar sill

Warm-water tongue

Warm, dense, very saline water

Western Mediterranean

Eastern Mediterranean

Gateways and waterfalls

Seafloor topography is varied and dramatic. The oceans are compartmentalized into great abyssal basins, with floors at quite different depths depending largely on the age of the ocean crust beneath, separated by immense submarine mountain ranges. As the conveyor belt grinds on, cold bottom waters pile up behind these topographic barriers until they reach the spill point. Typically, this is at a narrow gateway that cuts across the barrier, just as a mountain pass weaves its way through insurmountable peaks on land. As the huge mass of dense water funnels through the gateway, it is severely restricted in width and so accelerates. The bottom currents through such deep ocean passageways can be highly erosive, scouring away loose sediment and even grinding into bare rock. Bottom-current velocities of 1–2 meters per second (2.5–5 mph) are quite common—a rate that is strong and fast for the otherwise peaceful world of the deep sea.

As the narrow, high-velocity bottom current enters the adjacent basin, the dense water spreads out and cascades downslope. These features are called "submarine waterfalls" because of their immense height and power, although with such gentle slopes they are in reality more akin to the cataracts or rapids of terrestrial rivers. Whatever the gradient, their scale is awesome. Cold Antarctic Bottom Water, only a fraction of a degree above freezing point (at about 0.2°C/32.4°F) piles up behind the Rio Grande Rise in the South Atlantic and then cascades down into the Brazilian Basin—a drop of over 1,000m (3,300ft). With a temperature that is still below 1.4°C (34.5°F), this cold water mass eventually reaches the equator, where it meets a further obstacle, the Céara Rise. Here it pours down into the North Atlantic Basin, to eventually give way to North Atlantic Deep Water. But by far the most impressive submarine waterfall is located beneath the Denmark Strait, east of Greenland. Here, 5 million cubic meters (6.5 million cu yds) of water cascade downslope into the North Atlantic basin every second, tumbling and gurgling, generating giant eddies and turbulent whirlpools, all in the profound silence of the deep ocean. The fall drops a vertical distance of over 3.5km (2.2mi). The Denmark Strait waterfall truly dwarfs any similar feature on land, where the tallest waterfall, the Angel Falls in Venezuela, drops a mere 1km (0.6mi). The Guairá Falls on the Paraguay–Brazil border have the largest average flow rate, greater than the impressive Niagara Falls—but even they manage only a trifling 13,000 cubic meters (17,000 cu yds) per second, almost 400 times less than their submarine counterpart.

Abyssal storms

The true power of bottom currents has only recently been realized. In addition to their acceleration through oceanic gateways and down submarine waterfalls, they are affected by submarine topography as the rotation of the Earth and deflection by Coriolis force constrains slowly moving bottom-water masses against the western margins of ocean basins. Restricted and confined purely by forces at play in the deep sea rather than by any form of channeling, the bottom waters are transformed into narrow, high-energy bottom currents. These are known as western boundary undercurrents, and it is these that effect the bulk of deep-water transport—of water, heat, chemicals, and nutrients—through the global conveyor belt. Other weaker, broader return flows are found almost everywhere else across the floors of the abyss.

Strong bottom currents flow for many thousands of kilometers, transporting enormous volumes of the ocean's finest sediment in addition to a cocktail of dissolved chemicals. Sediments are deposited as giant, elongated drifts (or mounds), or else are molded into wonderfully regular sediment waves. Rather like ripples on a stream bed, except that their wavelength is measured in kilometers rather than centimeters, such waves cover tens of thousands of square kilometers of seafloor. Exactly how they are formed from bottom currents is still an unresolved challenge for ocean scientists.

Slowly we are gathering more data on these unseen rivers of the deep. Current meters and even seafloor laboratories have been left in their path for periods of at least a year. The great western boundary undercurrent that slides noiselessly beneath the Gulf Stream off eastern North America chugs lazily with its vast load at an average speed of only some 10–20cm per second (less than half a mile per hour), in a direction exactly opposing the surface current above. But the motion is far from steady—sometimes the flow is faster, sometimes slower, with eddies that develop and peel off, at times even yielding reverse flows.

One intriguing recent discovery that came directly from a long-term seafloor experiment off Nova Scotia was evidence that bottom currents are periodically subjected to much higher than normal turbulence, experiencing flow velocities averaging four or five times the mean. These abyssal storms are in some way linked to major instabilities at the ocean surface that are in turn coupled with atmospheric conditions. They can rage for days or weeks at a time and then disappear as suddenly as they came, leaving the seafloor strewn with debris and the sparse abyssal life struggling to recover.

This high resolution seismic profile (*below left*) reveals perfectly sculpted sediment waves of giant proportion lying deep on the seafloor. The distance between adjacent crests (wavelength) is about 1.5km (1mi). Clearly defined layers of sediment in the upper 60m (200ft) of seabed revealed by this section show a steady migration upslope (right of image) and upcurrent. The bottom current that created these remarkably regular profiles must have remained constant in the same location for the past 1 million years at least.

SEE ALSO

SALT, SUN, & SEA LEVEL
Heat, Light, and Sound   70–71
Ocean Layers   72–73

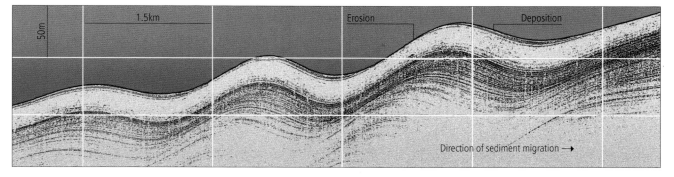

1.5km | Erosion | Deposition
50m
Direction of sediment migration →

Oceans and atmosphere are intricately linked—they drive the weather and buffer Earth's climate. But the pace of global warming today seems set to push the planet to extremes.

SILENT, SWIFT, & STRONG

Oceans and Climate

Oceans and atmosphere are closely coupled. Together they control and express both the daily drama of Earth's weather systems and the long-term changes in planetary climate. They respond as one to the influx of solar energy that bathes our planet in a never-ending supply of life-sustaining force. Winds drive the currents that redistribute heat from the equator to the poles, just as they are engaged themselves in the atmospheric heat engine. The transfer of heat energy every second of every day is staggering in its scale. Without such perpetual motion through atmosphere and ocean, the blistering heat delivered daily to the tropics would render them uninhabitable, while the rapid spread of ice sheets from north and south would exaggerate the already harsh environment of polar regions. A very different type of life would exist on planet Earth, most of it squeezed into a narrow band between these climatic extremes.

Thankfully for us, the actual situation is very different. The oceans operate as a giant thermostat that regulates global temperature and, working in partnership with the atmosphere, drives weather systems. The world's climatic zones are determined by the ocean's climatic zones—by the latent heat stored and transferred in the upper few hundred meters of our seas.

But the story of the weather is not all sweetness and light, as we all know only too well. Tropical cyclones are spawned where sea surface temperature reaches a critical point. As many as 100 of these violent storms are created every year in the tropics, each unleashing energy comparable to a massive volcanic eruption, with the potential to wreak havoc wherever they go. The El Niño phenomenon is a natural oscillation of the ocean–atmosphere system that profoundly affects the global climate over a period of a few years. Almost in passing, it so tampers with biological productivity in the Pacific Ocean that it can make or break the livelihood of South American fishermen.

There is no doubting the power and effect of the ocean–atmosphere couple on the daily weather we experience, for good as well as bad. But the system now has another player on the stage. Over the past two centuries or more, and most dramatically over the past few decades, the human species has been unwittingly contributing to climate change. By deforesting the land and creating urban sprawl, by digging up the Earth's precious store of fossil fuels and discharging their waste into the atmosphere as spent carbon dioxide, we are contributing to a unique and rapid phase of global warming. The causes and effects are still being debated. Whether we will take center stage or simply a bit part in this grand spectacle of ocean and climate has not yet been decided.

Stirring the atmosphere The envelope of gases forming the atmosphere extends an impressive 2,400km (1,500mi) above the Earth's surface, although it is the lowest and thinnest layer, only some 10–12km (6–7mi) thick, that contains 75 percent of the mass. This layer, known as the troposphere, comprises 78 percent nitrogen and 21 percent oxygen, as well as many other gases in minute quantities. It also contains virtually all the atmospheric moisture and is where the clouds, snow, rain, and winds occur. These are the vibrant parts of the atmospheric heat engine that work together with the oceans to moderate climate and create weather systems.

Three giant, spiraling circuits of wind dominate each hemisphere. They are set in motion by a differential influx of solar energy, directed by Earth's rotation and deflected by Coriolis

Gases in the atmosphere

Gas	Concentration (%)	Contribution to greenhouse effect (%)	Annual increase (%)
Nitrogen	78.084	—	—
Oxygen	20.946	—	—
Argon	0.934	—	—
Greenhouse gases			
Carbon dioxide	0.035	60	0.5
Methane	0.002	15	1
Chlorofluorocarbons	trace	12	4
Tropospheric ozone	trace	8	0.5
Nitrous oxide	trace	5	0.2

Monsoons and mountains

The location of continents within the Earth's blue oceans and the height of mountains across the continental interior have an added effect on the winds that stir the atmosphere. They bring about the seasonal reversal of winds known as monsoonal circulation, caused by the unequal heating of the low-latitude continents and tropical oceans as the Earth, with its tilted axis, makes its annual journey around the Sun. Recent research has shown that the monsoon circulation system, at least in its present intensity, was greatly enhanced within the past 15 million years by uplift of the Himalayas and the Tibetan Plateau.

During the northern hemisphere winter, while the sun is overhead in the south, the great continental landmass of Eurasia and North Africa cools dramatically. With little moisture to pick up and high pressures due to the cold, heavy air, the northeasterly winds that blow outward from the continental interior are cool and dry. As the winds blow across the ocean, they gather moisture and deliver torrential rains to southeast Asia, northern Australia, central Africa, and South America—the southern hemisphere monsoons.

A reversal of climatic conditions during the northern hemisphere summer leads to continental heating, especially over the Himalayas and the Tibetan Plateau, that in turn generates rising air masses. Moisture-laden winds are drawn in from the oceans, and the northern hemisphere monsoon rains pour down from May through September each year. They are particularly heavy in

A classic storm cloud—cumulonimbus—rains over a calm sea (*left*). Clouds form as warm, moist air rises from the ocean surface, then expands, cools, and condenses as rain.

The circulation of the atmosphere (*below*) is driven by heat from the Sun. Greater heating at the equator makes warm air rise and creates a low pressure zone below. The air sucked in to replace that lost causes the trade winds to blow across the tropics. The rising air races poleward, cooling rapidly before sinking once more at the horse latitudes—regions of permanent high pressure. A circulation cell forms at each side of the equator, with further pairs at mid-latitudes and at the poles.

force. The tropics experience persistent trade winds, towering cumulus clouds, and episodic hurricanes. Between the northeast and southeast trade winds lies the region known as the doldrums, their eerie calm and sultry humidity broken by frequent thunderstorms and squalls. The mid latitudes, from about 30° to 60° north and south of the equator, are buffeted alternately by mild westerlies, sweeping warm air away from the subtropical high-pressure zone toward the polar front, and harsh polar easterlies. Cold frontal depressions, also known as extra-tropical cyclones, are generated at the fluctuating interface between these two opposing wind systems. These are the storms familiar to people living at maritime mid-latitudes. They become stronger during the winter months when the pole–equator temperature gradient increases, driving the heat engine all the faster.

Above the great spiral wind cells of the troposphere are narrow, high-velocity jet streams, located between 10 and 15km (6–9mi) above the Earth's surface. Bands of wispy cirrus clouds race across the sky at speeds exceeding 160km/hr (100mph), streaked out by an invisible, tubelike core of very fast winds. Even at these speeds, the jet streams follow sinuous paths, and their meandering loops, diving and rising as they flow, help to create and direct cyclones below. In winter, speeds increase to as much as 500km/hr (over 300mph), the jet streams are straighter, and the storm winds stronger.

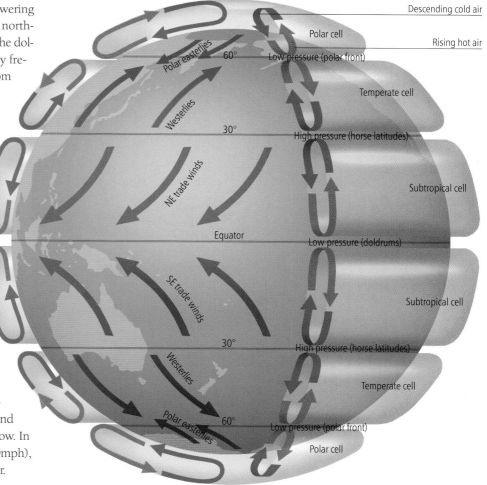

Descending cold air

Rising hot air

Polar cell

Low pressure (polar front)

Polar easterlies

60°

Temperate cell

Westerlies

30°

High pressure (horse latitudes)

NE trade winds

Subtropical cell

Equator

Low pressure (doldrums)

SE trade winds

Subtropical cell

30°

High pressure (horse latitudes)

Westerlies

Temperate cell

Polar easterlies

60°

Low pressure (polar front)

Polar cell

M onsoon rains arrive in Simla, India (above). This seasonal bounty from the Indian Ocean— freshwater from salt—dominates the climate of the subcontinent. Each summer the winds change from offshore northeasterlies to onshore southwesteries, bringing a months-long deluge.

A fully formed hurricane (or tropical cyclone) is the ultimate expression of energy exchange between the ocean and atmosphere. Summer heating of the tropics is intense and continuous. Heat and humidity build up over huge areas in all three of the major oceans that straddle the equator (map, *far right*). As the sea temperature reaches 27°C (80°F), circular winds develop around small areas of low pressure, forcing the warm air upward. In a complex dance of rising air and spiraling winds, a hurricane is born (*right*).

south and southeast Asia, where they often fall on lands that have been seriously drought-stricken through the winter months. The resulting floods are an inevitable and even welcome part of everyday life—rural activity surges into action, water and power supplies are replenished. But there is a very fine line between enough water and too little or too much. A failed monsoon may prolong drought for another year, whereas too much rain can spell equal disaster from floods, crop damage, and the spread of disease. It is salutary to recall that 3 billion people—half the world's population—depend on just the right monsoonal season for their survival.

Clouds and crystals

C louds are the most visible signs of the weather system, a reminder of the global heat engine that turns the water cycle in a never-ending display of power. They give clues to the temperature and humidity of the air, the direction of the wind, and the shape of the weather to come. They are composed of billions of minute water droplets or ice particles, so tiny that they float suspended in the atmosphere. These develop when the concentration of water vapor in the air, much of which has been incorporated from evaporation at sea, reaches its saturation point. The water vapor then condenses around a suitable nucleus—blown sea salt or dust particles, for example—to form a minute particle of liquid, or of ice where the temperature is well below freezing.

Warm air can transport more water vapor than cold, so condensation occurs and clouds form when moist, warm air is cooled. Such cooling occurs when the land heats up and hot air rises, or as winds blow from the sea to land and are forced up over hills and mountains. Night-time cooling of air at the land surface results in fog or mist (a ground-hugging cloud), and in the condensation of water vapor as dew. The thin streaks of cirrus cloud behind high-flying aircraft, known as contrails, are formed primarily by condensation from hot exhaust gases.

For the moisture in clouds to fall as rain, air currents swirling through a sizeable mass of cloud must cause many tiny droplets to collide and coalesce, eventually forming larger drops heavy

enough to escape through the base of the cloud. For snow, too, many individual crystals must join together to create heavier snowflakes, which usually arrive on land as tangled clumps with no particular pattern. High up in the cloud, however, the crystal lattice of ice ensures that snow crystals are hexagonal. Cold, dry conditions produce tiny needles and stubby prisms, whereas warmer, moist clouds allow an infinite variety of delicate snowflake stars to form.

In some cases, as falling crystals pass through moist clouds, further water droplets are scavenged and freeze around the surface, creating pellets of snow. Extreme turbulence in large snow clouds can lead the same pellet to be swept up and down many times, causing layer upon layer of ice glaze or rind to coalesce to form a hailstone, rather like the many skins of an onion. The stronger the snow storm upcurrents, the larger the hailstones, which on occasion can even attain the size of golf balls, oranges, or small melons. Such massive hailstones can cause serious damage and even loss of life when they fall to earth. They may also form around exotic nuclei that have been swept up into the sky, from coal slag to live frogs!

Tropical cyclones

T he largest, most violent storms the Earth's heat engine can unleash are known by various names around the world: in the Indian Ocean they are called cyclones, meaning coiled serpents; in the western Pacific they are typhoons, after the Chinese *tai fung*, "great wind"; and in the Atlantic and eastern Pacific they are hurricanes, named for Hurakan, a Caribbean storm god. It is little wonder that they are so respected, for the energy released by a single tropical cyclone in one day would be enough to power the entire industrial production of the United States for a year.

Tropical cyclones—as distinct from mid-latitude depressions—are born directly from overheating of the oceans. They develop over warm seas at tropical latitudes when sea surface temperatures exceed 27°C (80°F). As vast amounts of water evaporate from the ocean surface, the hot, moist air rises, and then condenses as it cools to form billowing cumulonimbus clouds. Air rushes in across the sea surface to fill the void left beneath them, evaporating more water as it whirls past and so encouraging more clouds to form. Heavy rainfall ensues, and spiraling bands of thunderstorms begin to rotate, counterclockwise in the northern hemisphere and clockwise in the southern.

As the storm grows, so it becomes self-sustaining, sapping heat energy from the ocean to drive the spinning winds ever faster. Wind speeds near the wall of the eye typically exceed 160km/hr (100mph), locally gusting to over 350km/hr (220mph). The calm eye region at the center is typically around 25–40km (15–25mi) across, and the most destructive winds are generally confined to a radius of 100km (60mi) around it. The whole, complex cyclone may extend over a circular area measuring as much as 500 to 1,500km (300–900mi) in diameter. The progress of the storm across the sea is relatively slow and erratic, making it extremely difficult to predict just where and when the storm will strike land.

The damage wreaked by winds of this force is much feared by vulnerable coastal communities. As the cyclone tracks inland, it slowly loses power as it becomes detached from the oceanic fuel supply. Even so, its potential for destruction lingers on. Thunderstorms deluge the land with torrents of water, leading to landslides, mudflows, and widespread flooding. Lightning from thunderstorms strikes the planet in total over 100 times every second of the day, the ultra-high-voltage sparks each carrying tens of thousands of amperes of electrical energy—enough to light a small city for several weeks. The surging current superheats a narrow column of air to 30,000°C (54,000°F), about five times as hot as the surface of the Sun. Although direct lightning strikes on people and animals are relatively rare, they are responsible for igniting many thousands of wildfires every year, especially in the dry forests and grasslands of Australia and North America. Thunderstorms on the outer fringes of cyclones spawn tornadoes or twisters, waterspouts at sea, and strong downbursts of wind (known as microbursts) that strike suddenly and with great destructive force.

It is little wonder, then, that cyclones can have dramatic effects even far inland, sometimes in places where they are least expected. In October 1987, warm tropical waters off Florida led to the development of Hurricane Floyd. Warm air rose high into the atmosphere, confronting the polar jet stream as it swept across the Atlantic, and the marked temperature gradient set up by the encounter accelerated the winds to some 320km/hr (200mph). These winds in turn transformed an ordinary depression into the worst storm to strike northwest Europe in nearly 300 years. Surface winds reached almost 200km/hr (125mph) along the English Channel, and were still only just shy of 160km/hr (100mph) in London. Great swathes of forest were flattened; between them, France and England lost 45 million trees. Property damage, power cuts, and even fatalities ensued—25 people lost their lives and over 120 were injured.

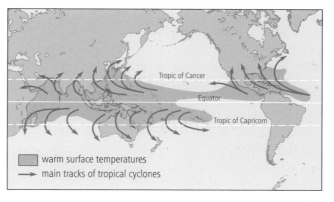

Tropical cyclones, such as Hurricane Katrina (1981, *above*), are spawned in regions where sea surface temperatures exceed 27 °C (80°F). Although the spiraling winds may reach speeds of over 200 km/hr (125mph), the cyclone itself tracks west and poleward much more slowly.

warm surface temperatures

→ main tracks of tropical cyclones

1 Warm, moist air is sucked toward low-pressure center.
2 Water vapor rises in convection towers up to 16km (10mi) high.
3 Towers of thunderstorms coalesce to form spiral rainbands as Coriolis force causes spin.
4 Air spirals in toward the center and out the top in the opposite direction.
5 Cool, dry air sinks into the cloud-free eye, a region of calm.
6 The eyewall; a dense wall of thunderstorms where the strongest winds occur.
7 The high-speed jet stream accelerates the exhaust winds spiraling out from the hurricane.
8 Trade winds steer a westerly path.

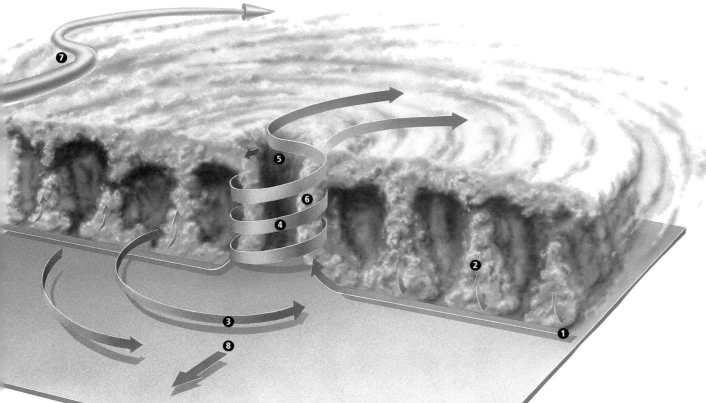

Colored satellite map of depleted ozone levels in the atmosphere over Antarctica in September 2001. At its annual maximum, the ozone hole (violet-blue) covered an area of 26 million km² (16 million mi²) that year. First seen in 1980, the hole is alarming evidence of the extent to which the ozone layer, which protects the planet against harmful ultraviolet radiation from the Sun, has been damaged by chemical pollutants.

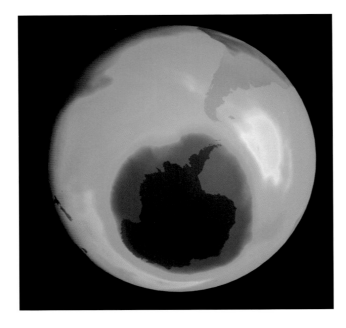

Holes in the sky

Far above the weather systems that are so apparent in our everyday lives lies the dry stratosphere, a further 40km (25mi) thick. Tiny amounts of water vapor sometimes escape from the troposphere to form thin swirls of nacreous (mother-of-pearl) clouds, with their shell-like iridescence. The stratosphere also contains large amounts of very highly dispersed ozone gas—a form of oxygen made up of three oxygen atoms rather than the two forming normal oxygen. It is this ozone layer that protects life on Earth from the harmful effects of the Sun's ultraviolet rays. It acts as an invisible sunscreen, as essential for our survival as the oxygen we breathe. Since oxygen first appeared in the atmosphere, building up to its present level about 2 billion years ago, this upper ozone layer has been kept in a state of dynamic balance by a complex array of chemical reactions. A dramatic rise in pollution since the 1960s, mainly involving chlorine gas from chlorofluorocarbons (CFCs) used in aerosols, refrigerators, and air conditioning systems, is tipping the balance toward ozone destruction.

The icebreaker *Kapitan Khlebnikov*, sailing near Baffin Island, is barely troubled by Arctic ice, melting due to global warming. The contributory causes, longterm trends, and broader effects of global warming are an ongoing debate and challenge.

A gaping "hole" in the ozone layer, representing a very marked thinning across an area about the size of North America, appears over Antarctica each austral spring. A smaller hole has now also appeared over the Arctic. The consequences of such ozone loss are very serious. A 10 percent reduction in ozone results in a 20 percent increase in the amount of ultraviolet radiation reaching the Earth's surface, and inevitably leads to an increase in skin cancer and eye cataracts, as well as a general suppression of human immune systems. Penetration of the ocean surface can damage plankton, with obvious repercussions higher up the marine food chain. Fortunately for all our futures, the world has now agreed to ban CFCs and is seeking to further reduce chlorine emissions.

Climate change

Earth's climate has not always been as it is today. There have been times when the Earth has experienced warm, humid conditions with little temperature differences between the equator and the poles. About 100 million years ago, in the Cretaceous era, palm trees and crocodiles lived as far north as 80°N—the equivalent of Arctic Canada or northern Greenland today. At other times, a far colder climate has gripped the planet; ice sheets and frozen tundra have pushed mercilessly outward from the poles, squeezing the diversity of climatic zones into a narrow, subtropical band.

As far as the evidence will allow, we can surmise that most of the geological past enjoyed warm, "greenhouse" conditions. These have been interrupted at irregular intervals by cold and glacial climates of altogether shorter duration—so-called "icehouse" periods. Intriguingly, we do not yet have a full scientific explanation for the causes and timing of such large-scale climatic fluctuations. We do know that the solar influx of heat, and moderation by the ocean–atmosphere coupling, are the key players involved. We know also that the nature of the Earth's surfaces—oceans, continents, ice-caps—as well as their relative positions, are important in regulating the amount of solar heat that is absorbed or reflected.

Substantially increased rates of volcanic activity in the past may have raised atmospheric levels of carbon dioxide—one of the principal greenhouse gases that help to keep Earth warm. Increased rates of seafloor spreading would lead to higher sea levels and the spread of water across an even greater proportion of the planet than today. In addition, greater reflectivity of solar radiation from such a large ocean surface would reduce temperatures. The drifting of continents across polar regions allows the greater build-up of ice caps, which then spread out from these areas as the planet cools. Uplift of the Himalayas and the Tibetan Plateau, and the complete ocean–atmosphere isolation of Antarctica, are both factors that are believed to have contributed to the onset of the most recent series of ice ages. There are numerous other scenarios and pretenders to the cause of climate change, but almost certainly the answer lies in a complex interplay of numerous variables. One of the key unknowns in all cases is the nature and timescale of ocean–atmosphere response to these external influences on the global heat budget. This is a key area of research in ocean science today.

Formally known as the El Niño Southern Oscillation (ENSO) phenomenon, El Niño is now recognized as a natural oscillation of the ocean–atmosphere system that occurs with an irregular periodicity of around 2 to 4 years. It is the single largest contributor to variation in Earth's climate at this timescale, and can result in changes to global surface temperatures of about 0.7°C (1.3°F) in just one year. This is similar to the amount by which Earth's average temperature has increased over the past century due to human-induced global warming. The short-term effects are numerous and dramatic, with some of the most severe impacts being felt along the coast of South America. In fact, in a mild form, El Niño occurs annually in this area around Christmas time—hence the name El Niño, given by Peruvian fishermen from the Spanish for "boy child," for the baby Jesus.

The equatorial waters off the coast of Peru normally experience a strong upwelling of cold, nutrient-rich waters from below, as warm surface water is blown offshore and across the Pacific by the strong trade winds. This upwelling supports very high plankton productivity, harvested by billions of anchovies and other fishes, and thereby also sustains large seabird colonies as well as a thriving local fishing industry. Toward Christmas each year, as the winds die down, the upwelling decreases and the water warms up. Periodically this effect is much more pronounced—warm waters spread right across the Pacific from Indonesia, reaching the coast of South America, where they completely suppress the upwelling. The plankton factory shuts down as nutrients decrease, fish die in the millions,

seabirds and marine mammals starve, and the whole fishing industry is brought to its knees.

The coupling of ENSO with climate is very marked. Under normal conditions, warm, moist air rises over the west equatorial Pacific, creating thunderstorms and heavy rainfall over Indonesia and northern Australia, whereas cold, dry air sinks in the east, leading to drought in South America. As ENSO intensifies, the conditions are reversed—heavy rains along the Peru–Chile coast results in floods and mudslides over the parched earth, while severe drought and wild-fires occur in Australia and Indonesia. Farther afield, droughts worsen in the Sahel region of Africa, the Indian monsoon weakens and may even fail completely so that famine becomes

inevitable, and mid-latitude storms are pushed further north, bringing drought to the central United States. In 1982–83 the world experienced the strongest El Niño in living memory, which played havoc with global weather patterns. It provided an important stimulus to atmosphere–ocean modeling, in an effort to better understand and predict such phenomena.

Slowly, we are beginning to understand the enormous complexities of the ENSO system. Evidence is emerging linking variations in its intensity with longer timescales—for example, the 11–22-year sunspot cycles that cause small changes in incident solar radiation. Even so, the links between ENSO and human-induced global warming are still not yet clear.

El Niño: an unwelcome Christmas present

Colored satellite image of the 1997 El Niño (red area along the equator, left) showing sea surface temperatures from well below normal (purple) to far above normal (red).

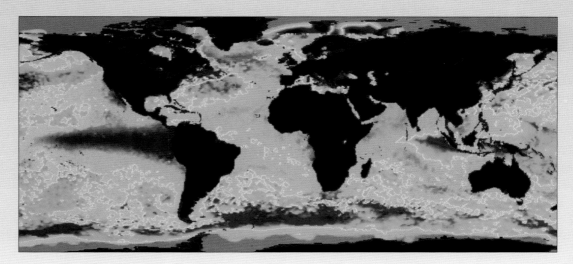

The planet heats up

Scientific evidence clearly shows that we are currently living through a short interglacial period in a world gripped by icehouse conditions. According to natural cycles of climatic change, we should be heading toward a full glacial period of much longer extent within the next 10–15,000 years. But now an altogether different force has come into play—direct human interference in the very delicate balance of natural greenhouse gases in the atmosphere.

The change has been brought about by the industrial and energy revolutions that have taken place in the course of the last few hundred years, coupled with an ever-increasing global population burning ever larger amounts of wood and fossil fuel. We are pumping huge amounts of greenhouse gases—carbon dioxide, methane, nitrous oxide, CFCs, water vapor, and others—into the atmosphere every second of every day. Global mean temperature—actually an incredibly difficult thing to measure—appears to have risen by around 0.5–1°C (0.9–1.8°F) in the past century. Our best estimates, based on projecting different trends into the future, suggest that temperatures could rise by a further 2–5°C (3.6–9°F) by the year 2100. An increase on that scale would make the Earth as hot or hotter than it has been during the previous interglacial periods of the last million years—not a very noticeable development for the planet, perhaps, but hugely significant for us.

Scientific understanding of the process of global warming and of its likely effects is still far from clear. The degree of warming will vary, with continental interiors and high latitudes being particularly strongly affected. Certain parts of the world will become wetter, and the growing season will lengthen; others will be driven to drought where once there was plenty. Global sea levels will continue to rise, probably a good deal faster than the current rate of 1–3cm (0.4–1.2in) per decade. Since many oceanic islands and large tracts of coastal lowland, including some of the world's major cities, are already perilously close to current sea level, the dangers of catastrophic flooding, and even of complete submergence, will inevitably increase. As the oceans become warmer, the frequency and scale of tropical cyclones are also likely to grow, and their effects will become still more devastating. The list of potential impacts from global warming and sea level rise, both positive and negative, is as long as it is hotly debated.

Few now doubt, however, that human-induced global warming is a reality, and that the world's nations should act collectively to reduce greenhouse-gas emissions. Because of the ocean's key role in affecting and moderating the world's climate, ocean scientists are perhaps best placed to gather data on past climate change, and to help build the models that will allow better prediction of future climates.

OCEAN
SYSTEMS

Far beneath the azure surface of an ever-changing ocean untold riches lie hidden in the silent deep. Over great tracts of ocean floor there are nodules, crusts, and bizarre but elegant spires of metallic minerals. But elsewhere the prize lies deeper still. Drilling many thousands of meters into thick piles of sediment below the seafloor, this oil rig in the Gulf of Mexico taps directly into pools of "black gold." Billions of tons of petroleum (oil and gas) are trapped within tiny pore spaces between grains of buried sand. Locating this resource in so hidden and remote a situation is a remarkable feat of technology and science.

HIDDEN RICHES OF THE OCEAN

CHAPTER 5

Energy and minerals from the sea

Thirty years ago, the mining industry was clamoring at the gateway to the deep oceans. Initially, interest focused on the manganese nodules discovered in abundance over many parts of the deep-sea floor. Then oil became the center of attention, at first in the relatively shallow North Sea waters. Now, however, eyes are turning again to the deep, for the Earth's last great wilderness is set to define the next generation of oil exploration.

Yet reaping the energy and mineral wealth of the seas is by no means a new occupation. Common salt has been extracted from evaporative pools since prehistoric times, and oil from natural seeps has been used for several millennia in China, the Middle East, and elsewhere. By the late Middle Ages, tidal mills for grinding wheat were in operation all along the coasts of Britain and Holland. Today, 12,000 carats of diamonds are taken each month from the Forbidden Coast off South Africa. More mundane, but of greater practical significance, offshore sand and gravel extraction exceeds that mined on land.

The sheer volume and diversity of resources available in the oceans will ensure their place in supplying an ever more demanding and crowded world. Their exploitation will undoubtedly also precipitate environmental change in a region we know less well than the surface of the Moon. To preserve the integrity of this fragile realm it is vital and urgent that we learn to manage ocean resources more successfully than we have those of the land. At stake are more than mere commodities; the riches of the ocean are also those of the heart.

New-found oil wealth in the deep seas will ensure that global reserves last well into the third millennium. But what will be the cost of exploiting this final frontier of the ocean environment?

HIDDEN RICHES OF THE OCEAN

Oil and Gas

Oil—"black gold"—has driven the pace of change through the 20th century. The exponential increase in the consumption of both oil and gas is linked with unprecedented development and profit for the countries and companies that have control of this liquid asset. The multinational oil giants have seen their annual turnover greatly exceed the Gross National Product of many developing nations. Less than 20 percent of the world's population, living in developed, industrialized countries, consume over two-thirds of the total energy supply. Not only has most of the world been left far behind in the oil race, this underprivileged majority has had to share the full environmental cost of such unchecked growth—the oil spills and acid rain, global warming and massive resource depletion.

Global oil consumption continues to rise with scant regard for dwindling reserves. The fastest growth in demand is expected to come from developing Asian nations including China, whose consumption may rival that of North America after 2020.

The Oil Age Although oil has been a part of the human story from the very earliest times, its popular and abundant use has been a remarkably recent affair. Lightning strikes on natural oil seepages are known to light wildfires that may burn continuously for years. Such natural conflagrations were quite possibly one of the ways that early hominids first came to harness fire, in China at least 400,000 years ago. By the time the Greek historian Herodotus was writing in 450 BCE, pitch and asphalt were being extracted from all over the Middle East and North Africa, the principal uses being for medicines, waterproofing, warfare, and oil lamps. But the main sources of energy before the Industrial Revolution came from the burning of wood, and from wheels and pumps driven by wind and water. From about the 12th century on, coal began to replace wood as local supplies became exhausted.

Coal then fired the Industrial Revolution that has shaped our modern world, its production escalating dramatically with increased mechanization, the spread of railways, and the generation of electricity. But still, at the start of the 20th century, oil and gas were hardly used at all; the oil boom was yet to come. The age of oil that followed saw the world shrink in a quite extraordinary fashion, first with the private automobile and then with passenger jets. The age of oil is also the age of plastics, and of countless other petrochemical products. It would be hard to imagine the world now without low-cost, petrol-fuelled transport, or without plastic. Such almost limitless demand is why the 21st century is set to drive exploration for oil and gas far into and beneath the global ocean.

Small beginnings Oil and gas, together with coal and peat, are fossil fuels based on the principal chemical elements common to all living matter—carbon and hydrogen. They are formed from the organic remains of dead plant material, through a process of decay and change over millions of years. Only a tiny fraction of the original living bulk is eventually preserved as a fuel resource. Humankind can burn in a matter of seconds what nature has taken countless millennia to produce. A reliable estimate is that, globally, we consume annually an amount of fossil fuel that took 1 million years to form.

In the case of hydrocarbons (oil and gas), the story begins in the sunlit surface waters of former seas and lakes. Billions upon billions of microscopic, unicellular plants, known as phytoplankton, harness the sun's energy and nutrients in the water to power their tiny life cycles. A steady rain of dead organisms filters slowly downward toward the dark, still depths. Much of this material is consumed by bacteria, either in the water column or

Global Oil Consumption

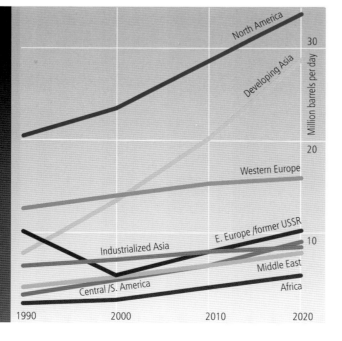

North America

Developing Asia

Million barrels per day

30

20

Western Europe

E. Europe /former USSR

Industrialized Asia

10

Middle East

Central /S. America

Africa

1990   2000   2010   2020

on the seafloor, but a small proportion finds its way into the sediment and is eventually buried and preserved.

There, locked within the sediment, usually in the form of a dark, organic-rich mud known as black shale, is the template for generating the particular hydrocarbon molecules that make up oil or gas. The remains of land plants tend to produce coal, but gas forms if these remains are more deeply buried, while marine plant material yields mainly oil, followed by gas at greater depths. The process is, however, immeasurably slow, like so much else that takes place in the geological realm. Slowly the organic matter is buried as younger sediment is deposited above it; slowly it becomes hotter as the depth increases.

There is a particular window of temperature, between about 50° and 100°C (120°–212°F), at which the very complex organic molecules are broken down into the long-chain hydrocarbons found in oil. At higher temperatures still (150°–250°C/ 300°–500°F), these are further cracked to yield the short-chain hydrocarbons, mainly methane, of natural gas. The typical thermal gradient, or rate of increase in temperature, is around 20°–30°C per kilometer of burial depth (60°–90°F per mile) beneath most parts of the ocean floor. This means that oil generation commonly occurs between 2 and 6km (1.2–4mi) below the seafloor, and gas at greater depths. It may take 10–20 million years of sediment accumulation before burial to these depths is achieved—a very long time to form such a precious resource, and that is just the beginning of the story.

Migration and entrapment Widely dispersed oil or gas in fine-grained, compressed black shale is very difficult to extract, and almost impossible when it is buried deep below the seafloor. However, the very great pressures that result from such deep burial force the mobile hydrocarbons, together with water still trapped in the sediment from the time of deposition, to migrate outward and upward to where the pressure is lower. It is estimated that as much as 90 percent of the oil and gas originally generated leaks out at the surface in natural seeps, part of the grand recycling of nutrient elements.

The remaining 10 percent is held back in the tiny pore spaces of sandstone and other sediments, or in minute cavities in some limestones, where shell or coral fragments have been partially dissolved; they fill up with oil and gas rather like a bath sponge with water. These thick accumulations of permeable rocks are the reservoirs that oil explorationists seek out. They can be deposited in many different ways, in oceanic or continental environments, but need to be of sufficient size in order to hold economically significant reserves. The structure of the reservoir rocks must form a trap into which hydrocarbons can migrate but from which an impermeable covering seal (or cap rock) prevents escape. Only then will the oil and gas continue to collect in these giant sponge pools deep below the surface.

Scientists and petroleum geologists work closely together in the quest for these riches, deeply buried and well hidden in sedimentary basins beneath land or sea. There is no pirate's map with an "X" marking the spot to drill, but there are a series of complex and related clues that need to be followed up if these modern treasure-hunters are to find the rewards they seek.

Small pockets of gas are generally found trapped above an oil reservoir. The gas is often simply flared off at the well site (*far left*) while only the oil is collected, causing both waste and atmospheric pollution.

The extreme pressure of burial squeezes petroleum out of the source rock. Passing through the network of tiny spaces in permeable rocks (as in this sandstone *below*) as well as larger faults and fractures, oil, gas, and water migrate up toward the surface.

Oil and gas are formed by the slow decomposition of plant material preserved and buried in sedimentary rocks (*below*). As the temperature increases with depth, the complex organic molecules break down into simpler hydrocarbons—a process called maturation. Marine algae tend to yield oil, while land plants generate gas.

As the fluids migrate upward, they may become trapped against an impermeable barrier —usually a fine-grained mudrock or shale. Oil traps form under a variety of structural and stratigraphic conditions (*below*).

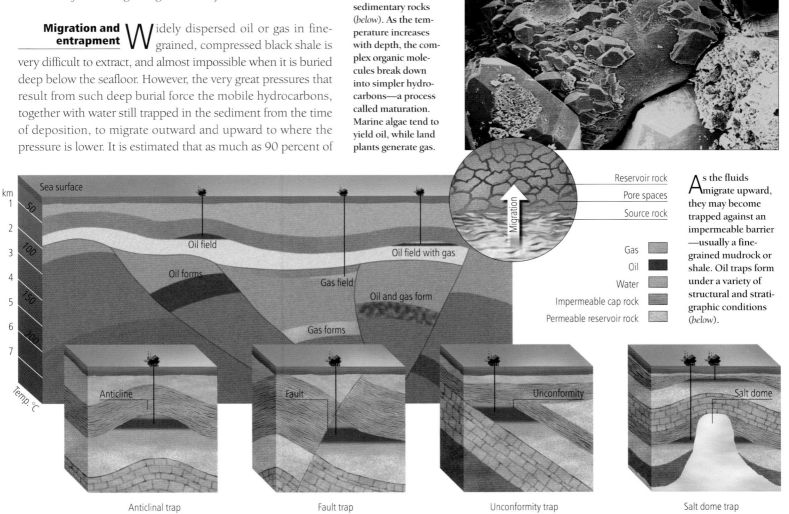

Gas	▨
Oil	▨
Water	▨
Impermeable cap rock	▨
Permeable reservoir rock	▨

Reservoir rock
Pore spaces
Source rock
Migration

Sea surface
Oil field
Oil field with gas
Oil forms
Gas field
Oil and gas form
Gas forms

km 1 / 2 / 3 / 4 / 5 / 6 / 7
50 / 100 / 150 / 200
Temp. °C

Anticline
Fault
Unconformity
Salt dome

Anticlinal trap
Fault trap
Unconformity trap
Salt dome trap

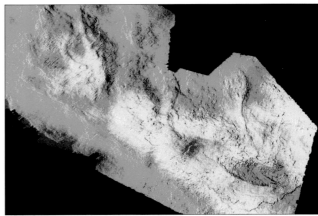

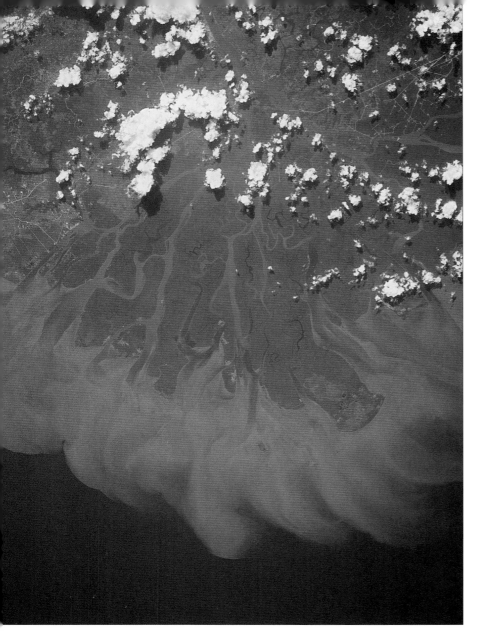

Hydrocarbons are only generated and trapped in parts of the world that have accumulated thick sedimentary successions over many millions of years. These areas are known as sedimentary basins, and many of them lie beneath major river deltas. This view of Borneo's Mahakam Delta from the space shuttle *Columbia* shows the dense plumes of sediment delivered daily by the Mahakam River. Roads to the numerous oil fields are clearly visible from space.

If we are to prospect for oil in the sediments of former ocean margins, for example, then we first need to understand the deep-water sedimentary system. Sands that will eventually form reservoir rocks are carried into the deep ocean by catastrophic events such as slides, debris flows, and giant turbidity currents. These are the submarine equivalents of the floods, avalanches, and landslides that periodically wreak such havoc on land. They are capable of transporting more than 400 cu km (96 cu mi) of material hundreds, even thousands, of kilometers into the deep sea and depositing it as a thick, very extensive sheet known as a turbidite—equivalent to removing a slab of material the area of New York City, transporting it the length of the United States' eastern seaboard, and spewing it out over the whole of Florida!

We need to understand how, when, and why such events occur, in part, of course, because they can be a hazard to exploration and to submarine installations. We are also aware that they must be repeated many times in the same region over a period of a million years or so in order to build up a reservoir of oil or gas big enough to be economically viable to exploit. We know that there can be several immediate triggers of these events—severe earthquakes at sea, a major collapse of island volcanoes, the sudden release of buried clathrates (frozen methane)

from beneath the seafloor, tropical storms, freak river floods. Many aspects of these processes, however, remain a mystery.

In order to locate potential reservoirs, it is critical that we discover the subsurface dimensions and architecture of such deep-water turbidites. Today most of the deep ocean is carpeted with fine muds and biogenic oozes, but careful surveying can reveal the input points, conduits, and depositional settings for the coarser sands and gravels that form oil reservoirs. In the recent geological past, during periods of greatly lowered sea level, sand input into the deep sea was significantly greater. Scientific drilling and giant piston coring, coupled with the latest industry techniques of seismic surveying (3-D seismics), can help elucidate such changes in sediment pattern.

Deserts, deltas, and shelf seas Deep-water turbidite reservoirs are one of many different types of hydrocarbon sources throughout the world, though all were discovered in the first instance by drilling on land. The very first discovery well was drilled in 1859 by Edwin Drake, who found oil at a depth of only 18m (59ft) at Oil Creek in Pennsylvania. Exploration soon spread to many other states, and the first tentative forays offshore were made from wooden piers in California and Texas by the early 1900s. Exploration in the Middle East oil crescent, which has long been the world's most prolific hydrocarbon region, began at about the same time. Most of the giant fields had been discovered by 1950, and many of these were hosted in limestone reservoirs. The limestone originally formed in warm, shallow seas that extended up to 2,000km (1,250mi) across a broad continental shelf in the course of the Mesozoic era, over 80 million years ago.

The existence of these plays (groups of oil fields) under the Caspian Sea and Arabian Gulf was known from geological prediction and test drilling near the beginning of the 20th century. With such vast onshore reservoirs, however, there was little need to develop the more complex and costly technology to drill at sea, in an unknown, and often hostile environment. Intensive offshore drilling, therefore, did not begin until the 1950s.

The immense, hidden riches of the North Sea sedimentary basin only began to be realized from the 1960s on, following similar geological prediction based on land exploration. In 1959, an onshore well near Groningen in the Netherlands

Mapping different attributes across many closely spaced seismic profiles can image potential oil reservoirs (*left*). The red and yellow colors show the likely distribution of deeply buried sands beneath the central North Sea.

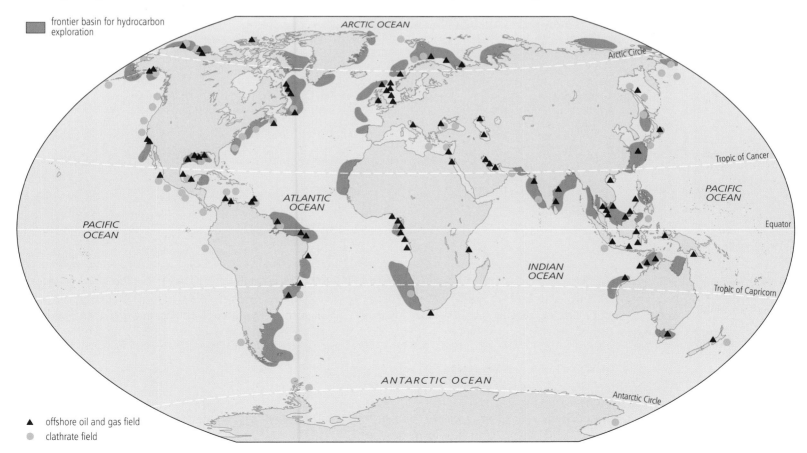

Shetland Islands — Brent Delta — Brent oil field — Viking Graben — NORWAY — Central Graben — BRITAIN — WESTERN EUROPE

Rivers flowing through the Viking Graben rift valley 165 million years ago carried huge volumes of sediment to what is now the northern North Sea. The sands deposited in the Brent Delta formed the reservoirs for one of the world's most prolific oil provinces.

struck gas about 3,000m (10,000ft) below the surface in thick sandstones originally deposited in a 250-million-year-old desert. This field has an estimated reserve of 1,500 trillion cubic meters (almost 2,000 trillion cu yds) of gas, and is still one of the largest known gas accumulations in the world. Geological exploration determined that the same reservoir formation, together with the underlying coal measures from which the gas was generated, lay beneath much of the southern North Sea. The first of many discoveries was made just five years later; this was the West Sole field, lying in relatively shallow waters off eastern England. There, 3,000m (10,000ft) below the seafloor, 150m (500ft) of highly porous desert sandstone held large reserves of gas.

It was another five years before the oil potential of the North Sea was discovered, this time when a wildcat exploration well was drilled in virgin territory in the extreme north of the basin, over 200km (125mi) northeast of the Shetland Islands. Good, light crude oil was found at a depth of 3,000m (10,000ft) in Jurassic-age sandstones of the Brent oil field. With an estimated 1.8 million metric tons of oil in place, the Brent field immediately became a giant on the world stage, and still ranks as the second largest oil pool of the North Sea.

It is now one of 20 similar fields in what has become known as the Brent Province, the most productive oil play in the North Sea. These sandstone reservoirs were formed some 160 million years ago at a time when earth movements had pushed the area that is now the central North Sea above sea level. As fast as the land rose, so erosion stripped bare the new uplands, and billions of tons of sand and mud were carried to the north in powerful rivers, to be deposited in what must have been one of the greatest deltas of the age. Coal-forming swamp forests colonized the muddy embayments, while tidal currents cleaned and reworked the sands until they became a near-perfect reservoir.

Offshore hydrocarbon resources (*below*). Beyond the known petroleum provinces, huge sedimentary basins remain untapped—frontier areas where exploration is just beginning. Exploitation of frozen methane (clathrates) is a potential for the more distant future.

frontier basin for hydrocarbon exploration

ARCTIC OCEAN — Arctic Circle — Tropic of Cancer — PACIFIC OCEAN — ATLANTIC OCEAN — PACIFIC OCEAN — Equator — INDIAN OCEAN — Tropic of Capricorn — ANTARCTIC OCEAN — Antarctic Circle

▲ offshore oil and gas field
● clathrate field

Everything about drilling operations at sea is large-scale and expensive. Exploration wells can take many months to drill and cost $50 million each—an expensive gamble, as many turn out to be failures! Silhouetted against a dawn sky off the coast of Louisiana, Penzoil's rig *Eugene Island 330* (*right*) is one that struck lucky.

Drilling crews meet the same challenges the world over, from the Sahara Desert in southern Libya to this oil production well in Bohai Bay, north-eastern China. Softer sediments near the surface are rapidly penetrated with giant, steel-knobbed drill bits. Harder reservoir rocks at depth may require narrow-diameter, diamond-studded coring bits.

Triumph and disaster The exploration and exploitation of the marine realm has truly been a triumph of technology and science. Gigantic drilling platforms are among the largest structures ever built. They are floated or carried by even larger transporter ships to wherever in the world they are needed, and then fixed in position for what may be several months of drilling or many years of production. Accommodation may be required for up to 300 workers at a time; personnel transport is usually by helicopter or boat, and a never-ending round of supplies must be brought in by ship.

The drilling process is an even more remarkable undertaking. Some 500 steel drill pipes, each 10m (33ft) long, must be joined one by one on the drilling platform to reach even the modest depth of 5,000m (16,500ft) below the seafloor. Reservoir targets

may be over 7,000m (23,000ft) deep. The holes drilled are like a collapsible telescope of decreasing diameter, from nearly 1m (39in) at the top to only 10cm (4in) at the base. They are drilled in stages, tested, and then cased in concrete to avoid collapse before drilling deeper. Giant, steel-knobbed drill bits are used to chew through the upper layers of sediment. Fine, diamond-studded bits are needed when the sediment becomes hard rock. At all times the advancing hole is cooled, lubricated, and weighted with a dense mixture of mud and water. Without this precaution, the pressure released on piercing a deeply buried reservoir would cause a blow-out, with devastating consequences.

Clathrates: frozen gas from beneath the waves

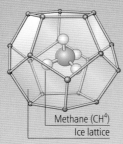

Methane (CH⁴)
Ice lattice

Clathrates, or gas hydrates, consist of molecules of gas, usually methane, surrounded by an icy "cage," or lattice, of water molecules.

In an energy-hungry world, the search for new supplies of oil and gas is an urgent one. On land, there are still enormous reserves of oil shales and tar sands that await either a more economic means of processing or a higher price per barrel to make their exploitation viable. At sea, there are several different options available. The first, and currently most productive, is the steady march of exploration into deeper and deeper waters, creeping further downslope with each new technological advance. Key target areas include the ocean margins off Scotland, West Africa, Brazil, Alaska, the Gulf of Mexico, and Borneo. The second is to move toward higher and harsher latitudes, drilling from artificial islands in the Arctic, off northern Norway and Svalbard, and around the Falkland Islands in the South Atlantic. Environmental restraint still maintains the Antarctic free from oil exploration, although the potential is very great.

The most recent and intriguing possibility, however, lies in the discovery of frozen methane gas at shallow depths below the seafloor. Vast fields of mixed gas and ice crystals, known as clathrates or gas hydrates, are now known to cover literally millions of square kilometers only a few meters below the seabed. The gas is formed during the very early stages of organic matter degradation, but freezes under conditions of high pressure and low temperature. Where the water pressure above is sufficiently great, at depths of over about 1,000m (3,300ft) and where the seafloor, swept by deep polar currents, is permanently cold, the gas remains in place as an icy meshwork within the sediment.

Gas hydrates are very difficult to detect by remote sensing techniques, although in some cases they show up on seismic profiles as bright spots that cut obliquely across other reflectors. As soon as they are penetrated by drilling and

brought to the surface in cores, the ice melts and gas escapes as pressure is released. However, an intensive program of investigation has begun to reveal something more of their anatomy and distribution. Physically, they appear as a white, powdery snow intermeshed with sediment, and occur in at least two forms. The first has up to 8 methane molecules within 46 water molecules, and can contain other gases such as ethane, hydrogen sulphide, and carbon dioxide. The second has a larger network of 136 water molecules and can contain more complex hydrocarbon gases—pentane and butane.

Initial estimates suggest that their potential as a future energy resource dwarfs existing stocks of other fossil fuels, although the technology and economic feasibility for their exploitation is not yet in place. Even if a more conservative view is taken of their volume and extent, we must be extremely cautious about unbridled exploita-

tion. They occur at very shallow depths within the sediment. The slightest operational accident could release to the oceans and atmosphere unknown amounts of one of the most potent greenhouse gases—and this at a time when the world is finally getting to grips with reductions in carbon dioxide emissions. We are fairly certain that natural release of gas hydrates has occurred in the past following major submarine slides. However, the reverse may also be true, with downslope mass movements being triggered by a substantial release of gas. Whichever comes first, the sudden escape of a giant bubble of gas through the water column could explain the mysterious disappearance of ships on an otherwise calm day in the open ocean. This is now a favorite explanation for disasters in the so-called Bermuda Triangle.

Professor Erwin Suess,
Geomar, Germany

Commercial drilling can now take place in water depths in excess of 3,000m (10,000ft), and anywhere from the ice-covered Arctic Ocean—drilling from artificial islands—to the sweltering tropics. Individual holes can be deviated from the vertical during drilling until they are advancing horizontally at a carefully preselected depth within the sediment pile.

Pushing at the frontiers of the possible inevitably brings failures as well as successes. A single wildcat exploration well in deep water can cost over $80 million, and the odds on striking oil first time in a virgin basin are as poor as one in 20 to one in 50. Our scientific detective work beneath the seafloor is still far from perfect. Even more unpredictable, however, are the ocean elements. Drilling rigs in the Gulf of Mexico are designed to withstand the seasonal battering by hurricane winds, an expected hazard in the region. However, nothing could be done when an entire platform suddenly and quietly slipped from view, to be lost with all hands beneath a calm sea, as the result of a huge submarine slide in the soft, unstable muds in front of the Mississippi Delta. Equally disturbingly, a fully occupied rig on the newly discovered Hibernia field on the Grand Banks of Newfoundland was destroyed by an iceberg.

Environment at risk Cooperation between science and industry is clearly bearing fruit as the march toward deep-ocean exploitation gathers momentum, but the conflict of interest with environmental concerns is also growing. The key players are not only oil giants and scientists, but also governments, environmental organizations, and a host of industries. Cable companies, for example, would like to know where best to site seafloor telegraph cables so as to avoid submarine hazards and costly cable breaks. The nuclear industry has long been looking to the deep sea as a solution for radioactive waste disposal. Several years of careful environmental study of the abyssal ocean plains off northwest Africa have helped to postpone this option, for the time being at least. But the experi-

ence of many years of oil transport across the oceans has seen the worst marine environmental disasters on record caused by super-tanker collisions and oil spillages. In 1989, the *Exxon Valdez* tanker ran aground at Bligh Island, Alaska. leaking 45 million liters (10 million gallons) of oil over an area in excess of 8,000 sq km (3,000 sq mi) and along several hundreds of kilometers of shoreline. The marine life that perished and the pristine coastal ecosystem that suffered tell a horrifying tale—and this is only one of many such disasters.

The environmental campaign group Greenpeace, which keeps an ever-watchful eye on the activities of the oil industry, would like to see a complete moratorium on deep-water exploration along the Atlantic Frontier (an area north-west of the British coast). They fear that seismic exploration, which uses submarine explosions and artificial sonar devices to map the seafloor, will adversely affect the lives and migration pathways of whales and dolphins. They are also concerned about the risk of pollution and the threat to delicate ecosystems, such as the recently discovered deep, coldwater coral communities that stretch for hundreds of kilometers along the Atlantic Frontier.

Added to these considerations are global concerns about over-reliance on non-renewable energy, the burning of fossil fuels, and the consequent emission of the greenhouse gas carbon dioxide into the atmosphere, with its possibly irreversible influence on climate. Environmental degradation, conservation, resource depletion, and energy diversification are arguably the most important issues of our time. Unravelling the marine frontier will undoubtedly create as many problems as solutions. What is essential is that the work of scientists should be used to realize a sound strategy of long-term environmental care and resource management throughout the global ocean.

Oil pollution on a Welsh beach (*below*). This slick originated in a spill of about 70,000 metric tons of light crude oil from the tanker *Sea Empress* off the coast of Milford Haven, Wales, in 1996. The oil is brown rather than black as it contains chemical dispersants. Although these detergents can reduce the size of the slick, they are also poisonous to wildlife.

A cleaner, sustainable alternative to fossil fuels is to harness the phenomenal energy locked up in waves, tides, wind, currents, and temperature differences within the oceans.

Renewable Marine Energy

KEY TERMS

FOSSIL FUEL

NUCLEAR FISSION

NUCLEAR FUSION

HYDROELECTRIC POWER

TURBINE

GEOTHERMAL ENERGY

SOLAR POWER

TIDAL BARRAGE

OCEAN THERMAL-ENERGY
CONVERSION (OTEC)

WIND FARM

Locked within the swirling blue waters that cover over two-thirds of planet Earth is a quite phenomenal store of solar and gravitational energy. It is an energy source that is clean and safe, without the threat of untold pollution and catastrophic disaster. It is a form of energy that is constantly renewed, daily and even hourly, as the Sun beats down on the ocean surface and the Moon drags the tides across the world in a never-ending cycle. We will not have to wait a million years for this fuel supply to collect and mature, hidden somewhere far beneath the ocean floor, for it is being replenished even as we use it.

This is the untamed energy of the oceans that beats as surf against the shoreline and keeps alive the creatures that make that unique intertidal world their home. It is the furious force that is unleashed with such terrifying might and unpredictability in the hurricane winds of the tropics, and it is the quiet, unseen power of ocean currents that act as a global thermostat, moderating between extremes of blisteringly hot lands and dark, ice-covered seas. On a modest scale, the use of marine energy is as old as human interaction with the sea. The first fishing rafts used sea currents to drift between islands, and all early voyages of discovery or trade were powered by wind-filled sails. Along the coastline, we soon developed the means to harness such energy more systematically—tidal mills and windmills became commonplace as human population and industry both expanded.

While fossil-fuel resources (oil and coal) have a limited future, renewable energy sources such as tidal and hydroelectric power promise constant returns, and the potential for nuclear fusion is almost limitless.

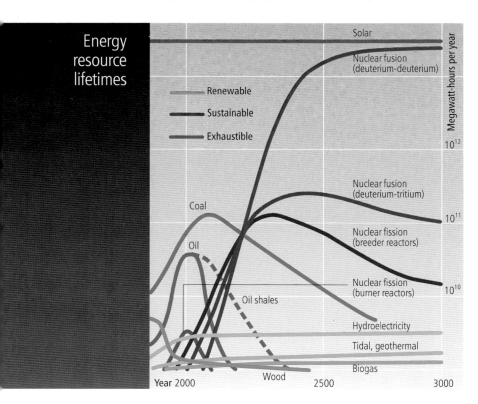

Energy resource lifetimes

Renewable
Sustainable
Exhaustible

Solar

Nuclear fusion
(deuterium-deuterium)

10^{12}

Coal

Nuclear fusion
(deuterium-tritium)

10^{11}

Oil

Nuclear fission
(breeder reactors)

Oil shales

Nuclear fission
(burner reactors)

10^{10}

Hydroelectricity

Tidal, geothermal

Biogas

Wood

Year 2000   2500   3000

Megawatt-hours per year

Global context Although copious volumes have been written on the subject, and very many pilot projects initiated, the great, untamed energy of the oceans is still very largely untapped. The environmental arguments in favor of moving to a greater use of renewable energy, both on land and from the sea, are incontrovertible. Yet it is the continuing low cost of oil and gas, coupled with their convenience for transport, storage, and use, that militate against progress toward sustainable options. In addition, coal is still very abundant throughout the world, and the technology for its exploitation and use is widely known.

Today, global energy use stands at around 80 percent from fossil fuels (oil, coal, gas, and minor amounts of peat), 5 percent from nuclear energy, and 15 percent from "renewable" energy. There has been much debate about the use of fossil fuels. They are finite natural resources, which we are currently consuming about 1 million times more rapidly than geological processes can replenish them. Their burning results in voluminous emissions of carbon dioxide and other pollutants into the atmosphere, contributing to global warming, health problems, and acid rain. Their accidental spillage can be ecologically devastating. So, why are they still used? The answer, of course, lies in the economic benefits that they bring to the richest 20 percent of the world's population, and the determination of developing nations to emulate the changes in lifestyle (such as vehicle-ownership) that are seen as concomitant with surplus energy.

Nuclear energy was once heralded as a panacea for the future, clean, safe, and limitless—"atoms for peace" was a slogan of the 1950s. Ambitious programs, often heavily subsidized by government, were introduced in a number of countries—France, Japan, the UK, the USA, and the Soviet Union among others—

Bathers swim in the steaming waters of Blue Lagoon, a runoff pond at the Svartsengi geothermal power plant in Iceland. Here, in the far northern Atlantic, a short segment of the global mid-ocean ridge system has been pushed up above sea level. The power released at mid-ocean ridges and subduction zones is essentially limitless, but it has only been tapped in a handful of countries around the world.

The solar alternative is widely recognized as the ultimate solution for harvesting on land or at sea, but developments are painfully slow in a world not short of other sources of supply.

The principal source of non-commercial energy is wood, which accounts for over 90 percent of total energy use in some countries. However, the global depletion of wood resources in this way presents humankind with one of our greatest environmental and energy challenges, for the current rate of usage is no longer sustainable. In addition, the wholescale destruction of rain forests for timber and to release land for agriculture presents a suite of other problems. It is as though a planetary cancer was eating away at the world's lungs, lessening the global capacity of plant life to consume carbon dioxide and to release oxygen into the atmosphere. The destruction of the rain forests is greatly accelerating the rate at which species are becoming extinct, fast making the present epoch the greatest mass-extinction event since the demise of the dinosaurs 65 million years ago.

Some of the answers lie in the oceans, in the great, untapped reserves of energy and in the gigantic phytoplankton forests that keep the world breathing.

and as much as 50 percent of energy supply in some nations is now nuclear. However, accidents such as that at Three Mile Island, Pennsylvania, in 1979, and the catastrophe of Chernobyl in the Soviet Ukraine in 1986, together with the enormous problem of the safe disposal of radioactive waste and the decommissioning of obsolete power stations, have meant that nuclear energy is fast falling from favor. The principal fuel required for the fission reaction of all nuclear reactors today is uranium, yet supplies of this mineral are severely limited. "Breeder" reactors, which are designed to produce more fuel than they consume, have not, on the whole, become a commercial option. The only long-term solution for nuclear power would seem to be the development of fusion reactors, using only "heavy" water (water enriched in deuterium or other heavy isotopes) as a fuel and generating vast amounts of energy with little or no harmful waste. However, this technology involves controlling power equivalent to a continued series of nuclear bomb blasts in a safe power station and releasing it slowly for peaceful use. We are still several decades away from resolving the necessary constraints.

Within the relatively large figure of 15 percent given for renewable energy sources many diverse programs can be found—both centralized schemes generating electricity for national grid systems and local ones for use by villagers in the developing world. The principal centralized form of renewable energy, accounting for over one-third of the total, is hydroelectric power. Clean, safe, and remarkably efficient at producing electricity directly from falling water on land, this is a widespread option with great future potential. Geothermal power schemes are fast developing in countries with active volcanoes and hot springs, such as Iceland, Italy, and New Zealand, and already over 15 percent of Central America's electricity is geothermal.

Tides and waves

Wherever tides are strong, their energy can be harnessed. Yet of the thousands of tidal mills that existed for hundreds of years, only one still exists—as a working museum in Toton, at the head of Southampton Water in southern England. The modern approach, however, is to think big. This means building a major dam or barrage across an estuary having a large tidal range, certainly in excess of 4m (13ft) and preferably over 8m (26ft). The high tide is temporarily held up in a great, enclosed bay behind the dam, and then let out through a series of large turbines, generating electricity directly in much the same way as a hydroelectric

There is no doubting the potential power locked in ocean waves. Harnessing just a fraction of their force could supply the global demand for energy—if it could be economically harvested. Coastal areas are, naturally, best placed to exploit wave energy. Following a successful 10-year pilot project, the *Limpet* commercial wave power station on Islay, a Scottish Hebridean island, was among the first to go into production. Its 500kW capacity provides electricity for 300 homes.

plant on a river dam. In fact, the generators can be used in either direction, yielding power from both flood and ebb tides.

The first large-scale commercial scheme was the Rance barrage near St. Malo in Brittany, a dam 1km (0.6mi) wide with 24 turbine generators constructed across an estuary in which the mean tidal range is 8.17m (27ft), peaking at 13.5m (44ft) during equinoxes. The project has been feeding almost half a million kilowatts into the French national grid system daily since it opened in 1967. Surprisingly, few other tidal barrages yet exist—those that do are in Norway, Canada, China, and the former Soviet Union—although detailed plans and costings have often been drawn up. There are plans for eight different estuaries around the UK, the largest and most advanced being the Severn tidal barrage, a multimillion-dollar construction 16km (10mi) wide that would tap into the world's second greatest tidal range. The proposal is for 200 turbines generating more than 7,000 megawatts of electrical energy, roughly equivalent to the supply from five conventional nuclear reactors.

Waves are another obvious and very visible source of energy. Their destructive capability is enormous, as is their slow but persistent erosion of the coastline. The power potential of an average wave per 1km (0.6mi) of beach has been estimated at about 40 megawatts, but the means of harnessing this potential are largely in the research and development phase. A small-scale but commercial wave-power plant, known as OSPREY, was constructed off northern Scotland in 1995. Wave motion entered into a submerged chamber open at the base, forcing air upward through turbines. The 2-megawatt facility fed the electricity generated to the shore some 300m (1,000ft) away via an underwater cable. Unfortunately, the plant was soon damaged by waves and later destroyed by a storm—testimony to the power of the sea, as well as to our underdeveloped technology.

Elusive flow Still more elusive is the power bound up in ocean currents. Since ancient times, mariners have utilized the flow of currents as best suited their course across the sea. The Gulf Stream, the fastest current in the world, was charted by Benjamin Franklin, former scientist and later US statesman, in an attempt to explain why mail sent from America to Britain arrived much faster than letters on the return route. Today, we know that this current sets out past Florida with a velocity of 5 knots (10 km/hr) and transports 30 million cubic meters (39 million cu yds) of water per second—five times more than the flow of all the world's rivers. An array of huge, slowly spinning rotary turbines submerged in the Gulf Stream as it emerges through the Florida Straits could, in principle, generate 1,000 megawatts of electricity every day—an amount roughly equivalent to the output of two nuclear power stations.

The tidal barrage on the Rance river estuary, at St. Malo in northern France, has been commercially operative since 1967. As the tide rises, six large sluice gates across its 1km (0.6mi) span are opened to fill the estuary, which acts as a reservoir. When the tide falls the gates are closed and the water is released through 24 turbine generators, producing a total of 240,000kW of power. Chromium-nickel steel is used for the turbine blades to prevent corrosion by aggressive seawater. Contrary to expectation, the estuary above the dam has not silted up, and the large artificial lagoon created there has led to a great expansion of water sports and leisure use.

In the days of windmills and sailing ships, the world had never heard of energy shortage and the air was still sparkling clean. Today, however, when both energy and environmental crises loom ever larger, wind power is one of our most underutilized renewable resources. Wind could still, as in earlier times, be made to yield an abundant, clean, inexhaustible, and non-polluting source of energy —supplied ultimately from the heat of the Sun.

Together, sun and wind play a constant and inseparable duet, providing between them a phenomenal display of untapped power. Each year the Sun gifts Earth with nearly a million times as much energy as is locked up in all the planet's oil reserves. A significant proportion of this solar energy, unevenly distributed as it is across a spinning world, is used to stir the planet's winds.

Slowly, we are learning again to use these winds to our collective advantage. Nearly 100 countries can currently claim to be putting wind power to use in some form, but generally the projects involved are local and small-scale, and almost all onshore.

The USA leads the way in total capacity; giant wind farms in California now produce some three-quarters of the planetary total of wind-generated power. But, in the world's most energy-hungry nation, the amount generated still accounts for a mere 1 percent of the country's annual needs.

Several European countries, notably Denmark, Germany, Italy, the Netherlands, and Greece, also have national targets for wind power, and are already making significant use of this resource. The UK has perhaps the greatest potential, since it has been estimated to command 40 percent of Europe's total realizable wind-power resources. So far, however, Britain's commitment has been negligible, in part at least as a result of issues of competition for land utilization, along with environmental concerns about the alleged "visual intrusion" of onshore wind farms on familiar and cherished landscapes.

However, less than a quarter of the European wind-power potential comes from onshore schemes. Most derives from offshore capacity, for stronger and more consistent winds are found at sea than on land. This fact is now beginning to turn attention toward the offshore prospects. The Republic of Ireland has already determined to tap into this resource, approving plans for a major offshore wind farm to be constructed on a sand bank in the Irish Sea, just south of Dublin.

An elegant row of wind turbines lines the Jutland coast of Denmark. Just as Norway, although richly endowed with offshore oil and gas, now produces 90 percent of its commercial energy needs from hydroelectric power, so Denmark has major plans for offshore wind energy.

Across the continent as a whole, however, Denmark is leading the way. The world's first commercial offshore wind farm is located just 1.5–3km (1–2mi) offshore from Vindeby, on Lolland Island in the Baltic Sea. There, 11 wind turbines sited in up to 5m (16.5ft) of water are generating 12 million kilowatt hours of electricity.

Still more ambitious are two further schemes currently under construction. At the Horns Rev site, 80 giant turbines are spaced at regular intervals across 20 sq km (8 sq mi) of the Danish continental shelf from 6 to 14m (20–45ft) deep. Completely out of sight of land, they rise 70m (230ft) above the water, with a rotor diameter of 80m (260ft).

Provided that these schemes are successful, Denmark will be well on the way toward meeting its target of supplying 40 percent of its total electricity requirement by 2030. Offshore wind power will, once again, have come of age.

Serious studies have been made, focusing on currents that are accelerated due to restricted flow through straits, around major headlands and peninsulas, or in the entrances to lochs and fjords. The calculations of the amount of potential energy that could be supplied by these hydroelectric hotspots are impressive. Pilot schemes have even been tested successfully utilizing both river flow (on the River Nile at Juba in 1980) and tidal flow (in the Kurashima Straits off Japan in 1988). Current power may well be a resource for the future, but is of limited immediate interest.

Hot and cold Closer to reality, and elegant in its simplicity, is the idea that we could tap into the almost infinite supply of thermal energy in the oceans, and convert this directly into electricity. The input of solar energy to the ocean surface is continuous and quite staggering in its extent. The weather systems already utilize this energy store in a variety of ways, including the generation of tropical cyclones: three or four average storms of this kind together draw enough power from the heat of the sea to fuel the whole world's industrial production for a year. Of course, we must ask whether we could do the same, and how.

The principle of ocean thermal-energy conversion (OTEC) is to use the difference in temperature between warm surface water and cold deep water to power a turbine and thus generate electricity. In tropical seas, surface waters at around 25°–30°C (77°–86°F) overlie cold waters of 4–7°C (39°–45°F) at depths of 500–1,000m (1,650–3,300ft). The warm water can be used to vaporize an intermediate fuel, such as ammonia, which in turn serves to power the turbine. Cold water pumped up from the depths is put to use to condense the ammonia back into liquid form, thereby creating a vacuum that causes the gas to circulate. An initial pilot scheme was built in Cuba in 1930, and the first mini-plant actually used to produce power was operated successfully off Hawaii in 1979. However, about 80 percent of the 50-kilowatt output was needed to pump the cold water up from depth, leaving a net output of only 10–15 kilowatts—obviously a rather low level of efficiency.

However, the potential is undoubtedly there. The Natural Energy Laboratory at Keahole Point in Hawaii has now developed an experimental 210-kilowatt open-cycle OTEC plant. It has been calculated that the energy locked up in such thermal gradients within the oceans is nearly 10,000 times that available from tides and waves combined.

The ocean is a vast storehouse of mineral wealth, from salt to diamonds,
fertilizers to valuable metal ores. Some we extract in abundance;
much more lies unknown or untouched.

**HIDDEN RICHES
OF THE OCEAN**

Minerals from the Sea

KEY TERMS

MINERAL

MANGANESE NODULE

ELEMENT

HYDROTHERMAL VENT

PLACER DEPOSIT

AGGREGATE

ARAGONITE

PHOSPHORITE

DESALINATION

Since their first discovery during the pioneering scientific voyage of HMS *Challenger*, which circumnavigated the world between 1872 and 1876, manganese nodules have remained something of an enigma. They occur as a bizarre, polka-dot carpet strewn across the seafloor, generally concentrated at depths of 4,000–5,000m (13,000–16,500ft). They are typically potato-shaped, ranging in size from about 2cm to 15cm (1–6 in), brownish-black in color, and strangely light to touch—in fact, about the same weight as a potato of similar size.

The nodules are mainly composed of manganese and iron oxides, abundant metals on land and of little economic interest. However, an attractive feature is their high content of copper, nickel, and cobalt, sometimes averaging 2–3 percent of their weight. Numerous other constituent metals, including zinc, lead, molybdenum, titanium, and vanadium, also add to their value. Pilot schemes for their mining, involving a giant vacuum-cleaner suction process, have proved successful, but economic conditions for their wholesale exploitation are not yet right.

The story does not end there, however, for the nodules pose many scientific questions. How, why, and where do they grow? Careful studies over the past decades have revealed some of the answers, but by no means all, and further scientific advance may have to wait on a renewed economic impetus.

Typically the nodules grow as fine, concentric bands around a tiny nucleus or core, usually a fragment of rock or shell, a shark's tooth, or a broken piece of an existing nodule. The source of the

Manganese nodules (*below*), rich in a variety of metals, are often highly concentrated on the ocean floor, yet their commercial extraction is far from straightforward. Currently they remain a scientific curiosity—nearly always at the sediment surface, rarely buried, and sometimes as much as 10 million years old.

metal elements concentrated in these layers is one of the unresolved issues of nodule study. Their ultimate provenance must be continental weathering, submarine volcanism, or cosmic debris—or, most likely, a combination of these sources. The more immediate supply, however, is probably chemicals either recycled from water held within the underlying sediment, or precipitated directly from the ocean bottom-water.

We have been able to measure the growth rates of nodules from many different sites, and almost invariably they are a matter of only a few millimeters per million years. This is far slower than the rate of sedimentation over much of the ocean floor, and yet they occur very largely at the sediment surface. The larger nodules must therefore have been growing slowly and imperceptibly in the same place for 10 to 15 million years, yet, paradoxically, without becoming buried in sediment. Three different mechanisms might explain their surface position: strong bottom currents may sweep away any sediment falling from above, or seafloor-dwelling organisms may remove any particles that do fall; also, as the nodules remove metals from the topmost sediment layers, a chemical gradient may form. Mobile elements would migrate along this gradient from the higher concentrations at depth toward the depleted sediment surface. Once there, they, too, would be precipitated onto the nodule in a continuous process.

Polymetallic vents Two more recent discoveries have opened our eyes once again to the untold metallic wealth of the deep ocean. In the early 1960s, hot brine pools were discovered in the Red Sea, the new ocean spreading zone along which Africa and Arabia are pulling apart from each other. Beneath these brines, formed by hot fluids leaching salt and other elements from the underlying rocks, are muds rich in metals—iron, manganese, zinc, copper, cadmium, lead, and silver. The Atlantis Deep is one of several narrow, elongated deposits, 200sq km (80sq mi) in area and containing some 3 million metric tons of zinc, 1 million tons of copper, 80,000 tons of lead, and 5,000 tons of silver. Of all the deep-sea metal deposits, these are the ones most likely to be mined first. Similar deposits are known to occur in the Guaymas Basin in the Gulf of California.

The second discovery, in 1975, was of similar hot, hydrothermal fluids discharging along the axial valley of the East Pacific Rise, associated with metal-rich deposits. Many such vent sites have now been found along the crest of the global mid-ocean ridge system, where seawater circulates through newly formed hot volcanic rocks, scavenging metals into solution and heating up in the process. At some sites, giant chimneys are formed of condensed metal sulphides (commonly copper, iron, and zinc), in places up to several hundred meters high. As hot (250°–350°C/480°–660°F) acidic fluids gush out of the summit and meet icy-cold ocean bottom water, metal sulphides precipitate rapidly, partly adding to the construction of the chimney and partly forming a dense, black, particulate smoke. These upwardly buoyant hydrothermal plumes, from which the vents derive the name "black smokers," spread out over a broader area, building sizeable mounds up to 200m (650ft) across. Although the costs of recovery are still some way from being economically viable, preliminary data show that such bodies contain several million metric tons of ore, and compare favorably with the largest massive sulphide deposits that are being mined on land.

Sparkling sands Far from the inaccessible metals of the ocean floor are the sands that adorn the world's coastlines. Yet these can also hold treasures. Those along the shores of Namaqualand (South Africa) and Namibia are rich in diamonds, and have been the site of mining operations for many decades. The jewels, formed at extreme pressures deep within the Earth, are brought to the surface in volcanic vents known as kimberlite pipes, which occur throughout southern Africa. Transported to the coast in rivers and then swept north by longshore currents, they now add an extra sparkle to the beach sands stretching over 100km (60mi) along the coastline.

More commonly, however, the mining carried out on beaches and in offshore sand deposits is for concentrations of heavy metals known as placers. The wash of waves, the pulse of tides, and the constant action of other marine currents in the nearshore zone sift and sort the sand and gravel, gradually concentrating those mineral grains with higher specific gravity—the heavier ones—into pockets and depressions along the coast.

The type of mineral found in these placer deposits depends on the local geology. Gold has been mined off Nova Scotia and Alaska, cassiterite (tin ore) off Malaysia, Indonesia, and Thailand, and chromite off the Oregon coast. Dredging operations now routinely extend down to water depths of 50m (165ft) in areas where concentrations of ore minerals for chromium, titanium, zirconium, tungsten, and the rare earth elements (lanthanides) have been found. Such deposits can be enormously valuable and are quite readily mined—the United States, for example, estimates the value of titanium in its coastal zone to be in excess of $35 billion.

Industrial wealth In a world where an ever-burgeoning population is steadily migrating from the countryside to the city, it is small wonder that the construction industry is booming. More houses, apartments, and high-rise business premises, more roads and airports, more hotels and tourist resorts all require copious amounts of the basic building materials—sand and gravel aggregates, together with cement. With an annual output of over 8 billion metric tons, the aggregate business is by both volume and value the second largest extraction industry in the world after hydrocarbons.

Marine aggregate resources, particularly in those countries that are highly industrialized and densely populated, are gradually replacing traditional supplies from rivers and other onshore areas. Beaches are the obvious sources of supply, but their use now presents a serious conflict of interest; with the galloping growth of tourism, their amenity value has come to supersede their value as construction materials. The next step is offshore. It is perhaps fortunate, therefore, that at least half the surface of the world's broad expanse of continental shelves is covered by

From the metal-encrusted "chimneys" (*far left*) of hydrothermal vents, hot, mineral-rich water discharges out of the ocean crust at mid-ocean ridges around the world. These deep-sea deposits remain a potential economic treasure for the future.

Diamond mining on the Namibian coast (*below*). Tens of thousands of diamonds, stripped by erosion from the interior of southern Africa, lie hidden in the sands.

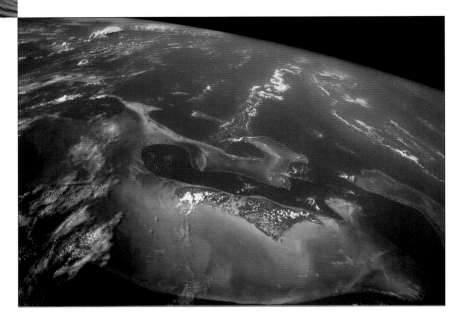

Below the clear blue waters of the Bahamas lie vast carbonate banks (turquoise areas, *above*), source of the mineral aragonite— "white gold." The islands' coral reefs and white sands also provide raw materials for the cement needed by the local tourist industry for the construction of a new airport, roads, and hotels.

region of the country, where 25 percent of resources now come from dredging in carefully licensed areas.

The other part of the construction equation, a source of lime for cement production, is still largely available from onshore limestone quarries. But island communities with no suitable supply must turn elsewhere. The brilliant white beach sands of coral islands, composed of broken reef fragments and shell debris, present an attractive alternative source. Tourist resorts built from these resources, however, may have traded short-term gain for long-term disaster, for natural coastal protection is all too easily removed by such operations, and the delicate reef ecosystem disturbed. It is not only coral islands, however, from which lime materials are gathered. Production occurs off Sri Lanka, Australia, and the United States. Even Iceland, with its dark volcanic rocks and complete absence of limestone, has turned to offshore dredging for shell-rich sand and gravel.

Other minerals are also obtained from the shallow marine environment. Very pure quartz (silica) sand is taken from beaches for the glass industry. A pure form of calcium carbonate (aragonite) is produced in large quantities from the warm seas of the Bahamas. Locally known as "white gold," it is an odorless, tasteless, dustless, non-toxic substance with multiple uses in the steel, glass, chemical, agricultural, and food industries; some estimates put the size of this resource at over 100 billion metric tons. Phosphorite is a sedimentary material containing various phosphate minerals, of prime importance in the manufacture of fertilizer. It is widespread as a recent deposit across continental shelf and upper slope regions in many parts of the world, generally at depths of less than 1,000m (3,300ft). Although onshore supplies have so far met global demand, the resource is there should future needs dictate.

ample supplies of sand and gravel. These are relict deposits from former rivers, beaches, and coastal dunes, formed during the past ice age when sea levels were much lower.

Japan is the world leader in offshore aggregate production, accounting now for almost 50 percent of the world total. Much of this comes from depths of less than 50m (165ft), but where the deposits are close to shore there is a danger in removing the natural offshore supplies for beaches and for coastal protection against major storms. Because of its location on a tectonically active plate margin, Japan has mainly narrow shelves, so that dredging operations are having to move into progressively deeper water. The UK is another major producer of offshore aggregates, especially to feed the construction-hungry southeastern

The principal mineral resources (apart from sand and gravel) currently available from the ocean realm include a wide range of metals and minerals extracted from placer concentrations in beach and shelf sands. Large-scale coastal operations have been developed to process millions of tons of sand for anything from iron ore to gold and diamonds. Huge metal resources farther out to sea await exploitation when economic conditions are more favorable. Offshore phosphorite deposits may soon substitute for onshore supplies of fertilizer minerals.

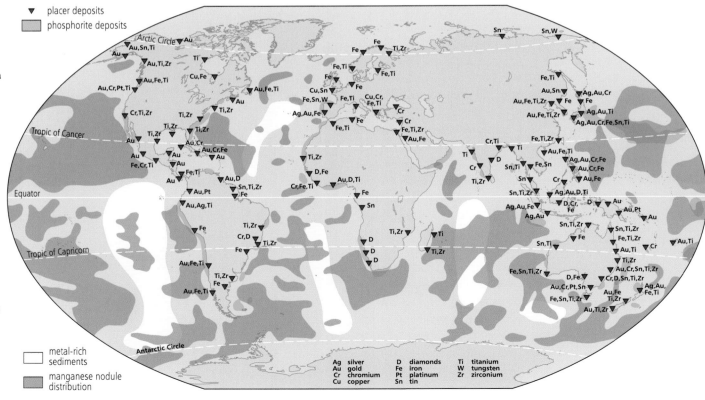

▼ placer deposits
▨ phosphorite deposits

☐ metal-rich sediments
▨ manganese nodule distribution

Ag	silver	D	diamonds	Ti	titanium
Au	gold	Fe	iron	W	tungsten
Cr	chromium	Pt	platinum	Zr	zirconium
Cu	copper	Sn	tin		

Of all the sea's mineral riches, salt has the longest history of use by humankind. Roman soldiers were even paid part of their wages as a salt allowance (Latin *salarium*), hence the term "salary." Through the ages the mineral has contin- ued to be produced in the same way, by the direct evaporation of seawater, as at this extraction plant on Lanzarote, Canary Islands. Although salt is still used exten- sively in the food and chemical industries, its economic impor- tance is now rivalled by a byproduct of the extraction process— fresh water. For the Arabian Gulf nations, the desalination of seawater is an essen- tial part of life.

Fresh water from salt The sea itself is also a vast storehouse of dissolved minerals. All 102 naturally occurring chemical elements can be found in seawater, derived ultimately from the erosion of continents and the input of fluids associated with mid-ocean-ridge volcanism. Most, however, are carried in such trace amounts that their extraction is uneconom- ic. Of the few that are sufficiently abundant as well as useful to be currently extracted on a commercial basis, by far the most common is sodium chloride, or common salt, which makes up 71 percent of the dissolved solids found in seawater.

Salt has been of prime importance for cooking and for trade for at least the past 5,000 years. In the past as now, the principal means of extraction was by direct evaporation of seawater, sever- al stages of crystallization being required to remove some of the magnesium, calcium, and iron compounds that otherwise make for a rather bitter taste. Today, evaporation accounts for about one-third of the world supply of salt, mainly from large-scale plants in India, Mexico, France, Spain, and Italy.

For many countries in arid regions, an even more important extract from saltwater is fresh water; desalination plants are a vital way of obtaining scarce supplies for both domestic and industrial purposes. Such plants are becoming increasingly common throughout the Middle East, the Mediterranean, and the North African region, and look to become still more impor- tant as an added supply for major coastal cities the world over.

The only other elements currently extracted commercially from seawater are bromine and magnesium. There is also con- siderable interest in the potential for extracting uranium for the nuclear industry; pilot research schemes have been successful, but they have not yet been followed through to full production. As for the other dissolved compounds, it is unlikely that we will move closer toward extraction for many years to come. Far more pressing are the tasks of understanding the environmental con- sequences of any extraction that does take place, and determin- ing the ocean's response to the enormous volume of waste prod- ucts currently dumped at sea.

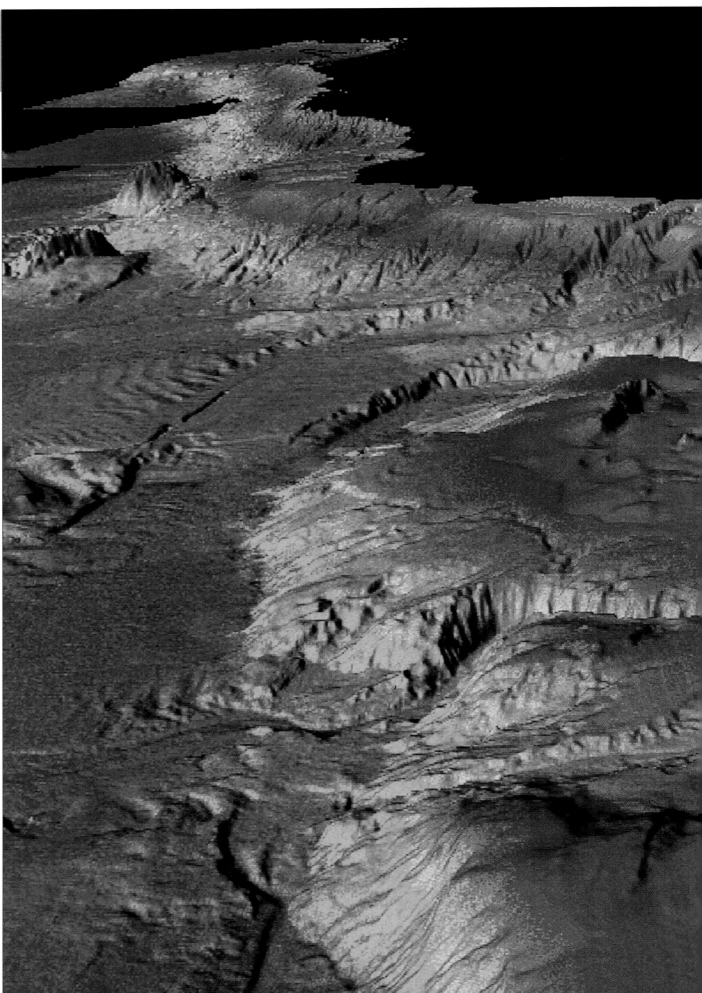

This high-definition image of the east Pacific seafloor was mapped using sonar (*s*ound *n*avigation *a*nd *r*anging) equipment. Pulses of sound emitted from a transmitter/receiver towed behind a research vessel bounce back off the seafloor and, with the help of computers, build up a mosaic image of near-photographic clarity. Artificial colors are used to emphasize detail. Here they show the depths of the ocean floor. The broad orange area is the continental shelf and upper slope off California (black), to a depth of 1,000m (3,300ft) below sea level. The continental slope is deeply incised by submarine canyons and plunges steeply to the deep-ocean basin (blue) at 3,000m (9,900ft).

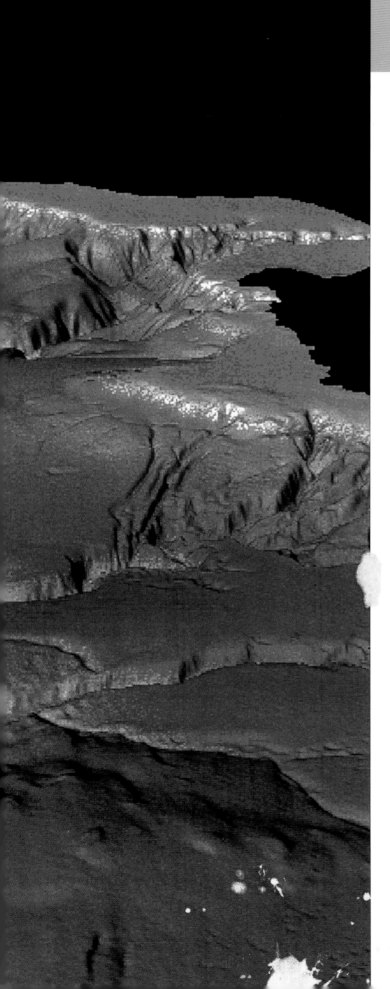

THE OCEAN FLOOR

Revealing a hidden landscape

For many years now, seafarers and scientists have plumbed the ocean depths, slowly compiling better and better maps of the seafloor. These are bathymetric maps, showing the depth of the bottom measured from the sea surface—the exact reverse of topographic contour maps, which show the hills, mountains, and valleys that course across the continents.

It was not long before oceanographers came to realize with astonishment that the seafloor bathymetry is every bit as varied and unique as topography on land. In the seas' depths there are submarine mountain chains many times longer than those on land and trenches that could swallow the Grand Canyon six times over. Not only are the mid-ocean ridges alive with earthquakes and volcanoes, but huge outpourings of lava, quite dwarfing India's Deccan Traps, lie quietly hidden beneath 3km (2mi) of water. In the pitch blackness of the ocean bottom, great abyssal plains spread out, monotonously featureless and more remote and hostile than the Siberian tundra or the Sahara Desert.

Explaining these new views of a world we still cannot see, as well as the magnetic, seismic, sedimentary, and other properties of the underwater landscape, has completely revolutionized ocean and earth sciences. But the mapping continues—from satellite observations, from deep-towed surveying craft, and from remotely operated submarine vessels that probe the ocean floor beneath polar sea ice. Scientists are awaiting with great excitement the fresh revelations that must inevitably result.

From icy fjordland to crocodile-infested mangrove swamps, from the stormy Wild Coast of South Africa to the Doldrums' unsettling calm, the Atlantic is truly the ocean of a thousand faces.

THE OCEAN FLOOR

The Atlantic Ocean

The broad, elongated S-shape of the Atlantic Ocean splits the world in two, stretching over 15,000km (9,000mi) from the Arctic in the north to the Southern Ocean that encircles the continent of Antarctica in the south. Although once no more than a narrow rift through the middle of the ancient supercontinent of Pangea, the Atlantic has very gradually widened over the past 200 million years, to as much as 5,000km (3,000mi) in places. Yet, as the landmasses to either side of the mid-ocean ridge system that runs down its center have drifted apart, a near-perfect jigsaw fit of the continental margins has been maintained. It was this close geometrical match that first inspired early notions of continental drift in the mid-19th century. Careful mapping of the ocean floor and its magnetic properties revealed that the crust became progressively older with distance from the rift valley, and ensured that seafloor spreading became firmly embedded in the paradigm of plate tectonics.

Ocean data

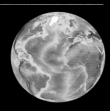

ATLANTIC

Area	82,000,000km² (31,660,000mi²)
Average depth	3,300m (10,830ft)
Maximum depth	8,605m (28,230ft) (Milwaukee Deep)
Volume	321,930,000km³ (77,231,000mi³)
Principal plates	African, Eurasian, N. American, S. American
Minor plates	Caribbean, Anatolian
Oldest ocean crust	175 million years (Middle Jurassic)

GULF OF MEXICO

Area	1,543,000km² (596,000mi²)
Average depth	1,512m (4,960ft)
Maximum depth	3,996m (13,110ft)
Volume	2,322,000km³ (560,000mi³)

CARIBBEAN SEA

Area	2,640,000km² (1,020,000mi²)
Average depth	2,520m (8,270ft)
Maximum depth	7,686m (25,220ft) (Cayman Trench)
Volume	6,652,800km³ (1,596,000mi³)

NORTH SEA

Area	575,000km² (222,000mi²)
Average depth	94m (310ft)
Maximum depth	700m (2,300ft) (Skagerrak)
Volume	54,000km³ (13,000mi³)

MEDITERRANEAN SEA

Area	2,966,000km² (1,145,000mi²)
Average depth	1,500m (4,920ft)
Maximum depth	4,982m (16,350ft) (Hellenic Trench)
Volume	4,449,000km³ (1,067,000mi³)

Hidden margins The Atlantic, including both the North Atlantic (this page) and South Atlantic (next page), now ranks as the world's second largest ocean, although with an area just less than half that of the Pacific. However, its drainage area is four times as large as the Pacific's, and it is the combined sediment influx from some of the muddiest rivers in the world that has helped shape the bathymetric features of its margins. In fact, these are classic divergent margins, with the continental shelf, slope, and rise draped in sediment. This has accumulated in parts to over 15km (9mi) in thickness,

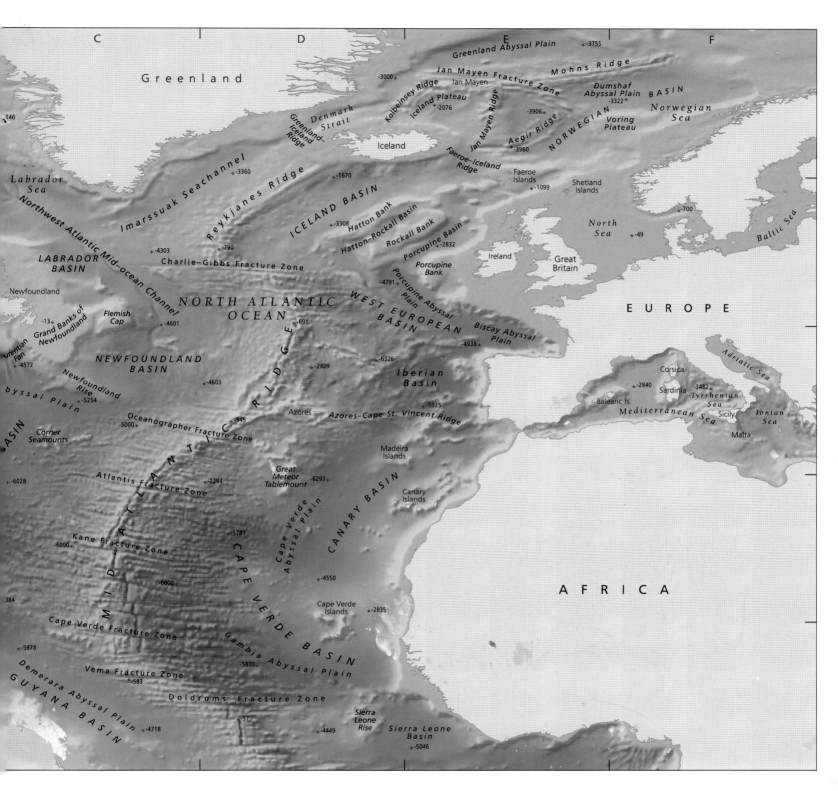

especially where major rivers —the Amazon, Niger, Congo, and Mississippi—have followed a similar route for many millions of years. Some of the continental shelves, such as the Grand Banks, the North Sea, and the southwest approaches of the United Kingdom, are over 250km (150mi) wide, whereas those off northeastern Brazil and much of west and northwest Africa are only some 20–40km (12–25mi) wide. The shelf edge is typically 100–200m (330–660ft) deep, and marks the true edge of the continents. It is underlain by continental crust, and is mostly exposed at times of lowered sea levels.

Steeper slopes (2°–5° gradient), coupled with gentler continental rises (less than 2°), extend for up to several hundred kilometers beyond the shelf edge. Somewhere deep beneath their thick sediment cover lies the transition between continental and oceanic crust—a fundamental boundary of still unknown complexity. Also trapped within these great sedimentary basins are hidden riches yet to be discovered—deep-water oil and gas resources we are just beginning to find. In places, the slopes are deeply incised by numerous submarine canyons and channels that effectively funnel sediment outward from the continental

The vast mountain range of the Mid-Atlantic Ridge dominates the topography of the Atlantic. The North Atlantic section began to open during the Jurassic period, around 175 million years ago, splitting North America from Gondwana and Europe.

shelves. Elsewhere they are smoothed by strong bottom currents or sculpted into elongate mounds, giant sediment waves, and long, linear furrows. Eventually the slopes give way to almost horizontal, thinly covered abyssal plains, which stretch across very large areas of ocean floor. These are truly some of the darkest, most remote, and least known expanses of the seabed.

Submerged mountain chain The topography of the axial region of the Atlantic Ocean is dominated by the very long and continuous submarine mountain range known as the Mid-Atlantic Ridge. This rises from abyssal depths to within 1.5-2.5km (1–1.5mi) of the sea surface, only becoming emergent at rare ocean islands, including Jan Mayen, Iceland, the Azores, Ascension, and Tristan da Cunha. The whole of the ridge is seismically and volcanically active; the youngest and freshest volcanic emanations can be found periodically bubbling to the surface, where they add incrementally to the ocean crust. The main zone of activity is within a narrow axial valley, between 25 and 50km (15–30mi) in width. Volcanic lava wells up and squeezes out like wads of toothpaste, forming elongate pillows on the seafloor. Vents of hot, metal-rich fluids (black smokers) occur sporadically along the ridge. Spreading rates at the present day range from 2–3 centimeters (0.8–1.2in) per year in the north to 4cm (1.5in) in the south—about the same rate as the growth of fingernails. A characteristic feature of this and all other mid-ocean ridges is their lengthwise segmentation into a series of much smaller sectors by numerous transform faults. These are expressed topographically by narrow ridges and deep linear clefts, which extend from tens to hundreds of kilometers on either side of the ridge axis.

The emergent islands of the mid-Atlantic have mostly resulted from very prolonged volcanic activity. These eruptions have led to the formation of areas of greatly thickened ocean crust, for example below the Greenland–Iceland–Faeroes Ridge in the north and the Walvis Ridge in the south. The Azores, together with a rather complex seafloor topography to the east, can be related to volcanic outpourings as the Eurasian and African plates slowly grind past each other. Islands closer to land are slivers of continent that have become temporarily detached.

Only two relatively small trench systems occur in the Atlantic. One is the Puerto Rico Trench, near 20°N, which has formed where a tiny part of the Atlantic is beginning to plunge beneath the Caribbean plate, and the other is the South Sandwich Trench, associated with localized subduction of the southernmost part of the South Atlantic plate. Mostly, though, the Atlantic is still growing at the expense of the Pacific Ocean, which is contracting. Although we now consider the Atlantic as a single ocean, its component parts were born at quite different times in geological history. The central North Atlantic dates back to the middle Jurassic, some 175 million years ago. Most of the South Atlantic first began to open 50 million years later, while the rift north of Iceland only began to separate in the early Tertiary period, 50–60 million years ago. We still do not understand the process that triggers the birth of an ocean, nor indeed the beginning of its subduction and destruction.

About 50 million years after the initial opening of the Atlantic took place in the northern hemisphere, rifts began to appear in Gondwana, the southern portion of the Pangea supercontinent. South America and Africa began to drift apart, their original jigsaw fit still apparent today. Between the sediment-covered continental margins and the bare rock of the Mid-Atlantic Ridge lie the vast abyssal plains.

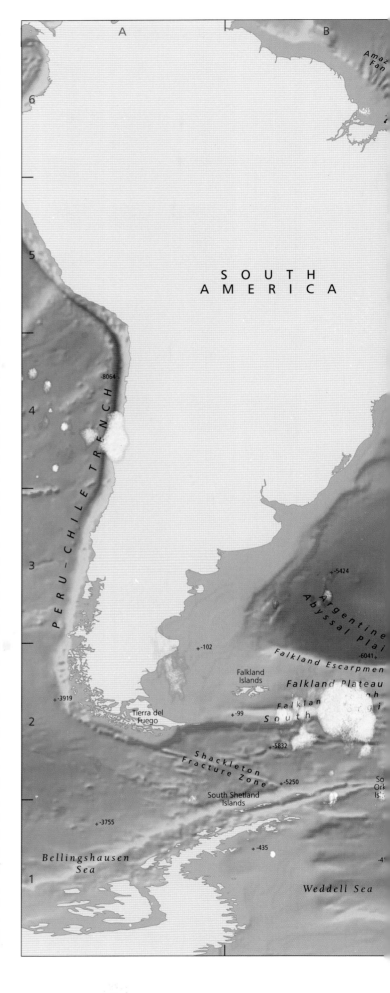

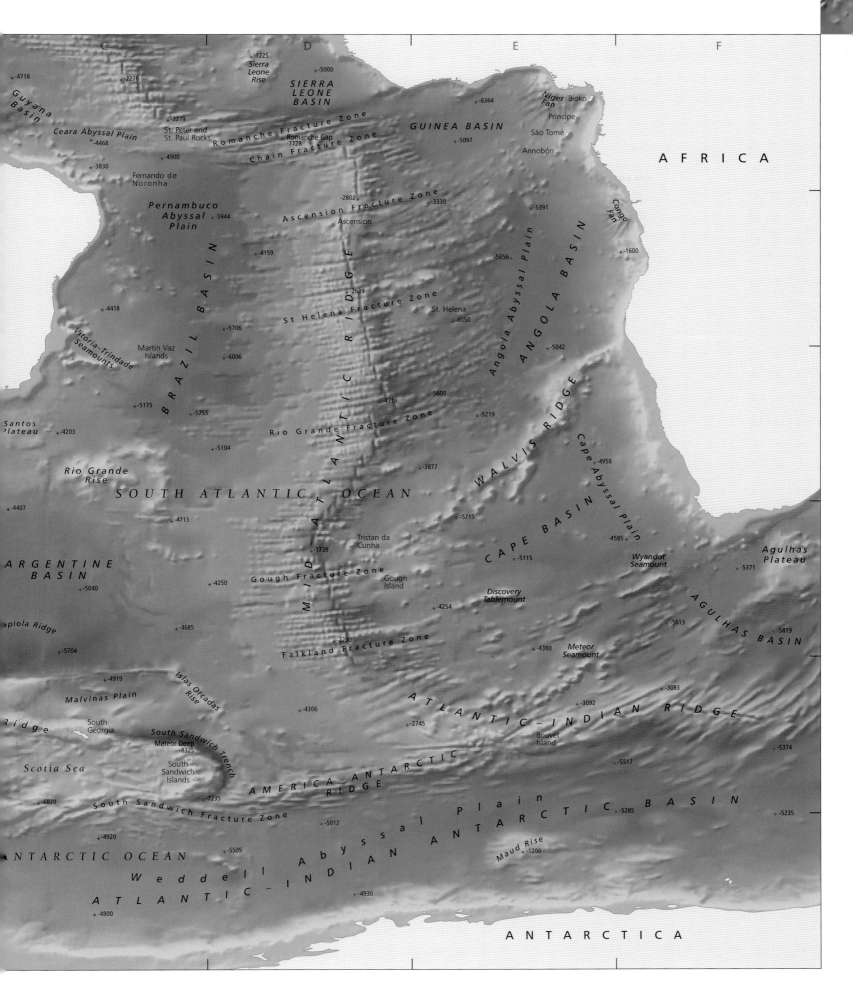

C
·-4718

·-2276

Guyana Basin

Ceara Abyssal Plain
·-4468

·-3279
St. Peter and
St. Paul Rocks

·-4900

·-3830

Fernando de
Noronha

D
·-1225
Sierra Leone Rise

·-5000

SIERRA LEONE BASIN

Romanche Fracture Zone
Romanche Gap
·-7728

Chain Fracture Zone

·-6364

E
Niger
Fan
Bioko

Príncipe

São Tomé

Annobón

GUINEA BASIN
·-5097

F

A F R I C A

Pernambuco Abyssal Plain
·-5944

·-4159

·-2802
Ascension Fracture Zone
·-3330
Ascension

·-5391

Congo Fan

·-1600

B R A Z I L   B A S I N

·-4418

·-5706

·-6006

Vitoria–Trindade Seamounts

Martin Vaz
Islands

·-5175

·-5755

·-2639

St Helena Fracture Zone

·-6050
St. Helena

·-4319

·-5600

·-5656

Angola Abyssal Plain

A N G O L A   B A S I N

·-5042

Santos Plateau
·-4203

·-4407

·-5104

Rio Grande Rise

S O U T H   A T L A N T I C   O C E A N

·-4713

·-3877

·-5219

Rio Grande Fracture Zone

M I D - A T L A N T I C   R I D G E

W A L V I S   R I D G E

·-4958
Cape Abyssal Plain

A R G E N T I N E   B A S I N
·-5040

·-4250

·-1739
Tristan da Cunha

·-5715

C A P E   B A S I N

·-4585

·-5115

Wyandot Seamount

Agulhas Plateau

·-5371

apiola Ridge

·-4685

Gough Fracture Zone

·-4254

Gough Island

Discovery Tablemount

·-5613

A G U L H A S   B A S I N

·-5819

·-5704

·-4380

Meteor Seamount

·-4919

Islas Orcadas Rise

·-2290
Falkland Fracture Zone

·-3083

·-3092

Malvinas Plain

·-4306

·-2745

A T L A N T I C – I N D I A N   R I D G E

Bouvet Island

·-5517

·-5374

Ridge

South Georgia

South Sandwich Trench
Meteor Deep
·-8325

South Sandwich Islands

A M E R I C A – A N T A R C T I C   R I D G E

Scotia Sea

·-6820

·-7235
South Sandwich Fracture Zone

·-5012

W e d d e l l   A b y s s a l   P l a i n

·-5285

A T L A N T I C – I N D I A N   B A S I N

·-5235

·-4920

·-5505

N T A R C T I C   O C E A N

A T L A N T I C – I N D I A N   A N T A R C T I C

Maud Rise
·-1200

·-4930

·-4800

A N T A R C T I C A

Atlantic marginal seas Several interesting and very different marginal seas and small oceanic basins extend to the east and west of the North Atlantic, forming irregular indentations into and between the surrounding continents. Some are no more than broad, shallow shelf seas that have temporarily inundated a low-lying portion of continent. Others are deep, tectonically active mini-oceans—ideal laboratories in which to view large-scale ocean processes.

The Gulf of Mexico and the Caribbean Sea, although neighbors, have very different characteristics and origins. Rifting of North from South America stretched the continental crust and allowed a shallow, sediment-filled embayment to develop in the Gulf area. As the two continents moved together again, accreting the material that would form the Panamanian isthmus, a "tongue" of Pacific lithosphere was squeezed eastwards to form the Caribbean plate.

The geological complexity of the Caribbean region is reflected in its bathymetry of deep, separated basins, broad shelves, narrow trenches, and a wealth of submarine ridges and mounts, many capped by the islands that make the area so popular with vacationers. While the ocean crust beneath the Gulf of Mexico and the Yucatan Basin is thought to consist of semi-isolated portions of the much larger North American plate, the Caribbean Sea covers what is probably an ancient fragment of the Cocos plate. It became isolated from the Pacific by the formation of the Middle America Trench along the west side of the Panamanian isthmus, and now forms a separate, small Caribbean plate, completely enclosed by zones of intense plate interaction. Destructive plate margins lie to the east, where Atlantic crust is consumed into the Puerto Rico Trench, and to the west, where the Cocos plate dives beneath the Panamanian isthmus. A series of

complex faults lie at the southern boundary, where South America is gradually moving westward, while to the north, a giant fault zone absorbs the lateral movement of the North American plate. The Cayman Trench, which lies along this fault, reaches a depth of 7,686m (25,220ft). Although the islands that fringe these tropical waters appear idylls of tranquillity, plate movements regularly lead to some of the most violent earthquakes and volcanic eruptions anywhere in the Atlantic realm.

On the opposite side of the Atlantic, the Mediterranean is an even larger, very shallow, and almost completely landlocked marginal sea. Its tranquil surface hides an extremely complex structure and history, resulting from the collision between the African and Eurasian plates and the closure of the once-great Tethys Ocean. A series of microplates and partial plates now jostle for position and survival, leading to seismic activity throughout its central and eastern regions. Restricted circulation caused by the sea's near isolation has periodically led to basin stagnation, especially in the eastern Mediterranean and the adjacent Black Sea, and the preservation of thin layers of black, organic-rich oozes in the sediment pile. Complete isolation due to the closure of the Gibraltar gateway led to the whole region temporarily drying out some 6 million years ago, when it became little more than a blistering white salt pan.

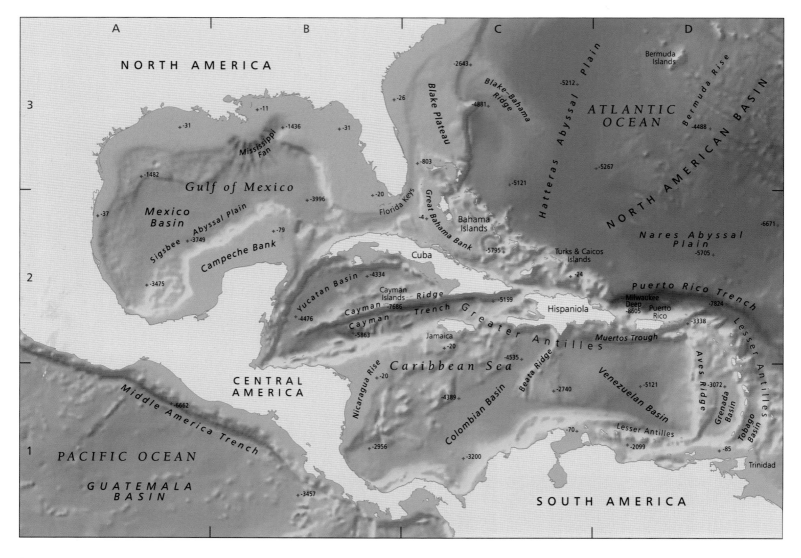

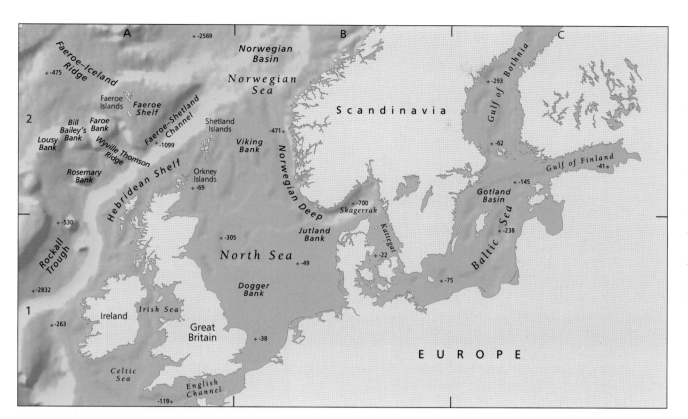

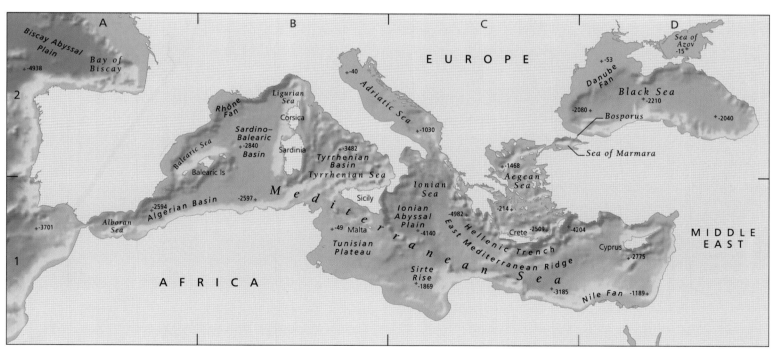

The North and Baltic
Seas are relative
backwaters, formed
mainly by the inunda-
tion of a large area of
the European conti-
nental shelf. The tec-
tonic activity that
opened the northern
arms of the Atlantic
around 55 million
years ago caused old
rifts in the crust
around Britain to be
re-activated, but no
new ocean developed.
The rift between
Greenland and North
America is now inac-
tive, but the eastern
rift between Green-
land and Europe con-
tinues to widen.

The seafloor topography beneath the shallow shelf waters of the North Sea to some extent belies its history. Although, apart from the Norwegian Deep, there are few obvious relief features, a deep rift valley, or graben, lies hidden beneath several thousand meters of sediment in the center of the basin. Rifting began in Permian times, around 285 million years ago, with a later phase during the Jurassic and Cretaceous as the North Atlantic began to open out. The North Sea rifts failed to become a new ocean, but subsidence of the rift valley over millions of years, together with changing land and sea environments, led to the accumulation of thick and varied sediments; these have become host to one of the world's richest oil and gas provinces.

During the Permian period, the North Sea was a desert plain with inland seas and salt lakes. High sea levels during the Cretaceous inundated the basin, but with the coming of the last Ice Age the region became dry once more as water was locked up in immense ice sheets. Glacier tongues cut deep fjords and sea lochs and deposited huge lobes of debris (moraine) such as the Dogger Bank. The eventual melting of the ice sheets again filled the basin and created the cold, shallow Baltic Sea.

The narrow Strait of Sicily links the Mediterranean's two halves. The Balearic Basin to the west is a smooth abyssal plain, while the eastern basin is dominated by the Mediterranean Ridge system. This is not a spreading ridge, but an area of compressed crust.

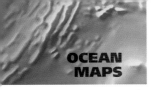
The mid-ocean ridges that dominate the bed of the Indian Ocean have run up against the northern impasse of the Himalayas. Sediments stripped from the mountains wash into the ocean as huge submarine fans.

THE OCEAN FLOOR

The Indian Ocean

KEY TERMS

SEDIMENT

SUBMARINE FAN

SEAMOUNT

The Indian Ocean is the third largest on the planet, and is relatively younger than either the Pacific or Atlantic. It began to develop about 120 million years ago from two long, narrow rifts in the supercontinent of Pangea. Spreading ridges began to form within these rift zones and effectively severed the landmass that was to become the Indian subcontinent from Africa to the west and Antarctica–Australia to the south and east.

As India moved north, so the Indian Ocean opened in its wake. Today, the active spreading ridges (the Carlsberg, Southwest, Southeast, and Mid-Indian Ridges) that dominate the ocean are elevated some 2km (1.2mi) above the adjacent abyssal plains. Together they form a giant inverted Y, the arms of which link to the circumpolar ridge system. The spreading rate is now 1.7–3.7cm (0.7–1.5in) per year for much of the ocean, as the continued northward movement of India is significantly impeded by continent–continent collision and mountain uplift.

As the mountains have grown, eroded material has fed sediment into two of the world's major drainage systems. The Indus and Ganges–Brahmaputra rivers together empty close to 1.3 billion metric tons annually into the ocean. Over millions of years, this material has spread south, draping and smoothing an irregular seafloor and building up two gigantic submarine fans—the Indus and Bengal Fans. Long, sinuous channels across the surface of these fans act as express highways transporting sediment up to 2,500km (1,500mi) into the Central Indian Basin.

An unusual feature of the Indian Ocean is the large number of shallow submarine plateaus and ridges, which locally emerge as tiny, isolated islands. Whereas the Seychelles are clearly fragments of continental crust broken off from Africa and Madagascar long ago, the Mascarene Plateau, together with Mauritius and Réunion to the south, are dominated by volcanic material. Thousands of tiny, coral-fringed islands—the Maldives—are now found capping the volcanic seamounts of the Chagos–Laccadive Ridge south and west of India. Those found over the Ninety East Ridge are more complex, and some, such as the Andaman Islands, may include fragments of off-scraped and uplifted sediment from the deep Bengal Fan.

The mid-ocean ridge system of the Indian Ocean began to form around 120 million years ago when the ancient southern landmass of Gondwana started to fragment. Madagascar and India split off from Africa, India slowly drifting north to eventually fuse with Asia, a collision that is still uplifting the Himalaya mountains today. Rifting processes also continue, with the gradually widening Red Sea a new ocean in the making.

Ocean data

INDIAN

Area	73,600,000km² (28,400,000mi²)
Average depth	3,890m (12,760ft)
Maximum depth	7,450m (24,440ft) (Java Trench)
Volume	292,131,000km³ (70,082,000mi³)
Principal plates	African, Indo–Australian, Antarctic
Minor plates	Arabian
Oldest ocean crust	120 million years (Early Cretaceous)

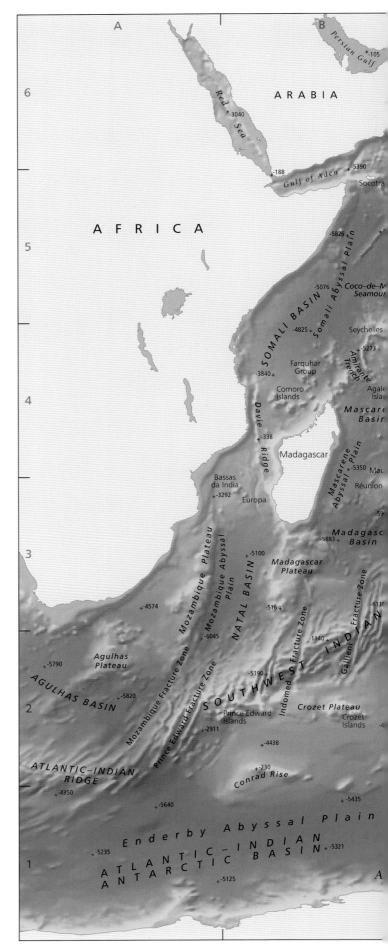

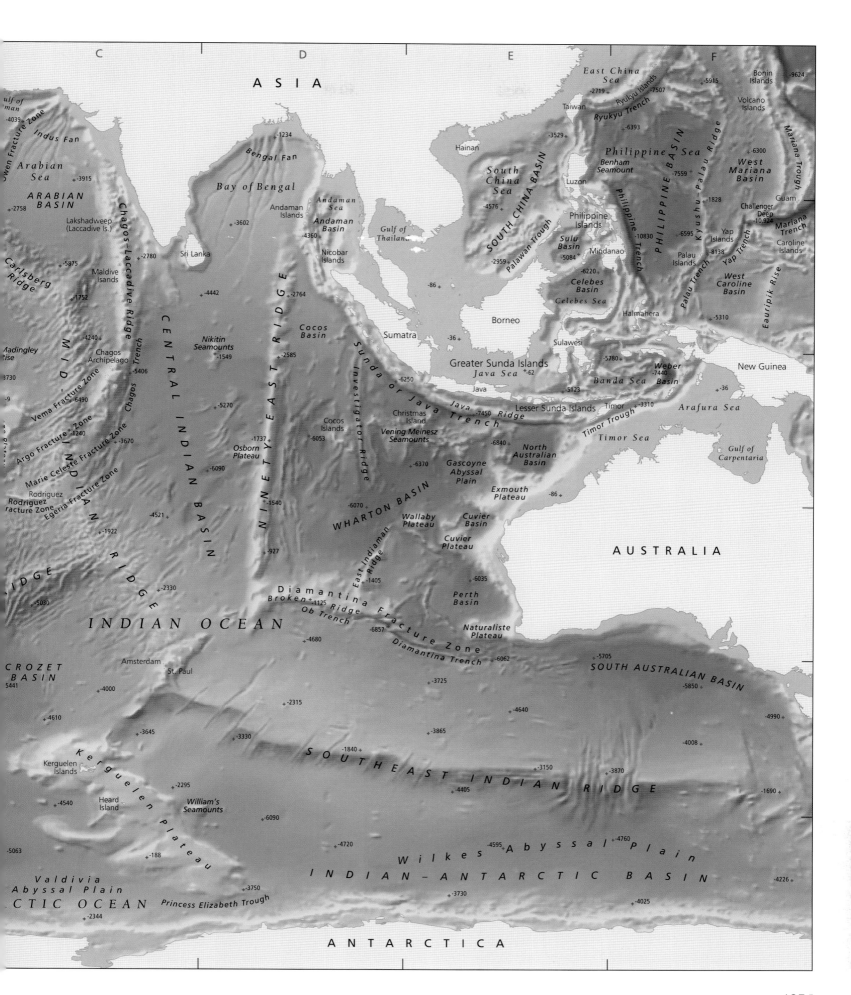

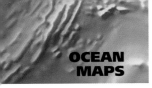
Boasting the deepest point on Earth as well as the greatest mountain, ringed by active volcanoes and earthquake zones, the Pacific, although unrivalled in size, is slowly shrinking.

THE OCEAN FLOOR

The Pacific Ocean

The Pacific Ocean dominates the globe, stretching from the Arctic in the north to the Antarctic in the south, and from the Americas to Asia and Australia. It is what remains of Panthalassa, the "world ocean" that surrounded Earth's prehistoric continents and that still covers one-third of its surface. At its widest point, the Pacific is over 17,700km (11,000mi) across. It is also the deepest ocean, averaging 4,280m (14,040ft), and it contains more than half the world's seawater.

Equally impressive is the seafloor, with its highly variable topography, more dramatic in some ways than that of any of the continents. Apart from its steep rise to the Antarctic landmass, the Pacific Basin is entirely surrounded by deep trenches. These drop far below the level of the adjacent seafloor, plummeting to as much as 10,924m (35,840ft) in the case of the Challenger Deep in the Mariana Trench—the deepest place on Earth. The trenches represent zones of subduction, along which ocean crust is being lost at a rate of over 2.5 sq km (1 sq mi) annually. They are also regions of intense seismic activity, related to great friction between the moving plates.

The Pacific is distinctive for its many hundreds of islands, from tropical to icebound and heavily populated to deserted, but particularly striking are the innumerable submerged seamounts that lie up to 4,000m (13,000 ft) below the surface. Most of these originated as volcanic peaks over hotspots and formerly active parts of mid-ocean ridges, in some cases building up above sea level only to subside after becoming extinct. Mauna Kea on the island of Hawaii is one of the most active volcanoes on Earth and, taken from its base on the ocean floor to its peak on land, is taller than Mount Everest, at 9,200m (30,200ft).

A still longer chain of mostly submerged mountains forms the East Pacific Rise and the Pacific–Antarctic Ridge. These are

KEY TERMS

SUBDUCTION ZONE
SEISMIC
HOTSPOT
SPREADING CENTER
BACK-ARC BASIN
FRACTURE ZONE
TRANSFORM FAULT
FAULT SCARP
MANTLE
OCEANIC CRUST
ISLAND ARC

The main evolutionary pattern of the Pacific Ocean consists of expansion of the Pacific plate (west of the East Pacific Rise) and destruction of plates around the margin by subduction. The prominent fracture zones to the northeast are all that remain of earlier seafloor spreading. The western Pacific is distinct in having numerous volcanic chains and several large plateaus, also of volcanic origin.

Ocean data
PACIFIC

SOUTH CHINA SEA

Area	166,000,000km² (64,000,000mi²)
Average depth	4,280m (14,040ft)
Maximum depth	10,924m (35,840ft) (Challenger Deep)
Volume	723,700,000km³ (173,615,000mi³)
Principal plates	Pacific, Nazca, Antarctic, Indo–Australian
Minor plates	Philippine, Gorda, Cocos
Oldest ocean crust	150 million years (Late Jurassic)
Area	2,318,000km² (895,000mi²)
Average depth	520m (1,700ft)
Maximum depth	4,576m (15,010ft)
Volume	1,274,900km³ (305,850mi³)

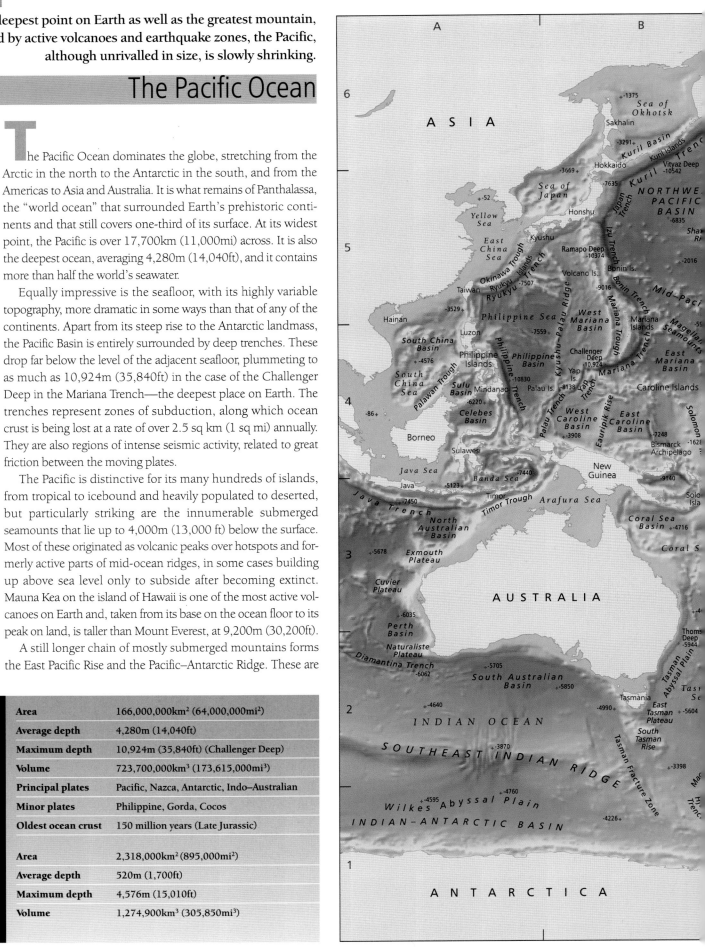

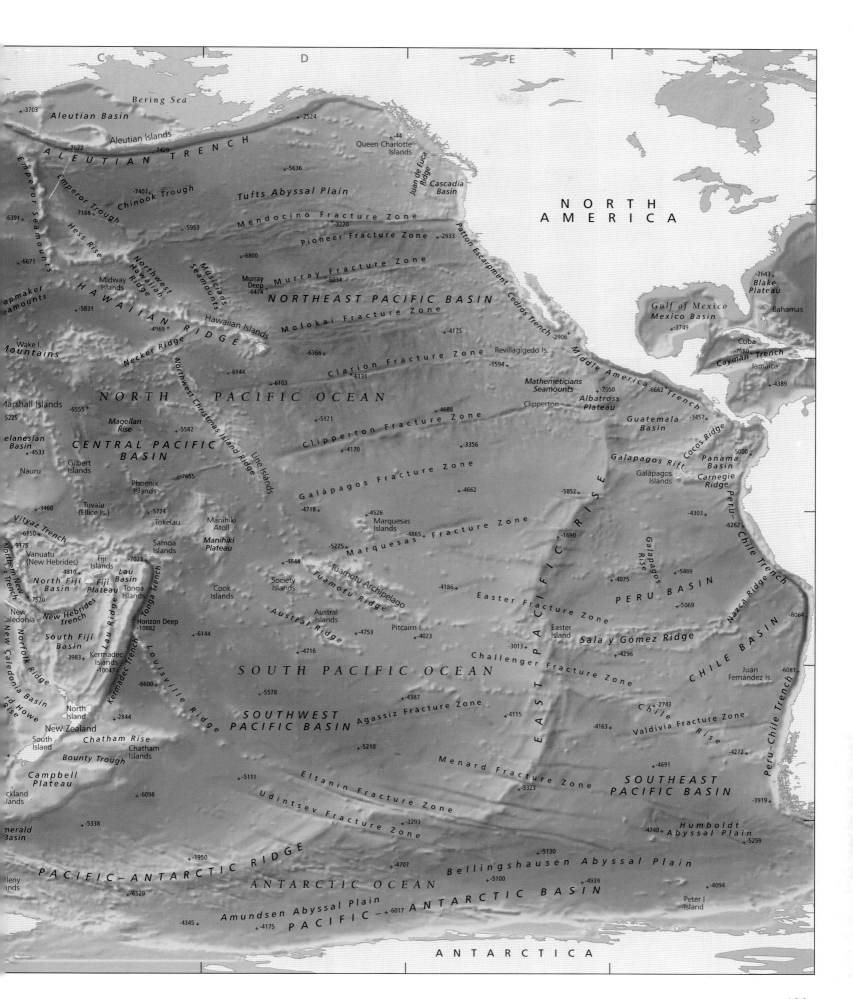

C   D   E   F

Bering Sea
+-3703
Aleutian Basin
Aleutian Islands   +-2524
-7922+   +-7429
+-7407   Tufts Abyssal Plain   Queen Charlotte
Chinook Trough   Islands   +-44
-7168+
+-5636   Juan de Fuca
Emperor Trough   Ridge   Cascadia
-6391+   Basin
+-5953   -3220
Mendocino Fracture Zone   **NORTH**
+-6671   Pioneer Fracture Zone   +-2933   **AMERICA**
+-6800   Murray Fracture Zone   Patton Escarpment
Midway   Murray Deep   -6034
Islands   +-5831   Murray   **NORTHEAST PACIFIC BASIN**   -2643+
apmaker   Deep   Cedros Trench   Blake
eamounts   -6474   Plateau
Hawaiian Islands   Molokai Fracture Zone   **Gulf of Mexico**   Bahamas
+-5831   +-4175   -2906+   **Mexico Basin**
-4969+   +-3749
Wake I.   +-6366   Revillagigedo Is.   Cuba
Mountains   Necker Ridge   +-6144   -1594+   -7680+ **Cayman Trench**
+-6144   Clarion Fracture Zone   +-4389   Jamaica
+-6103   -6131   Mathematicians
NORTH   **PACIFIC OCEAN**   +-4688   Seamounts -2050+
-6555+   +-5121   Clipperton Albatross   +-6662
Marshall Islands Magellan   Plateau   Guatemala
5225   Rise   +-5582   Clipperton Fracture Zone   Basin   -3457+
CENTRAL PACIFIC   +-4170   +-5000
elanesian **BASIN**   -7655   +-3356   Galapagos Rift Panama
Basin   Gilbert   Galápagos   Basin
+-4533   Islands   Galápagos Fracture Zone   Islands   Carnegie
Nauru   +-4662   Ridge
Phoenix   +-4718   +-4303
+-3460   Islands   +-4526   Marquesas Fracture Zone   -6262+
Tuvalu   -5724   Manihiki   Marquesas
Vityaz Trench (Ellice Is.) Tokelau   Atoll   Islands -4865+   +-1690
-6150+   Samoa   Manihiki   -5275+   Easter Fracture Zone   -5469+
-9175+   Vanuatu   Islands   Plateau   -4844   +-4075
Northern New (New Hebrides) -7023+   Tuamotu Archipelago   +-1690   **PERU BASIN**
Trench   -4810+   Fiji Lau   Cook   Society Tuamotu -4186+   -5069+
North Fiji   Islands Basin Islands   Islands Ridge   Galápagos
Basin   Fiji Lau   Austral Austral   +-4186   Easter   Rise
-7570+   Plateau Ridge Tonga   Ridge Islands Pitcairn I.   Island
New   New Hebrides Islands Horizon Deep   +-4753 -4023+   -3013+ Sala y Gómez Ridge +-4296
Caledonia   Trench   Tonga Trench -10882   +-4716   **Challenger Fracture Zone**
South Fiji   New Caledonia   +-6144   Juan   -6081+
Basin   Ridge   Kermadec   +-6144   **SOUTH PACIFIC OCEAN**   Fernández Is.
-3983+   Islands   **CHILE BASIN**
rd Howe   -10047+ Louisville Ridge -5578   +-4387   +-2743
Rise   -6600+   +-4387 Agassiz Fracture Zone +-4115   Chile
North   **SOUTHWEST**   -4163+ Rise
Island   -2844+   **PACIFIC BASIN**   +-5210   Valdivia Fracture Zone -4272+
New Zealand   +-4691
South   Chatham Rise   +-5130   Menard Fracture Zone   **SOUTHEAST**
Island   Chatham   +-3323   **PACIFIC BASIN**
Bounty Trough Islands   +-5111   +-4691   -3919+
Campbell   Eltanin Fracture Zone
Plateau   +-5111   +-4115
ckland   -6098   Udintsev Fracture Zone   -3293+   +-4740 **Humboldt**
lands   Abyssal Plain -5259+
nerald   -5338   +-5130
Basin   +-1950   +-4707 Bellingshausen Abyssal Plain
lleny   **PACIFIC-ANTARCTIC RIDGE**   **ANTARCTIC OCEAN**   +-5100   +-4939   +-4094
ands   -6529+   **PACIFIC-ANTARCTIC BASIN**   Peter I
Amundsen Abyssal Plain   +-6017   Island
-4345+   +-4175

ANTARCTICA

129

One of the features of the western Pacific are back-arc seas, such as the Philippine Sea, which stretches for 1,600km (1,000mi) between the Mariana and Ryukyu island arcs. The elements of a typical island arc system are a deep trench down which the subducting oceanic plate descends, with a parallel chain of volcanic islands lying about 100km (60mi) closer to the continent, and an adjacent back-arc basin (or marginal sea) between the island arc and the mainland. The tectonic history of this region is complex, with several stages of development, as new spreading centers and subduction trenches began and then, at a later point, ceased activity, leaving behind inactive "fossil" formations such as the Kyushu–Palau Ridge.

the mid-ocean ridges at which both the Pacific and Nazca crustal plates were formed, and along which a constant upwelling of volcanic material pushes apart existing seafloor and creates new ocean crust, at a rate only just less than it is being consumed at the trenches. This is the fastest spreading center on Earth, with the rate of separation along the crest reaching 13–16cm (5–6in) per year. Two smaller, less active ridges—the Galápagos Rift and the Chile Rise—branch eastward from it. Another minor branch is in the process of separating Baja California from the rest of North America. Among the most prominent features of the eastern Pacific are the great, linear, east–west fracture zones, made up of steep fault scarps and volcanic ridges. These are the inactive scars of transform faulting across the mid-ocean ridge.

Pacific marginal seas The Pacific Ocean has a special relationship with the continents that surround it. For the most part, the tectonic plates that floor the ocean plunge beneath the continental lithosphere, subducted into the underlying mantle along a series of deep trenches. The many volcanoes and earthquakes that rumble and shake along the plate boundaries girdle the Pacific with an immense "Ring of Fire." Against the Asian and Australian continents, former trench systems have been abandoned and new ones formed, building up curving chains of volcanic islands (island arcs) as the ocean crust is melted, as well as associated back-arc basins.

The many marginal seas created in this way are varied and unique in character. Cold and remote in the far north, the Bering Sea touches fingers with the icy Arctic, while the Sea of Okhotsk laps the frozen shores of eastern Siberia. At the other extreme, the Arafura and Coral Seas north of Australia are quintessentially tropical—palm-fringed, with coral lagoons and a rich array of marine life. In between, the Sea of Japan has fertile, temperate waters that are growing slowly wider as tectonic forces stretch the ocean floor. The basins of the Yellow, East China, and South China seas and the Gulf of Thailand are slowly being filled; sediment-laden rivers pour into them, carrying huge quantities of material stripped from the Tibetan Plateau, the high Himalayas, and the plains of China and southeast Asia.

Sprawling across the ocean from Asia to Australia, separating the Pacific from the Indian Ocean, are countless islands, all with slightly different geological histories. They are deeply weathered, and the traces of former rivers that flowed from their volcanic interior are still etched across shallow shelf seas that flooded when sea level rose at the end of the last ice age.

In other parts of the region, the seafloor plunges to oceanic depths—over 6,000m (19,700ft) in the Celebes Basin. Pacific Deep Water finds its way through a series of deep, narrow gateways between the island archipelagos, flowing out into the Indian Ocean. The whole area is swept by tropical cyclones, drenched by monsoon rains, encircled by volcanoes and earthquake zones—and yet its scattered islands are home to over 270 million people, and its seas teem with a rich biodiversity that mirrors that on land. Still baffling in their complexity, the seas of south and east Asia present a serious challenge for ocean scientists attempting to understand the mighty Pacific.

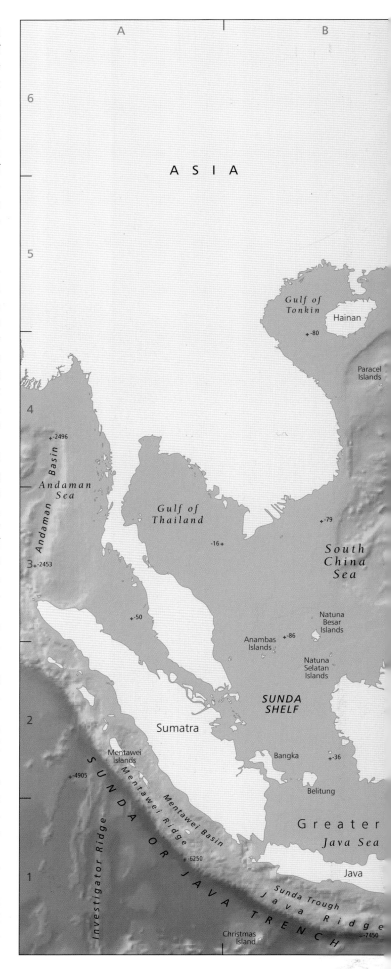

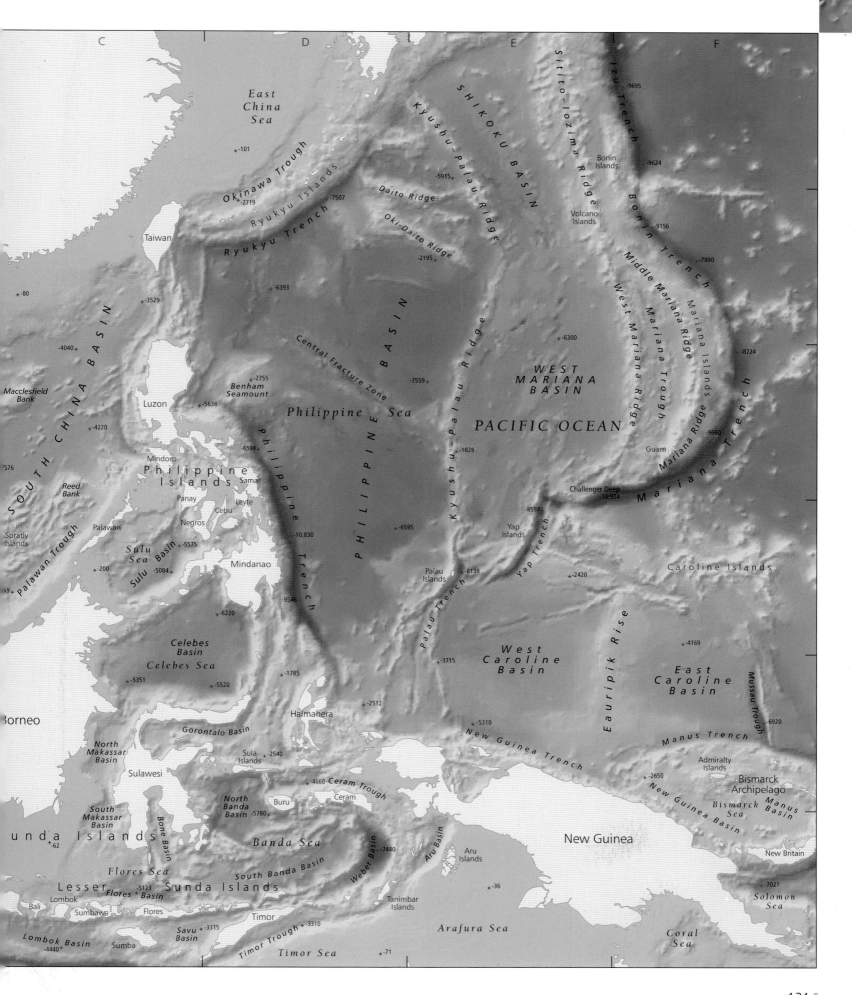

C   D   E   F

East
China
Sea

+ -101

Okinawa Trough
+ -2719

Ryukyu Islands

Ryukyu Trench

Taiwan

+ -3529

+ -80

Macclesfield
Bank

SOUTH CHINA BASIN

+ -4040

+ 576

Reed
Bank

+ -4220

Luzon

+ -5638

Benham
Seamount
+ -2755

Mindoro

+ -8594

Samar

Philippine
Islands

Panay

Cebu
Leyte

Negros

Palawan

Sulu
Sea
Basin + -5575

Spratly
Islands

+ -200

Sulu
Sea
-5084

Mindanao

-9 + Palawan Trough

+ -10,830

+ -6220

Celebes
Basin

Celebes Sea

+ -5351

+ -5520

Borneo

+ -1785

North
Makassar
Basin

Sulawesi

Gorontalo Basin

Sula
Islands + -2540

South
Makassar
Basin

Bone Basin

North
Banda
Basin -5780

Buru

Ceram

+ -4660 Ceram Trough

Halmahera

unda Islands

+ -62

Banda Sea

Weber Basin + -7440

Flores Sea

Lesser + -5123 Sunda Islands
Flores + Basin

Bali

Lombok

Sumbawa

Flores

Timor

Savu + -3315
Basin

South Banda Basin

Lombok Basin
-4440 +

Sumba

Timor Trough + -3310

Timor Sea

+ -71

Kyushu-Palau Ridge

Daito Ridge

+ -5915

+ -7507

Oki-Daito Ridge

-2195 +

+ -6393

Central Fracture Zone

Philippine Sea

PHILIPPINE BASIN

+ -7559

Philippine Trench

+ -9546

+ -1828

Kyushu-Palau Ridge

+ -6595

Palau
Islands

Palau Trench

+ -8138

-2512

+ -3715

-8597

Yap
Islands

Yap Trench

SHIKOKU BASIN

Sitito-Iozima Ridge

Izu Trench

+ -9695

Bonin
Islands

+ -9624

Volcano
Islands

+ -9156

Bonin Trench

+ -7890

West Mariana Ridge

Middle Mariana Ridge

Mariana Trough

+ -6300

WEST
MARIANA
BASIN

PACIFIC OCEAN

Mariana Islands

Mariana Ridge

Mariana Trench

+ -8724

Guam

+ -9660

Challenger Deep
-10,924

Caroline Islands

West
Caroline
Basin

+ -2420

Eauripik Rise

+ -5310

New Guinea Trench

+ -4169

East
Caroline
Basin

Mussau Trough

+ -6920

Manus Trench

+ -2650

New Guinea Basin

Admiralty
Islands

Bismarck
Archipelago

Bismarck
Sea

Manus
Basin

New Guinea

New Britain

Solomon
Sea

+ -7021

+ -36

Tanimbar
Islands

Aru
Basin

Aru
Islands

Arafura Sea

Coral
Sea

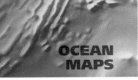
The North Pole sports an ice-covered ocean enclosed by land, the South Pole an ice-capped continent surrounded by water. These are the polar oceans—the Earth's least known, and still greatly feared by sailors.

THE OCEAN FLOOR

The Polar Oceans

Icebound, ice-cold, and inhospitable—such is the nature of the polar seas. Water temperatures are close to the edge of survival for marine life. Winters are so severe that the sea itself freezes and huge tracts of ocean are covered by ice. Summers see only partial melting, calving icebergs from glaciers and ice shelves. But there the similarity between the two polar oceans ends, for in all other respects, they are markedly different.

Arctic Ocean The Arctic Ocean is the smallest by far of the world's oceans and is almost completely landlocked. The only deep-water connection with the global ocean is through the Fram Strait between northern Greenland and the Svalbard archipelago. This narrow gateway marks where the northern extension of the Mid-Atlantic Ridge is offset along the Nansen Fracture Zone and connects to the presently active Arctic Mid-Ocean Ridge. Two other ridges cross the deep ocean—the Lomonosov Ridge and the Alpha-Mendeleyev Ridge—effectively dividing it into several abyssal plains, the deepest of which lies directly beneath the North Pole.

The huge continental shelf areas that surround these basin plains are a unique feature, underlying almost one-third of the total area. Off northern Asia and Scandinavia the shelves

Three major submarine ridges divide the Arctic Basin, with the Lomonosov Ridge rising some 3,000m (10,000ft) above the Pole Abyssal Plain. This ridge is composed of continental crust, separated from the vast continental shelf off northern Asia by the opening, during the early Cenozoic, of the Arctic Mid-Ocean Ridge.

extend from over 500km (300mi) to as much as 1,600km (1,000mi) in places—the widest shelf seas in the world. They are dotted with some of the most remote and inhospitable islands on the planet; the Frans Josef Land and Svalbard clusters on the edge of the deep Barents Sea shelf represent some of the closest solid land to the North Pole.

The other unique feature of the Arctic Ocean is its perennial cover of floating ice. In the depths of winter, the ice is up to 4m (13ft) thick, and no more than 5–10 percent of the ocean is open water—a figure that roughly doubles in the short summer months, when 24 hours of daylight help the annual thinning and melting process.

Ocean data

ARCTIC

Area	12,173,000km² (4,700,000mi²)
Average depth	990m (3,250ft)
Maximum depth	5,608m (18,400ft) (Molloy Deep)
Volume	321,930,000km³ (77,231,000mi³)
Principal plates	North American, Eurasian
Oldest ocean crust	55 million years (Paleogene)

ANTARCTIC

Area	35,000,000km² (13,514,000mi²)
Average depth	3,350m (11,000ft)
Maximum depth	7,235m (23,740ft) (South Sandwich Trench)
Volume	117,250,000km³ (28,130,000mi³)
Principal plates	Antarctic
Oldest ocean crust	120 million years (Early Cretaceous)

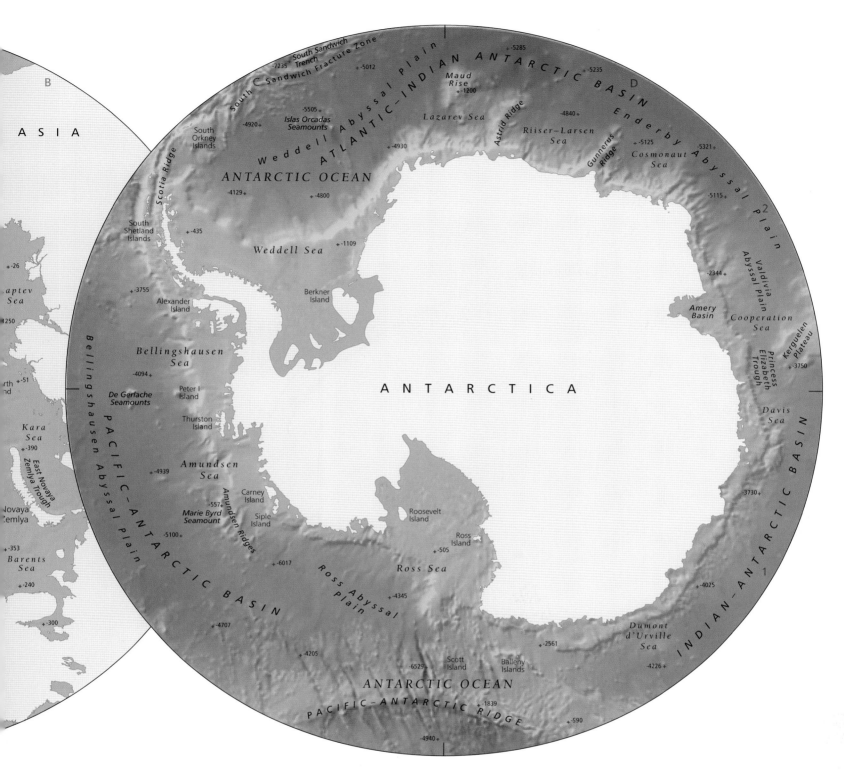

Map labels (clockwise/around the map):

South Sandwich Trench · 7235 · 5012 · ·5285 · ·5235
South Sandwich Fracture Zone
Weddell Abyssal Plain · ATLANTIC–INDIAN ANTARCTIC BASIN
Maud Rise · ·1200
Lazarev Sea
Astrid Ridge
Riiser–Larsen Sea · ·4840
Enderby Abyssal Plain
·5505 · ·4920
Islas Orcadas Seamounts
South Orkney Islands
Scotia Ridge
ANTARCTIC OCEAN
·4930 · Gunnerus Ridge · ·5125 · ·5321
Cosmonaut Sea
·4129 · ·4800 · ·5115
South Shetland Islands
·435
Valdivia Abyssal Plain
·1109
Weddell Sea
·2344
Amery Basin
Alexander Island · ·3755
Berkner Island
Cooperation Sea
Kerguelen Plateau
Princess Elizabeth Trough · ·3750
Bellingshausen Sea
ANTARCTICA
·4094
Davis Sea
Peter I Island
De Gerlache Seamounts
Thurston Island
PACIFIC ABYSSAL PLAIN
Bellingshausen Abyssal Plain
Amundsen Sea · ·4939
·3730
Carney Island
Amundsen Ridges
·557
Marie Byrd Seamount
Siple Island
Roosevelt Island
Ross Island
·5100
·505
·6017
INDIAN–ANTARCTIC BASIN
PACIFIC–ANTARCTIC BASIN
Ross Sea
·4025
Ross Abyssal Plain
·4345
Dumont d'Urville Sea
·4707
·2561
·4205
Scott Island
Balleny Islands
·6529
·4226
ANTARCTIC OCEAN
PACIFIC–ANTARCTIC RIDGE · ·1839
·590
·4940

Left-hand (partial) map labels:
B · ASIA
·26
·250
aptev Sea
Kara Sea · ·390
East Novaya Zemlya Trough
rth · ·51
nd
Novaya Zemlya
·353
Barents Sea
·240
·300

Antarctic Ocean

The Antarctic, or Southern, Ocean encircles Antarctica and is fully open to the global ocean system. At around 55°S there is a sudden rise in surface water temperature of 2–3°C (3.6–5.4°F) caused by the sinking of Antarctic surface water beneath warmer sub-Antarctic water. This border, known as the Antarctic Convergence, is generally taken as the ocean's northern limit. It is associated with high plankton productivity, and hence with a rich marine life.

The continental shelves that enclose Antarctica vary in width from less than 25km (15mi) to over 500km (300mi), and are generally much deeper than those found in other oceans. They have formed through crustal downwarping caused by the huge load of ice, which was even greater during the last ice age. A thick layer of sediment has built up beneath the continental slope over the years, shaped and molded by strong bottom currents. Beyond the slope lie deep basins and flat abyssal plains.

Some 90 percent of the Earth's freshwater is locked up in Antarctic ice, which is up to 2km (1.2mi) thick over parts of the continent and covers more than half the ocean in the winter months. Each year, the spring and summer melt reduces the sea ice by over 80 percent, so that only some 4 million sq km (1.5 million sq mi) are permanently frozen. The large variations in ice cover have a profound effect on the transfer of heat and moisture to the atmosphere and hence on the world's climate.

Although the Antarctic Plate is a discrete entity, defined by a ring of mid-ocean ridges, the Antarctic Ocean itself is really the southernmost extension of the Atlantic, Indian, and Pacific Oceans. The continent became isolated over the pole 30–40 million years ago, when Australia drifted north.

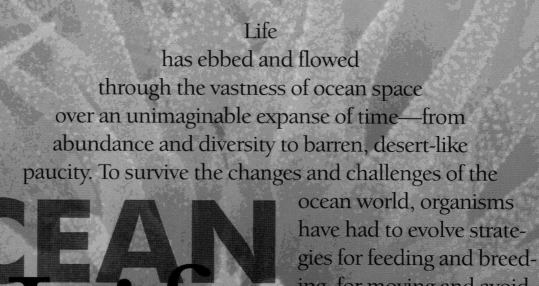

Ocean Life

Life
has ebbed and flowed
through the vastness of ocean space
over an unimaginable expanse of time—from
abundance and diversity to barren, desert-like
paucity. To survive the changes and challenges of the
ocean world, organisms
have had to evolve strategies for feeding and breeding, for moving and avoiding predation, that almost
defy belief. From sophisticated weaponry to elabo-
rate camouflage, from
lethal poisons to peaceful symbiosis, from remarkable
fecundity to inspiring migrations, the oceans are host
to a rich and intricate web of existence. Our task
is to learn how to use and enjoy this wonderful vitality without causing its
eventual demise.

The coelacanth— the only living relative of a group of fishes that flourished during the Devonian period (417–354 million years ago)—was first discovered by surprised fishermen near Madagascar in 1938. This rare creature, with its distinctive mottled markings (unique to each individual), inhabits the deeper waters of the southwest Indian Ocean. Heralded as one of the most significant zoological finds of the past century, the coelacanth was identified as a member of the primitive lobefinned fishes (Sarcopterygii). Its Devonian relatives were ancestors of the amphibians that first crawled out of the oceans, and therefore also of all terrestrial vertebrates, including humans.

EVOLUTION & EXTINCTION

CHAPTER 6

The flux of life on the blue planet

Life is inextricably bound to the oceans, from its humble origins to its evolution through eons of time and its blossoming into the rich variety we know today. Nearly 4 billion years ago, soon after the oceans themselves had condensed from the toxic atmosphere surrounding the early Earth, the first organic molecules came into being. Several theories compete to explain this remarkable occurrence, but most involve the early ocean. There life remained for over 3 billion years before first plants and then animals began to colonize the land.

Evolution was painstakingly slow—from single-celled to multi-celled organisms, from asexual to sexual reproduction, from soft-bodied creatures to those with protective skeletons. Explosions of diversity followed some novel colonization of a new ecological niche, only to be followed by catastrophic phases of mass extinction, in one of which up to 80 percent of the planet's species were wiped out. Yet somehow the thread of existence prevailed and proliferated, so that life on our blue planet is now unbelievably rich and diverse.

The march of life through the eras of geological history has always been closely linked to the natural environment. The changing nature of Earth's atmosphere and the salinity of its oceans, the rise and fall of global sea level, the drifting of continents and fluctuations of climate are all keenly felt. Humans' capacity to influence the natural environment is now well documented, but our success at riding the changes we wreak remains a great unknown.

Life began in the oceans over 3.5 billion years ago. For much of that time, before sexual reproduction provided a means to combine genetic material, the process of evolution was immeasurably slow.

EVOLUTION & EXTINCTION

Life Unfolds

KEY TERMS

AMINO ACID

PROTEIN

RNA

DNA

NUCLEIC ACID

CHROMOSOME

GENE

BASE PAIR

CYANOBACTERIA

FOSSIL

STROMATOLITE

CELL

PROKARYOTE

EUKARYOTE

PROTIST

The fecundity of our ocean world today is the direct result of eons of evolution and extinction. Even the most cursory examination of ocean life reveals its remarkable complexity and diversity. The Blue whale is probably the largest animal ever to have existed on Earth, greater even than the super-dinosaurs known from the Jurassic era, even though it lives off tiny krill, which at their biggest are no more than 5cm (2in) long. Killer whales must rank as some of the most efficient killing machines ever to have preyed in our seas, more effective even than *Tyrannosaurus rex* or the powerful saber-toothed cat were on land. Dolphins are certainly among the most intelligent species to have lived in the ocean. But the process by which this grandeur developed from the simplest and most unremarkable of single-celled organisms seems almost miraculous—although we think we now know at least part of that story. The origin of life itself on planet Earth is one of the scientific questions for which we do not expect to ever achieve a definitive answer.

First origins Geologists refer to the first 500 million years of the Earth's existence as the Hadean era, evocatively named from the Greek word for "hell." In the bubbling inferno of molten rock and erupting volcanoes that was the early Earth, battered from outer space by meteorites and asteroids, life almost certainly could not have survived. Nothing now is left of that era; it remains a blank about which we can only speculate. However, the Earth eventually cooled and the bombardment lessened. Water vapor in that early atmosphere of hydrogen, ammonia, and methane condensed and fell as rain, and the seas and oceans slowly filled.

It was at this stage that amino acids first appeared. These are water-soluble organic compounds comprising carbon, oxygen, hydrogen, and nitrogen, and are the structural components of proteins, essential to all living organisms. A series of classic laboratory experiments carried out in 1953 by Stanley Miller and Harold Urey at the University of Chicago showed for the first time that, given the right conditions, life could arise spontaneously. They demonstrated that an electrical spark discharged through a mixture of gases thought to be equivalent to those of the early atmosphere, coupled with exposure to ultra-violet radiation, eventually yielded amino acids, the building blocks of life.

Perpetual lightning strikes during the storms that relentlessly lashed our primitive Earth could have created a veritable soup of organic molecules, from simple to increasingly complex, in those early seas. After millions of years of trial molecular associations and separations, two very special chemicals were created—the nucleic acids, first RNA (ribonucleic acid) and then DNA (deoxyribonucleic acid). This last is a long-chain organic molecule shaped like a twisted rope ladder—the famous double-helix structure.

The unique and significant properties of DNA are twofold. First, it can act as a blueprint for the assembly of amino acids into proteins; secondly, it has the ability to replicate itself. It achieves this feat by unzipping down the middle and then combining again with other split molecules to form a new double helix. It seems the fundamental essence of life—reproduction—developed by purely chemical means.

Primordial soup All life on Earth depends on this same structural material, so it seems likely that these chemicals developed very early. An oceanic origin is supported by the fact that all living organisms are dominated by water—in some cases up to 80 percent of their body mass is made up of the water molecule. But just which part of the ocean was involved is less clear. The first suggestion following the Miller–Urey experiments was that the oceans in general formed a primordial chemical soup in which life first emerged. Other scientists consider that volcanic hot springs and geysers, bubbling through metal-rich mud pools, may have provided a more ideal setting, shielding newly formed organic matter from the

Mats of thermophilic (heat-loving) bacteria and metal-rich mud surround a volcanic hot spring. Life may have originated at the ocean fringe in this kind of hostile environment.

harmful effects of intense ultraviolet radiation. The tiny pore spaces within fine clay sediments, in shallow estuaries or deep beneath the seafloor, could have provided similar protection as well as an electrically charged environment in which molecular concentration and exchange would have more readily occurred.

Since the discovery in the 1970s of black smokers emanating from submarine vents, a new theory has been put forward that the first organic materials of life appeared in such environments. Water superheated up to 350°C (660°F) is expelled from these mid-ocean-ridge sites together with sulfides and other metallic ions. Primitive bacteria may have been able to metabolize sulfur from this hot chemical effluent some 3.5 billion years ago just as they continue to do today. There is even evidence of layered sequences of metal sulfide deposits from rocks of that age—the highly metamorphosed Precambrian sequence known as Archean. However, such evidence is by no means definitive—it simply hints at the possibility of an alternative site for the origin of life in the oceans.

Today it seems very probable that life may also have originated in many other parts of what appears to be an infinite universe. Where it might be found and what forms it might take remain completely unknown; resolving such questions is a goal of space exploration. We do know, however, that some meteorites contain carbon compounds similar to amino acids, so that an extra-terrestrial origin of life's building blocks is also possible.

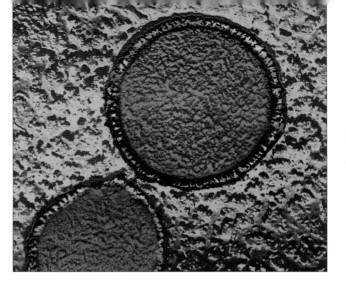

Micrograph of a section through the sulfur-metabolizing microorganism *Staphylothermus marinus*. This is a hyperthermophilic species (liking very high temperatures) found in deep-ocean vents. Once thought to be a bacterium, it belongs to a quite separate group, the Archaea.

Evolution underway Several different templates for early life may have originated in these steaming worlds at the dawn of history, somewhere between 3.8 and 3.6 billion years ago. Replication and multiplication of early life-forms proceeded according to the blueprints locked in their RNA and DNA. Occasionally, however, a small mistake was made, and two strands of the double helix joined up imperfectly. Perhaps a rung in the molecular ladder was missed out or added, or there was an incorrect pairing of two amino acids. Such copying errors, known as mutations, sometimes produce changes in an organism's protein structure. These changes may be inherited and can gradually accumulate within a population or species, eventually even contributing to the evolution of new species.

DNA (*below*) is a remarkable and complex molecule that holds the key to heredity and evolution. Within the nucleus of a cell, long spirals of DNA are found in the form of chromosomes. These carry a series of specific segments known as genes, and hence the whole genetic code for replication of each individual and species.

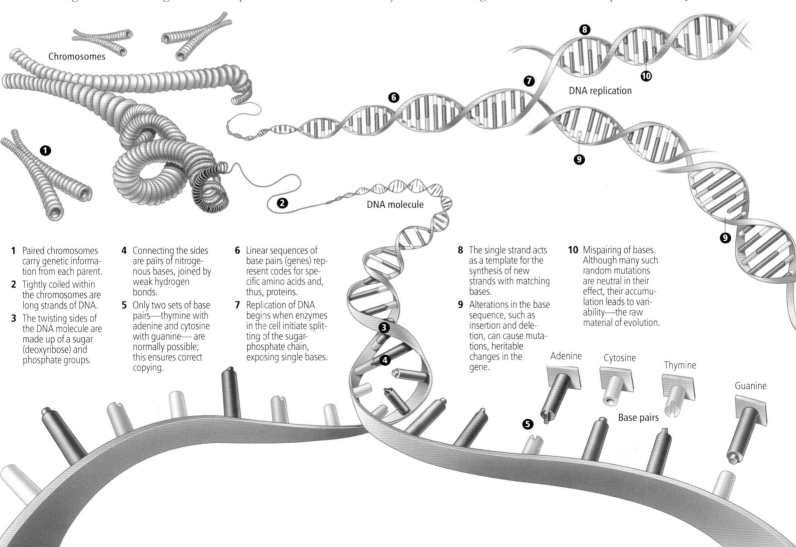

Chromosomes

DNA molecule

DNA replication

1 Paired chromosomes carry genetic information from each parent.

2 Tightly coiled within the chromosomes are long strands of DNA.

3 The twisting sides of the DNA molecule are made up of a sugar (deoxyribose) and phosphate groups.

4 Connecting the sides are pairs of nitrogenous bases, joined by weak hydrogen bonds.

5 Only two sets of base pairs—thymine with adenine and cytosine with guanine— are normally possible; this ensures correct copying.

6 Linear sequences of base pairs (genes) represent codes for specific amino acids and, thus, proteins.

7 Replication of DNA begins when enzymes in the cell initiate splitting of the sugar-phosphate chain, exposing single bases.

8 The single strand acts as a template for the synthesis of new strands with matching bases.

9 Alterations in the base sequence, such as insertion and deletion, can cause mutations, heritable changes in the gene.

10 Mispairing of bases. Although many such random mutations are neutral in their effect, their accumulation leads to variability—the raw material of evolution.

Adenine   Cytosine   Thymine   Guanine

Base pairs

Stromatolites crowd the warm, shallow waters of Shark Bay, Western Australia. These bizarre stony cushions were constructed layer upon layer by matlike colonies of cyanobacteria, in a cementation process that has remained unchanged for over 3 billion years. Stromatolite fossils have been found in some of the oldest rocks on Earth, clearly indicating that blue-greens acquired the means of photosynthesis near the very beginning of evolution.

When one particular type or species evolved that was better adapted to its environment, or that reproduced more rapidly and successfully, then that form became dominant—a process known as natural selection. The template for building more of the same was transferred between generations by the codes, or genes, locked within the DNA molecule. Over countless millennia, the rich variety of life evolved through genetic variation and natural selection. Today there are between 5 and 10 million species in existence. Many more have come and gone, many of which left behind no fossil record of their existence.

Wherever in the ocean life began, it was undoubtedly a hostile environment. The first organisms were microscopic, single-celled prokaryotes—bacteria and bacteria-like archaea ("ancient ones"). They endured a harsh existence, surviving not just the heat of Earth's interior but also damaging ultraviolet radiation from the Sun, and meteorite bombardment. The oldest known fossils, preserved in rocks called cherts (silica-rich sediments like flint), come from the ancient continental cores of Australia and Africa, which have been radiometrically dated at around 3.5 billion years. Although primitive, the several distinct lifeforms present are evidence that evolution was already under way.

Different bacteria and archaea evolved to take advantage of alternative sources of energy. Some utilized methane in the early atmosphere, some the hydrogen sulfide from volcanic emanations. Before long, new forms arose that were able to release hydrogen from water. These were rather more complex than the first bacteria; formerly called blue-green algae, they are now known as cyanobacteria, or simply "blue-greens." They contained the chemical compound chlorophyll within their cells, which gave them an ability to harness the Sun's energy directly, by the process we call photosynthesis. By dissociating hydrogen from water molecules, they released oxygen into the atmosphere.

This breakthrough was a clear evolutionary advantage for the blue-greens, because of the abundance of both water and sunlight. It was also a development that was to have a profound effect on all life that followed. For the first time the levels of free oxygen in the atmosphere began to rise. Some of the oxygen took the form of the ozone molecule, which provides a protective screen cutting out harmful ultraviolet rays from reaching the Earth's surface. In time a new, oxygen-breathing lifestyle would evolve, which would eventually include almost all modern animals and plants. But this step was still eons into the future.

For the first 3 billion years of life on Earth, evolution and advance were interminably slow. Blue-greens flourished then as they continue to do today, wherever there is constant moisture and sunlight. Green-colored mats blanketed the bottom of shallow waters. Some species, by binding particles of calcium carbonate (lime) between their filaments, formed irregularly shaped stony cushions known as stromatolites. These are among the earliest fossils but, still more remarkably, they have survived almost unchanged until the present day. To gaze out across their spectacular clusters in Shark Bay, on the tropical shore of northwest Australia, is to go back in time over 3 billion years.

Sexual reproduction Life continued for over 1 billion years exclusively in the form of single-celled organisms, having no clear differentiation of parts within the cell and reproducing by a simple process of splitting in two. The first major change came with the development of cells having a nucleus—a core of material containing all the DNA instructions for life and reproduction. These organisms, known as eukaryotes, gave rise to all other forms of life.

But the real step forward in the process of evolution occurred around 1.8 billion years ago, when reproduction became modified to involve the mixing of gene sets (contained in single-celled gametes, or sex cells) from two parents of different sexes rather than simple transfer from one. This development eventually led to the production of a large, comparatively immobile cell (the egg) by the female parent and a smaller, active one (the sperm), driven by a flagellum, by the male. The combination of egg and sperm cells is the essence of sexual reproduction, leading to the creation of unique organisms with new characters. By increasing the potential for genetic variation, it accelerated the rate at which evolution could proceed. The principal mechanism for evolutionary change was (and remains) genetic variability due to the mixing of genes from two parents.

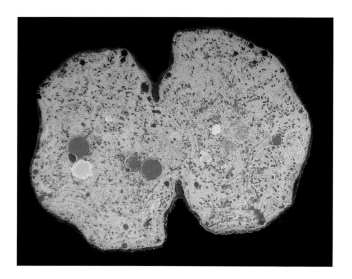

A dividing cell of the cyanobacterium *Microcystis acruginosa*, formerly considered a primitive alga. Cyanobacteria are prokaryotes, containing no membrane-bound nucleus. The nuclear material, including DNA, is dispersed throughout the cell.

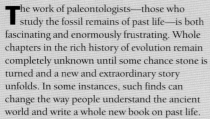

The work of paleontologists—those who study the fossil remains of past life—is both fascinating and enormously frustrating. Whole chapters in the rich history of evolution remain completely unknown until some chance stone is turned and a new and extraordinary story unfolds. In some instances, such finds can change the way people understand the ancient world and write a whole new book on past life.

Just such a discovery was made in the 1940s by geologists working in the Flinders Ranges of South Australia. There, in a region known as the Ediacaran Hills, they noticed odd but regular shapes and distinctive patterns in the sandstones, fuelling a lively debate as to their organic or inorganic origin.

The find soon turned out to open a window onto a remarkable, thriving ocean world, a fossil realm that helped bridge the gap between the simple algae, archaea, and bacteria that apparently reigned supreme through most of the Proterozoic era, and the Cambrian explosion into higher forms of life. The rocks were known to be extremely old, and have since been dated radiometrically to around 640 million years ago. The well-rounded sand grains, ripplemarks, and other sedimentary structures all indicated deposition on a beach or other coastal environment.

Thanks to decades of careful research, over 1,400 of these Ediacaran fossils have now been collected—all of them, quite amazingly,

preserving the delicate impressions and molds of a host of soft-bodied multicellular organisms that lived all those eons ago. At least 15 different species of creatures resembling jellyfish have been recognized, including some quite large specimens with floats up to 12.5cm (5in) in diameter, similar, although on a smaller scale, to the great Portuguese man-of-war, much feared in the oceans today. There can be little doubt that these Ediacaran forms were preceded by many millions of years of evolution.

Today, such creatures have a fair degree of complexity in evolutionary terms. They consist of two layers of cells separated by a jellylike substance that offers some rigidity to withstand the buffeting of waves. They have a primitive nervous system, simple muscles, and rows of stinging cells. They release eggs and sperm into the sea, with fertilization starting a life cycle that in some cases involves a quite different, sessile, flowerlike organism called a polyp. Eventually the polyps bud into new, tiny free-swimming jellyfish (medusae) that will grow into adults. Of course, we do not know

whether the Ediacaran creatures behaved in the same way, but certainly they were much more advanced than anything that had gone before.

In addition to the "jellyfish," the Ediacaran sandstones contain fossils of sluglike animals and a variety of feeding trails across the sand, as well as crescent-headed annelid worms up to 10cm (4in) in length with a clearly defined head and 40 body segments. Among the most spectacular organisms are the soft corals, colonial forms with delicate, featherlike fronds that were home to thousands of tiny coral polyps, wafting in the shoals and firmly anchored to the substrate. Some of the simple, circular impressions found nearby are interpreted as the holdfasts of these already highly evolved species.

Ediacara: soft-bodied wonders

1 *Spriggina* (annelid worm)
2 Sea anemone-like organism
3 Holdfast impressions
4 *Charniodiscus* (soft coral)
5 Jellyfish
6 *Dickinsonia* (annelid worm)
7 Burrowing worm

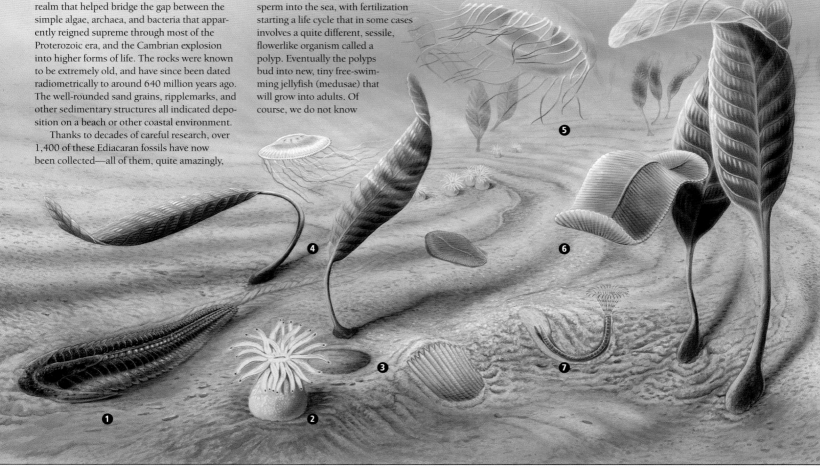

The stage was then set for a great experimentation in life-forms, of which sadly we have no fossil evidence. Yet the variety of single-celled organisms that evolved during this period is perhaps mirrored in part by the 25,000 types of protists that exist today. These simple organisms range from microscopic dots to tiny globes 1mm (0.04in) across, from irregular amoebas with bulging fingers of matter to highly coordinated cells covered by a mat of flailing threads, from some with packets of chlorophyll-like tiny plants to others that feed on them. In today's ocean world, many have developed delicate and beautiful skeletons of silica or lime—but this innovation was yet to arise.

Colonies and clusters There was a limit, however, to the evolutionary development possible in single-celled organisms. So new directions were taken, probably around 1.4 billion years ago in the first instance. One route was to form a colony of constituent cells with a coordinated life plan; surviving examples include the sponges, which are still only collections of single, quasi-independent cells. Another, more successful development was to unite single cells into true multicellular organisms. For the first time there was the potential to differentiate groups of cells according to function. From that remarkable day on, life on Earth was profoundly changed.

Life burst forth with breathtaking innovation around 500 million years ago, filling the seas with myriad new forms. Ever since, the oceans have teemed with color and profusion.

March through Time

The first 3 billion years of life on Earth are shrouded in mystery. The whole of this period is sometimes called the Cryptozoic eon, a word meaning "hidden life." Then, suddenly, after this long phase of early development, from which very few fossils remain, a wide range of previously unknown forms appear in the fossil record. This emergence happened around 545 million years ago, marking the beginning of the Phanerozoic ("visible life") eon. Each subsequent period through the whole of the Phanerozoic is marked by a new stage of evolution.

The world at the start of the Cambrian—the first geological period of the Paleozoic ("ancient life") era—had about 15 percent less continental landmass than today. The Late Proterozoic supercontinent Rodinia had broken into many isolated fragments, mostly aligned along the equator. The climate became much warmer and sea levels rose steadily, creating extensive shallow basins with a wide variety of new habitats.

Evolutionary timescale

	Geological period	Date (million years ago)	Characteristic life
CENOZOIC	QUATERNARY	0—1.8	Mammoths; saber-toothed cats; modern humans
	NEOGENE	1.8—24	Grasses; hoofed mammals; rodents; snakes; first hominids
	PALEOGENE	24—65	First whales; coral reefs; early ungulates; primates
MESOZOIC	CRETACEOUS	65—144	Calcareous plankton; flowering plants; placental mammals
	JURASSIC	144—205	Modern fishes; early mammals; first birds
	TRIASSIC	205—248	Scleratinian (stony) corals; reptiles; first dinosaurs
PALEOZOIC	PERMIAN	248—295	Sponge–bryozoan reefs; advanced mammal-like reptiles
	PENNSYLVANIAN (late Carboniferous)	295—324	Conifers; winged insects; first mammal-like reptile
	MISSISSIPPIAN (early Carboniferous)	324—354	Seed-bearing plants; giant land-scorpions; first reptile
	DEVONIAN	354—417	Large land plants; ammonoids; lungfishes; sharks; amphibians
	SILURIAN	417—443	Vascular plants; jawed fish; early land animals
	ORDOVICIAN	443—490	Tabulate corals; jawless fishes; spores of land plants
	CAMBRIAN	490—545	Arthropods; Burgess Shale fauna; first chordate
PHANEROZOIC / **CRYPTOZOIC**	PROTEROZOIC	545—2,500	Multicellular organisms; Ediacara fauna
	ARCHEAN	2,500—4,550	Archaea; bacteria; cyano-bacteria; stromatolites

Skeletons for survival Within a very short time, a diversified and relatively advanced flora and fauna had colonized the Cambrian seas. This event is referred to as the "Cambrian explosion," and it witnessed the appearance on the scene of reef-building calcareous algae, foraminifers, corals, bivalves, gastropods, echinoderms, and arthropods, all with a newly evolved feature—a hard shell or exoskeleton. Now extinct but then highly successful groups such as trilobites and graptolites swam, crawled, or floated throughout the marine world. New species, ever more complex and exotic, came and went with a hitherto unknown rapidity.

There are various possible explanations for the development of hard parts. Shells and external skeletons would have given protection against predators and against fierce ultraviolet radiation, especially near the ocean surface and in shallow waters, and would have prevented rapid drying out in intertidal environments. Skeletons also gave support and improved mobility and dexterity. Whatever the reasons, the change not only profoundly affected the development of life but also eased the task of modern-day paleontologists, for mineralized skeletons are 1,000 times more easily preserved as fossils than is soft tissue. The skeletons were variously made of calcium carbonate, opaline silica, or chitin (a hard organic polysaccharide), but not phosphatic bone, which had yet to appear.

Early sea worlds From the Cambrian on, therefore, we have a much clearer image of how evolution proceeded. We can recreate a catalog of past oceans and can trace the ancestral pathways of the wonderfully diverse life that inhabits today's marine world. Indeed, for many millions of years life remained firmly in the sea. But always there was change, and often groups or species that had once flourished subsequently vanished without trace or descendants.

Trilobites were the most important group of arthropods ("jointed-legged" animals) for over 100 million years but they eventually died out completely, leaving no successors. They looked rather like wood lice, with long feelers at the front and many legs, together with bulging compound eyes. They evolved rapidly and adopted a range of different lifestyles: some fed by plowing through the muddy seabed, others were more or less sessile filter-feeders, while others again were effective hunters, swimming, crawling, or floating after their prey. Another strange group that evolved similarly rapidly and has long since vanished was the graptolites ("written stone"). These were in fact free-floating colonies of animals, made up of branching strings of little cups in which individuals lived, filter-feeding on the algal plankton that floated on the ocean surface.

Microscopic in size but with profound longterm effects on the nature of seawater, innumerable different forms of zooplankton evolved to vie with the graptolites for food. These new, free-swimming organisms, tiny crustaceans and their relatives, systematically ingested everything of an appropriate size, organic or inorganic, hoping for something edible. The indigestible waste was later excreted in the form of membrane-coated fecal pellets, which sank rapidly through the water column. The effects of this

Mt. Stephen is a beautiful and tranquil location in the Canadian Rockies of British Columbia. It is the site of one of the most amazing and significant fossil discoveries of all time. In 1909 the American geologist Charles Walcott began a series of expeditions to the area during which he collected over 40,000 specimens from a relatively small and localized Cambrian outcrop. He named the formation the Burgess Shale, and proceeded to describe in great detail the wealth of its fossils, which are now dated at around 530 million years old.

Besides the sheer numbers of fossils that represent so well the first great explosion in life that took place at the dawn of the Phanerozoic, the Shale's unique importance lies in the almost perfectly preserved soft bodyparts of many of the organisms. This kind of preservation is both rare and very important in that it enables paleontologists to infer the details of primitive metabolisms. It occurs rarely because conditions must be just right. The Burgess Shale was laid down as a fine-grained mud deposit near the foot of a steep slope, in water around 150m (500ft) deep. Few animals lived in those dark, stagnant waters. But once in a while, muddy undersea currents (turbidity currents) swept organisms living on the seafloor swiftly toward a

deep-water grave. Many kinds of small sea creatures were snatched from their shallower habitats, while others were picked up by the powerful flow as it careered downslope. Because they were buried quickly in a tomb of soft, fine material, they were removed from the aggressive action of decomposing bacteria and disturbance by bottom scavengers. In this way even some of their soft organic tissues became mineralized during the process of fossilization. Much later, great earth-moving forces folded and elevated these now lithified sediments, pushing them upward as part of the Rocky Mountain chain. Miraculously, the small patch of Burgess Shale remained largely unscathed.

This fossil treasure chest contains over 120 genera. Some are primitive forms of familiar groups, while others have completely unknown affinities. Evidence of an early attempt at reeflike communities is there in the shape of algae, cyanobacteria, sponges, and crinoids (sea lilies). These early formations were populated by all kinds of walking, swimming, scavenging, and burrowing creatures. Arthropods (segmented invertebrates with jointed limbs) were especially well represented by over 40 different species. Some of the trilobite fossils show extraordinary details of the paired, articulated legs attached to

each body segment, of feathery gills and delicate feelers, of the muscle fibers that enabled the animals to roll up into a defensive ball, and of an alimentary canal running the length of the body. Other, equally detailed fossil forms test our current understanding: *Hallucigenia* (an aptly, almost despairingly named creature) with seven pairs of soft, retractable legs and seven pairs of sharp spines, was, until 1991, reconstructed upside-down; the fearsome predator *Anomalocaris*, with the rounded circlet of plates that form its mouth and its spiny, segmented appendages, also took many years of puzzlement and study to reconstruct. A dozen different annelid worms have been identified; also *Pikaia*, a wormlike organism with a primitive backbone that may have been the ancestor of fishes. It seems as though there was a great deal of early experimentation with animal design. Perhaps most of the attempts represented in the Shale never made it through the test of time and in the face of increasing competition. Others evolved slowly into creatures that become increasingly familiar as fossils in the long procession through succeeding ages.

Professor Simon Conway Morris,
Cambridge University, UK

Burgess Shale: fossil treasure chest

1 *Anomalocaris* (anomalocaridid)
2 *Marrella* (arthropod)
3 *Aysheaia* (lobopod)
4 *Olenoides* (trilobite)
5 *Wiwaxia* (halkieriid)
6 *Habelia* (arthropod)
7 *Dinomischus* (unknown affinity)
8 *Odaria* (arthropod)
9 Sponges
10 *Odontogriphus* (lophophorate)
11 *Hallucigenia* (lobopod)
12 *Nectocaris* (arthropod)
13 *Sidneyia* (arthropod)
14 *Ottoia* (worm)
15 *Pikaia* (chordate)

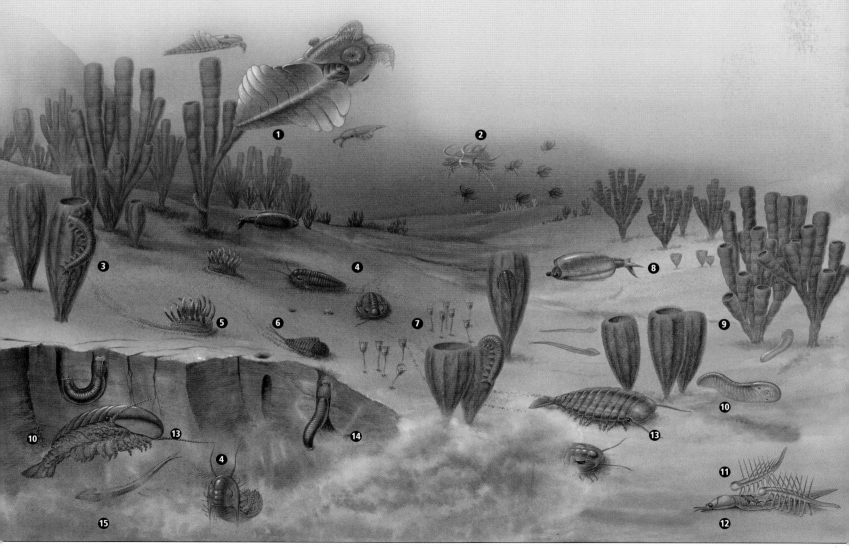

Graptolites were free-floating colonies of tiny organisms. They lived from the Middle Cambrian to Early Carboniferous and are commonly found as flattened, carbonized fossils in dark shales. Their wide distribution and varied forms are important in dating Paleozoic rocks.

Contemporaries of the dinosaurs, ammonites reached their zenith during the Mesozoic era. Over 10,000 species have so far been described—evidence of their rapid evolutionary radiation into almost all marine environments. They died out suddenly during the end-Cretaceous extinction event, leaving no direct descendants.

during this period. The Silurian, Devonian, and Carboniferous reefs were massive constructions; their remains now form some of the world's finest limestone scenery, while polished slabs of their fossil-packed stone are still widely used in buildings.

Through time, the changing reef environments spawned a plethora of new species. Although we can never trace evolutionary change to its actual birthplace, it is very probable that the first true fishes inhabited the colorful submarine world of the Silurian seas some 425 million years ago. There was already a great diversity of early, jawless "fishes," many armored with a bony array of shields, tubercles, and scales. The development that made their (and our) existence possible had occurred very quietly during mid-Cambrian times—the evolution of a backbone. The primitive creature that displayed the first evidence of such a structure was *Pikaia,* a flattened, wormlike animal found in the Burgess Shale. It swam in a series of S-shaped bends, rather as fishes do, using the stiffening rod (notochord) and paired, segmented muscle blocks common to all vertebrates. Even more closely related to the first vertebrates were the sharp-toothed, jawless, eel-shaped conodonts, fast-swimming predators of the Ordovician seas. Other accomplished hunters were the tentacled cephalopods, giant sea scorpions, and, by Devonian times, primitive sharks.

The innovation that marked out the true fishes from their precursors was the development of hinged jaws and paired fins. With greater maneuverability and a genuinely predatory mode of life, these aquatic gymnasts were set to conquer. There are places in the world where we can still walk across the seabed of 350 million years ago. In the unlikely setting of a cattle station in the northwestern Australian desert stands the Gogo Formation—an imposing limestone range cast from the fallen glory of an ancient coral reef. Among the debris are superbly preserved remains of over 20 species of Devonian fish—ray-fins and lobefins, lungfish, armor-plated placoderms, large bottom-dwelling scavengers and smaller free-swimming hunters.

particular chance turn of evolution were dramatic and immediately apparent. The surface waters were eaten clean, and organic detritus was discharged to the seafloor. A host of planktonic bacteria and fungi that had formerly fed on this material could no longer survive, allowing oxygen levels to build up and a variety of new life forms to develop. The continuous rain of organic waste provided a boost in food supplies for a growing variety of burrowing benthic (bottom-dwelling) fauna, but spelled disaster for delicate filter feeders, weakly anchored in the substrate.

The rapid demise of large numbers of these early species marked the first mass extinction event of which we have evidence. It was closely followed by a great radiation of new groups. This pattern of boom and bust is as much a part of species diversity over time as it is a contemporary economic reality.

The Devonian period is known as "The Age of Fishes," because most of the major groups had evolved by then. The jawless fishes (agnathans) that had dominated in the Ordovician and Silurian were overtaken by armored placoderms, spiny acanthodians, primitive sharks, lobefins, and lungfish. Many of these fish lived in freshwater lakes (*right*), but by the end of the period most species were extinct. Most modern fishes are ray-fins (Actinopterygii), which diversified during the Mesozoic.

Fishes evolve A series of major extinctions brought the Cambrian to a close. The falling sea levels that accompanied them did not last, however, and soon warm shelf seas again offered a host of welcoming habitats, refuges, and opportunities to new life forms. New reef communities developed and proliferated. Early Cambrian reefs had been, in the main, composed of stromatolites and archaeocyaths (primitive sponges). The major radiation of gastropods (snails) during the Ordovician brought about a decline in stromatolites as the snails grazed on the algae. The stromatolites were replaced by stromatoporoids—another type of calcified sponge—that remained the dominant reef-builder until the end of the Devonian. The first reef-building corals also came onto the scene

1 *Coccosteus*
   (placoderm)

2 *Dipterus*
   (primitive lungfish)

3 *Pterichthyodes*
   (placoderm)

4 School of
   Palaeospondylus

5 *Glyptolepis*
   (primitive lobefin)

6 Primitive vascular
   plants

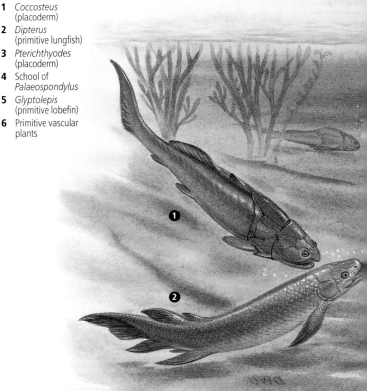

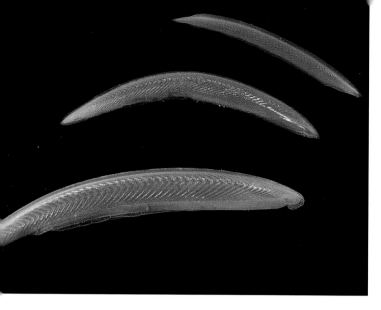

came into existence following the accumulation of leaf litter and other plant debris on a substrate broken down by increased levels of weathering. Plant roots stabilized the soil, allowing more sophisticated vegetation to develop. Nutrient cycles, once confined to the oceans, changed forever.

Once a bridgehead had been established on land by plants, pioneers of the animal world were also able to venture forth. As mosses, liverworts, and miniature forests of primitive vascular plants such as clubmosses spread inland from the edges of estuaries and rivers, the first animal colonists crept forward in their wake—distant ancestors of the millipedes, centipedes, scorpions, and spiders, followed, some 360 million years ago, by the first amphibians. The ancestor of these remarkable creatures was the lobefin fish, its muscular fins the forerunner of the limbs all tetrapod ("four-footed") creatures possess. These vertebrate pioneers onto land eventually gave rise to reptiles, birds, and mammals, but their amphibian descendants still inhabit the in-between world of swamps and lagoons, lakes, and rivers, the true meeting places of continent and ocean.

As the Paleozoic drew to a close, life was flourishing as never before across a whole spectrum of environments. The great Carboniferous and Permian coal-forming swamp forests rivalled the spectacular limestone reefs in scale and diversity. The remains of a Permian reef in Texas, which once rose 600m (2,000ft) from the seafloor to the water's surface, now preserves on land more than 360 species of marine fossils.

The lancelet *Branchiostoma* (*left*) is a small, soft-bodied marine chordate similar to the Cambrian fossils that are believed to be ancestral to all vertebrates. Chordates possess both a dorsal nerve chord and a strong but flexible notochord. It is this feature that later evolved into a true backbone.

Escape from the sea

It was at about this time too that the singular dominance of the marine realm as home to life on Earth began to be challenged. For as many as 200–300 million years, bacteria, algae, lichens, and fungi had been coping with increased exposure to air, and had already begun to colonize the once-barren land. Seaweeds developed a tough envelope, which allowed some rigidity without the support of water as they fanned out along the shoreline fringe of an otherwise desolate landscape. Before long the continents were as green as the seas were blue. The increasing colonization of the land by plants was to have a major and irreversible effect on the whole global environment.

The brave new world of terrestrial plant life meant that much of the carbon dioxide in the atmosphere was rapidly converted to oxygen. Eventually a more or less steady-state oxygen level was reached that allowed animals with lungs to live on land. A symbiotic relationship with bacteria gave some plants the ability to fix nitrogen directly from the air. The development of the unique system of evapotranspiration in vascular plants, whereby water is transported from the soil to the very top of the plant and thence into the atmosphere, greatly influenced patterns of rainfall, global temperature, and atmospheric circulation. Soil only

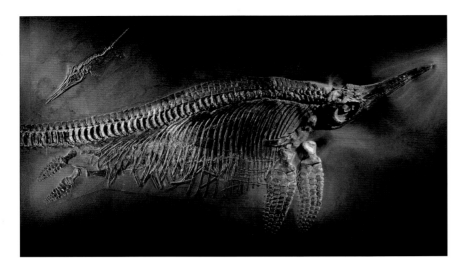

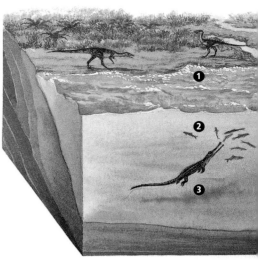

A remarkable fossil find—an ichthyosaur mother with infant and five unborn babies. This imposing predator was one of the best-adapted marine reptiles of the Jurassic seas. Its streamlined, dolphin-like form and rows of razor-sharp teeth were to return much later on an altogether different branch of the evolutionary tree—that of marine mammals.

Fur seals (*right*), sea lions, true seals, and walruses form a group of marine mammals, known as pinnipeds ("wing footed"), that are closely related to dogs and bears on land. Sometime in the latter part of the Paleogene period, a branch of the vulpavine (dog-like) carnivore lineage turned to the sea in a successful bid to escape from increasing terrestrial competition.

Reptiles and revolution The Mesozoic ("middle life") era is commonly referred to as the "Age of Reptiles," best known for the dominance of dinosaurs on land. The same is not true, however, of the oceans, for although reptiles played an important part in the unfolding story of marine life, they were never at center stage. Following the elimination of up to 95 percent of all species in the end-Permian mass extinction, when the greatest death toll was at sea, the oceans offered a world of opportunity for new colonists. The fossil evidence from these times is even more abundant and remarkable than that which came before. From the primary producers at the base of the food chain to the top predators at its peak, new forms appeared and flourished.

New types of microscopic plant life took hold of the phytoplankton world. Coccolithophores and diatoms bloomed in countless abundance in the clear surface waters, building their tiny calcium-carbonate and silica skeletons in a beautiful and varied array of intricate shapes. Dinoflagellates, one of the principal primary producers in warm waters today, also became common at this time. There was a parallel explosive radiation of zooplankton to feed upon the bounty, with lime-secreting foraminiferans in the lead. The brilliant white cliffs and upland areas of soft white chalk that are so common throughout the world today were built, over many millions of years, from the skeletal remains of these tiny organisms. Sharp, black nodules of flint, ranged in regular bands through the chalk layers or plucked out, rounded, and sea-worn by waves along the shore, owe their presence to an endless rain of diatom and radiolarian skeletons.

More coveted by the amateur fossil collector are the remains of organisms that scavenged on this plenty. Marine invertebrates capitalized on the ocean rejuvenation that was now underway with an exuberant radiation in form. Mollusks proliferated, from the oysters and clams that burrowed into or attached themselves to the substrate to the free-swimming cephalopods with their chambered shells. Among these were a magnificent variety of coiled ammonites, some growing to as much as 1m (3.3ft) in diameter, which were some of the most common animals in the Mesozoic ocean. New types of corals and bryozoans appeared in the warm, shallow seas. But everywhere the seafloor was dominated by voracious predators—meat-eating snails and crabs and lobsters able to crack open even the hardest of shells. Burrowing became a common avoidance technique.

Crowning what had now become a highly complex food web were some of the more advanced fishes, such as sharks, and the marine reptiles. Many of these were fearful predators indeed. There were porpoise-like ichthyosaurs, short-necked pliosaurs, and long-necked plesiosaurs, some up to 15m (50ft) in length, with rows of sharp teeth. The mosasaurs are a less well-known group, of similar size and ferocity, related to today's Komodo dragon. Crocodiles and turtles also appeared for the first time during the Mesozoic era.

Alongside all this activity at sea, several key evolutionary developments that were to greatly influence the nature of terrestrial life took place quietly on land. First, during the late Triassic period, came the evolution of mammals from the cynodont ("dog-toothed") group of advanced, mammal-like reptiles. At the time, however, the mammals remained small and kept a low profile beneath the overwhelming dominance of reptiles. Second, in the early Jurassic, was the evolution of birds, such as the renowned *Archaeopteryx*, from small, meat-eating dinosaur ancestors. And third was the appearance of angiosperms (flowering plants) in the early Cretaceous period.

The ascent of mammals Recovery from another mass extinction, famous for the complete elimination of dinosaurs from the planet, was the now familiar story played out at the start of the Cenozoic ("recent life") era. From top to

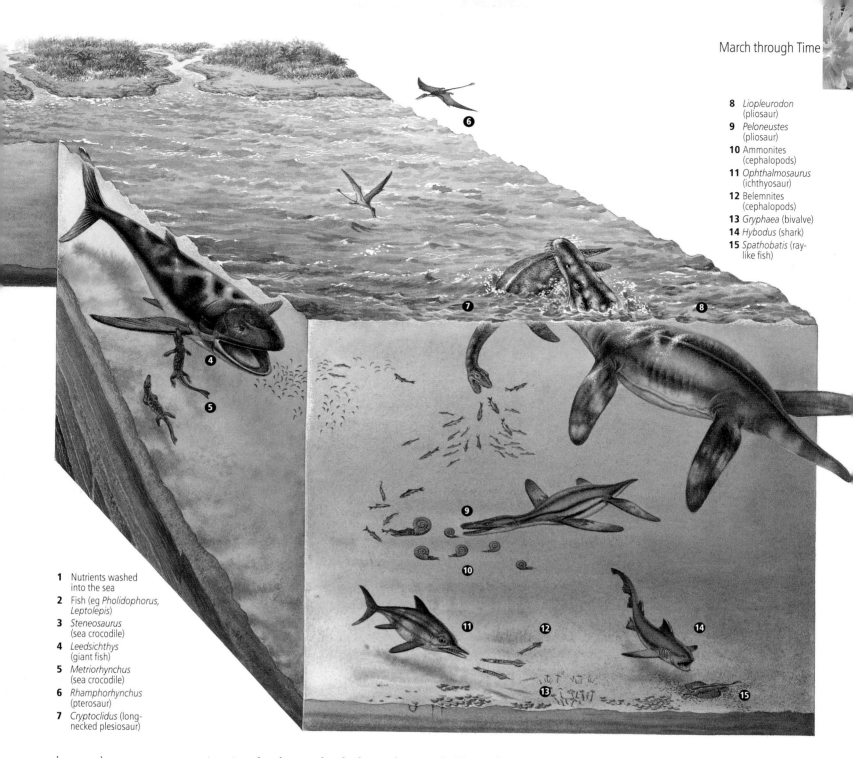

1 Nutrients washed
into the sea

2 Fish (eg *Pholidophorus,
Leptolepis*)

3 *Steneosaurus*
(sea crocodile)

4 *Leedsichthys*
(giant fish)

5 *Metriorhynchus*
(sea crocodile)

6 *Rhamphorhynchus*
(pterosaur)

7 *Cryptoclidus* (long-
necked plesiosaur)

bottom the seas were once again stripped and repopulated. The template for our modern oceans evolved really quite rapidly, so that a great many of the fossils we find from this era are almost indistinguishable from their living relatives. The phytoplankton factory regenerated at the base of yet more complex food webs. Coral reefs assumed the breathtaking beauty that modern reefs have today. Great numbers of marine animals adapted to almost every known habitat in the oceans, from the surface to the deep abyss and from the ice-cold polar seas to superheated, toxic waters emanating from deep-sea vents.

The great Tethys Ocean that girdled the world during the Paleogene period was very much a modern ocean. Although it was lost forever as the Earth's tectonic plates moved slowly to their present-day position, the fossil glimpses that remain show an ocean every bit as fascinating and varied as those of the tropi-cal world today. Some aspects were, of course, different, such as

the great shell banks formed of giant *Nummulites* foraminiferans that lined Tethyan shores between 40 and 50 million years ago, the likes of which have never been seen since.

Also along the shores of the Tethys, another extraordinary evolutionary development took place. The marine reptiles of the Mesozoic had accompanied the dinosaurs into extinction, but whereas on land the vacant niches were soon filled by a prolifer-ation of new mammal species, in the sea, at first, only the bony fishes expanded to fill the void. It was some 10–15 million years later that a group of terrestrial mammals began their return to the ocean. Possibly as a result of the disappearance of the huge marine predators, or to escape competition on land, the curious bear/hyena-like ancestors of today's whales and dolphins began to live in the nearshore realm, eventually adopting a fully marine lifestyle. A range of fossils found in the shoreline sediments of the ancient Tethys documents this remarkable transition.

Extinction is a natural part of evolution, and so, apparently, are the extraordinary periods of mass extinction. The causes of these events are almost certainly complex rather than simple.

EVOLUTION & EXTINCTION

Mass Extinction

KEY TERMS

EVOLUTION

EXTINCTION

MASS EXTINCTION

SPECIES

GENUS

"SHOCKED" QUARTZ

BIODIVERSITY

Evolution and extinction have always been closely intertwined and interdependent. The passage of species through geological time—their appearance, development, and demise—weaves an intricate pattern. At certain times and places conditions have favored adaptation and change, great radiations of new creatures such as the one that occurred in the Cambrian explosion of life. Other eras have witnessed the rapid and simultaneous demise of large numbers of species, and even of whole families of related genera. These are the mass extinction events that have engaged the imagination of both the public and the scientific community alike. But such fascination has encouraged wild speculation and simplistic explanations of these complex phenomena. The truth is undoubtedly more complicated.

Extinction events

Little has been preserved of the forms of life prior to that great Cambrian radiation, around 550 million years ago. The ensuing Phanerozoic eon is altogether much better known. It is divided almost in two by the largest extinction event known, that which occurred at the end of the Permian period 250 million years ago. Four other major mass extinction events, as well as about ten smaller ones, are clearly marked in the fossil record. Arguably we are now in the midst of another, mainly of our own making. There is almost an uncanny cyclicity in the timing of these events, at least since the great Permian extinction, with one occurring every 26–30 million years. But the spacing of cycles is not perfect, and the potential causes are hotly disputed.

The end-Permian event was quite unprecedented in its scale and effect; an estimated 50 percent of all known families disappeared, which translates into an astonishing 95 percent of all species, still mostly marine at that time. Two major environ-

The pattern of extinctions in marine invertebrate genera over the past 600 million years (*below*) shows several major peaks (mass extinctions). There is also a gradual decline in the number of extinctions; this is possibly an artificial result reflecting the difficulty of classifying the more ancient fossil groups, or an indication of improving ocean conditions.

mental factors conspired with dramatic effect to bring about the extinction. The first was the slow but immensely influential fusing together of continents to form the single great landmass known as Pangea. The consequences were manifold. Sea levels fell to an all-time Paleozoic low, reducing habitat diversity and promoting ecological instability. The escape of gas hydrates from beneath the ocean floor, and the oxidation of carbonaceous deposits, also a result of lowered sea levels, released large quantities of carbon dioxide into the atmosphere, causing global warming. Continental climates were extreme and harsh, severely stressing plant and animal life on land.

The second factor was an intense period of volcanic activity, particularly in the Ural Mountains and the Siberian Traps, where vast fields of lava were deposited. Immense clouds of volcanic ash would have blotted out sunlight, temporarily cooling the planet. Emissions of carbon dioxide and other gases would then have led to global warming and anoxic marine conditions, and probably also poisoned the atmosphere and oceans with fluorine, acid rain, and trace metals.

Pushed to their limits, species saw their rates of natural decline increase many times over. The loss of certain plants affected a host of animals that fed on them, and their passing in turn impinged on predators further up the food chain. Competition inevitably increased as favorable habitats diminished. Diseases may have spread more aggressively between animal groups as once-isolated landmasses came together. In the final analysis, the Permian extinction was an unparalleled biological disaster from which the world took many millennia to recover. Almost 5 million years were to pass before plant life returned to anything like its former glory.

90	Percent extinction of marine genera

Chart axis labels: 90, 80, 70, 60, 50, 40, 30, 20, 10, 0

Chart vertical labels: Mass extinction 443 mya; Terminal Paleozoic extinction 248 mya; Mass extinction 205 mya; Terminal Mesozoic extinction 65 mya

Chart horizontal labels: Cambrian | Ordovician | Sil. | Dev. | Carb. | Perm. | Tri. | Jurassic | Cretaceous | Tertiary

diameter, requiring an object of 10km (6mi) diameter to produce it. Concentrations of iridium, an element common in meteorites, have been found dating from this time at different places around the world, together with grains of "shocked" quartz, displaying the fractures that can only form under intense stress. But, in reality, the evidence supporting this view is limited. Such catastrophist theories have difficulty explaining the fact that many plants and animals were barely affected; most land plants survived, and so did most mollusks, sharks, bony fishes, placental mammals, and all amphibians.

There is abundant evidence, however, for a combination of environmental agents at this time. There were large outpourings of lava, as evidenced by the Deccan Traps in India which cover 500,000sq km (200,000sq mi); such an eruption would have had much the same effect as an asteroid impact. The high sea levels and warm climate enjoyed throughout the Cretaceous came to an abrupt end, a feature we can measure and even quantify from the oxygen-isotope record in fossil shells. As coastal habitats diminished and land bridges opened up, the spread of diseases between animal groups, already subject to enhanced competition, would have increased. There is further evidence of longterm decline in many groups of dinosaurs, and in the extinction of certain marine fauna over a period of several millions of years. Detailed microfossil records in marine sediments drilled beneath the oceans reveal a rapid but stepped species extinction. Catastrophist and gradualist theories are not mutually exclusive, but it is difficult to avoid concluding that the demise of the dinosaurs was due to a complex combination of environmental factors that triggered extreme biological stress.

Present-day extinctions About 95 percent of all the creatures that have at some time existed on the Earth are now extinct. The fossil record is full of species that have appeared, lived their brief day, and vanished. In addition to the biotic crises of mass extinctions, there has been a steady rate of "background extinction" of about one species per year over the past 3.5 billion years.

If we compare the present situation with past extinctions, the result is wholly frightening in its implications. The most conservative estimate for extinction rates during the 1990s is a staggering 27,000 species per year—a figure close to the maximum rate witnessed during the great Permian mass extinction. Within the next three decades, that figure could increase more than tenfold, rising to several hundred species per day.

There seems little doubt that we are currently witnessing a mass extinction event of our own making. Many larger animals have been hunted to extinction; others are disappearing as we encroach upon their habitats and pollute their environment with our activities. Today there are over 11,000 threatened species of plants and animals listed, but only a small proportion of recognized species have been evaluated, and many more quietly disappear without ever having been discovered at all. Loss of biodiversity threatens human existence, too. Our food supplies, our search for medicines, the water we drink, and the air we breathe all depend upon an undamaged web of life.

Bones from Blue whales lie scattered on a beach in the South Shetland Isles, remnants of an old Antarctic whaling factory (*left*). The very tenuous international moratorium on whaling is the only reason why these magnificent creatures have returned from the brink of extinction. The current unprecedented rate of species extinction worldwide is mostly a result of human activity. Despite the ban on killing Blue whales, deterioration of their habitat has depressed their recovery and it is still doubtful whether a viable population exists.

The Chicxulub crater (*left*) is a ringlike structure that extends northward from the Yucatán Peninsula (coastline shown as white) into the Gulf of Mexico. The outer rim, seen here as the larger colored arc is about 180km (110mi) in diameter. It is thought to be evidence of an asteroid impact, which some scientists believe contributed to the extinction of the dinosaurs.

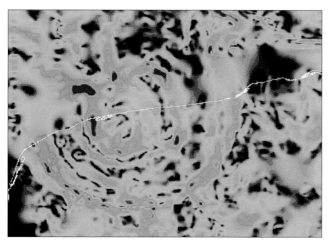

Death of the dinosaurs The mass extinction that marked the end of the Cretaceous period was arguably the second greatest in its overall effects. For about 180 million years dinosaurs and their close relatives had ruled the land, terrorized the seas, and fought for prowess in the sky. But some 65 million years ago the Age of Reptiles came to an end. With them went 75 percent of the birds and marsupial mammals, and roughly a quarter of all species of crocodiles, turtles, and fishes. Altogether about 20 percent of known families and an estimated 50 percent of all species became extinct.

Perhaps the most popular current explanation for this catastrophe is the idea that a large asteroid struck the Earth somewhere in the Gulf of Mexico region. A deeply buried ring-like structure, named the Chicxulub crater, has been found off the coast of the Yucatán peninsula; it measures 180 km (110mi) in

The astonishing productivity and diversity associated with coral reefs is a fine example of a highly complex series of feeding relationships. The primary producers include reef bacteria and algae, seagrass, phytoplankton, and, also, a host of tiny symbiotic organisms that live within the tissues of coral polyps. Together these support the primary consumers—herbivorous fishes, brittle stars, gastropods, and sea urchins. In their turn, these creatures are prey for those higher up the food chain, the many species of carnivorous fishes that bring such dynamism and color to the reef community. Among the major predators are barracudas, sharks, groupers, and octopuses.

THE WEB OF LIFE

CHAPTER 7

Phytoplankton, zooplankton, and the ocean food chain

Richness and diversity are the very essence of ocean life. The six great kingdoms—archaea, bacteria, protists, fungi, plants, and animals—into which we classify all living things are each present and profuse in the seas. From the restless surface waters to the dark and obscure ocean depths, there are oases of plenty along with marine deserts of eerie calm. Life is nowhere static. It presents an interwoven series of complex chains and webs, a colorful, constantly changing pattern contrived from the endless, inevitable quest for food.

The sunlit world of the photosynthesizing plankton that drift across the ocean swell is the single most important food source in the sea. Over two trillion metric tons of new phytoplankton grow every year, utilizing only solar energy and a handful of nutrient elements present in seawater. These bountiful pastures of microscopic organisms are grazed continually by equally tiny creatures with startling shapes and intricate shells—the zooplankton. Even the giant kelp forests and rich shorelines festooned with colorful seaweeds cannot compete with this scale of primary food production.

Primary producers such as plankton have fueled the entire history of life on Earth. So great were the quantities involved that the buried plankton of past oceans have yielded the offshore oil and gas fields we are busy exploiting today. Most primitive and constant of all are the billions upon billions of bacteria that decompose and recycle the ocean's wealth of nutrients.

The kaleidoscope of life sparkles and turns through every last corner of the ocean, and yet scientists estimate that 90 percent of the Earth's sea-dwelling species have still to be discovered.

The Diversity of Life

KEY TERMS

TAXONOMY
DOMAIN
KINGDOM
PHYLUM
CLASS
ORDER
FAMILY
GENUS
ARCHAEA
BACTERIA
PROKARYOTE
EUKARYOTE
NUCLEUS
ORGANELLE
MITOCHONDRIA
PROTIST
METAZOAN

A bewildering variety of life confronts even the most casual observer of the world's oceans. There are extremes of size and form, of impressive antiquity and evolutionary perfection, ranging from microscopic bacteria that have shown little change in lifestyle since they witnessed the dawn of life over 3.5 billion years ago to the superbly efficient, streamlined bulk of one of today's top marine predators, the Great white shark. The world's lightest fish, *Schindleria*, weighs only 2 milligrams, while the oceans are also home to the largest animal ever to have graced our planet, the Blue whale.

Order out of chaos As on land, we can only begin to comprehend the dazzling variety of life encountered in the oceans by arranging the countless different forms into groupings (taxa) that share similar traits. The most general yet most fundamental taxonomic level commonly used is that of the kingdom, while at the other end of the scale the ultimate division is into individual species, defined as groups of potentially interbreeding organisms. In between are several standard levels of classification—phylum, class, order, family, and genus.

The concept of kingdoms dates back to the 18th century, when the Swedish biologist Carl von Linné (Linnaeus) proposed the first division into animals and plants. By the 1950s the kingdoms had been expanded to five: animals, plants, fungi, bacteria, and the unicellular protists. Now, modern molecular analysis has once more redrawn the tree of life. In the 1970s Carl Woese, of the University of Illinois, observed fundamental differences between organisms that had previously all been called "bacteria." A new term was coined—"archaea"—and a new kingdom added to the tree. Some biologists, including Woese, recognize just three major divisions, or domains—Archaea, Bacteria, and Eukarya.

The tree of life has its roots more than 3.5 billion years back in time, when the first pioneering microbial life appeared. From these "archaea" (ancient ones) all the more complex, nucleated organisms—the eukaryotes—gradually evolved, branching off into their diverse kingdoms—protists, plants, fungi, and animals. The Eukarya, together with the Archaea and the Bacteria, constitute the three great domains of life.

Plants are multi-celled eukaryotes with an ability to produce their own food through photosynthesis—using the energy from sunlight to convert inorganic chemical nutrients into organic molecules. They are known as autotrophs (self-feeders).

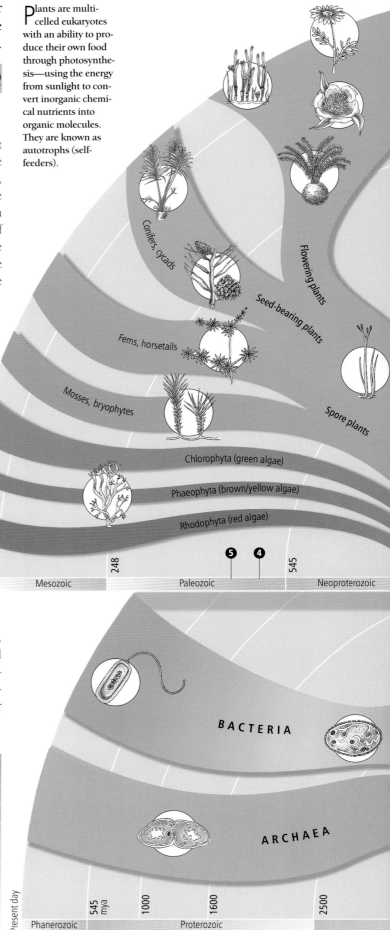

Taxonomic classification

	Diatom	Eelgrass	Sperm whale
KINGDOM	Protista	Plantae	Animalia
PHYLUM	Chrysophyta	Tracheophyta	Chordata
CLASS	Bacillariophyceae	Angiospermae	Mammalia
ORDER	Pennales	Najadales	Cetacea
FAMILY	Monoraphidineae	Zosteraceae	Physeteridae
GENUS	*Achnanthes*	*Zostera*	*Physeter*
SPECIES	*A. marginata*	*Z. marina*	*P. catodon*

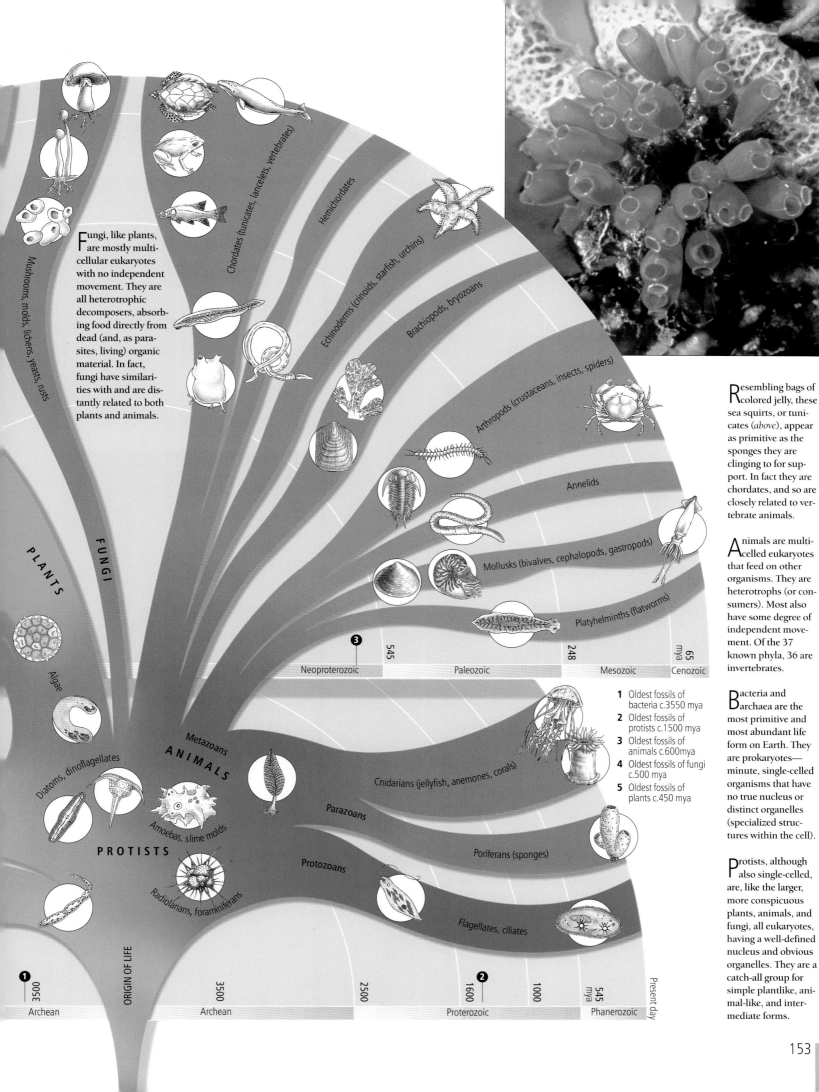

Fungi, like plants, are mostly multi-cellular eukaryotes with no independent movement. They are all heterotrophic decomposers, absorbing food directly from dead (and, as parasites, living) organic material. In fact, fungi have similarities with and are distantly related to both plants and animals.

PLANTS

FUNGI

Mushrooms, molds, lichens, yeasts, rusts

Algae

Metazoans
ANIMALS

Diatoms, dinoflagellates

Amoebas, slime molds

PROTISTS

Radiolarians, foraminiferans

ORIGIN OF LIFE

Chordates (tunicates, lancelets, vertebrates)

Hemichordates

Echinoderms (crinoids, starfish, urchins)

Brachiopods, bryozoans

Arthropods (crustaceans, insects, spiders)

Annelids

Mollusks (bivalves, cephalopods, gastropods)

Platyhelminths (flatworms)

3

545
Neoproterozoic

248
Paleozoic

65 mya
Mesozoic | Cenozoic

1 Oldest fossils of bacteria c.3550 mya
2 Oldest fossils of protists c.1500 mya
3 Oldest fossils of animals c.600mya
4 Oldest fossils of fungi c.500 mya
5 Oldest fossils of plants c.450 mya

Cnidarians (jellyfish, anemones, corals)

Parazoans

Poriferans (sponges)

Protozoans

Flagellates, ciliates

1
3500
Archean

3500
Archean

2500

2
1600
Proterozoic

1000

545 mya

Present day

Phanerozoic

Resembling bags of colored jelly, these sea squirts, or tunicates (*above*), appear as primitive as the sponges they are clinging to for support. In fact they are chordates, and so are closely related to vertebrate animals.

Animals are multi-celled eukaryotes that feed on other organisms. They are heterotrophs (or consumers). Most also have some degree of independent movement. Of the 37 known phyla, 36 are invertebrates.

Bacteria and Archaea are the most primitive and most abundant life form on Earth. They are prokaryotes—minute, single-celled organisms that have no true nucleus or distinct organelles (specialized structures within the cell).

Protists, although also single-celled, are, like the larger, more conspicuous plants, animals, and fungi, all eukaryotes, having a well-defined nucleus and obvious organelles. They are a catch-all group for simple plantlike, animal-like, and intermediate forms.

153

The Sperm whale (*right*), which can measure up to 20m (66ft) in length, is one of the largest animals ever to have lived on Earth. This remarkable blunt-nosed giant is an aggressive predator at the top of the food chain; it is particularly partial to squid and cuttlefish. On the whole, the higher up a food chain an individual organism operates, the physically larger it is likely to be. However, despite its size, this species is now rare; puny humans, occupying the role of top predator, have hunted it to the verge of extinction.

The six kingdoms

Each of the six kingdoms is represented in the marine realm, from which all life once originated. The oldest of all organisms are the bacteria and archaea; for the first billion or more years of life on Earth, in the earliest Archean times, these microbes were the only life forms on the planet. They have stood the harsh test of time, surviving dramatic environmental changes and sudden mass extinctions. Even today, they are the most abundant, widespread, and various of all living things. Both are simple prokaryotes, with only a single cell, no nucleus, and no separate organelles (functional units such as mitochondria), yet they are still critically important to life in the oceans.

Many bacteria are decomposers, secreting powerful enzymes that convert dead tissue into its component elements and releasing vital nutrients, such as carbon, back into the environment. Others manufacture their own food, either by harnessing the energy in sunlight, as in the case of cyanobacteria, or by the oxidization of sulfur, nitrogen, and other elements. They are tolerant of a wide range of environments: some depend on gaseous oxygen for their survival and growth, while others require the anaerobic conditions of deep underwater sediments; some live embedded in the Arctic ice, while others thrive in the boiling temperatures of hot springs or deep-sea hydrothermal vents.

Kingdom Protista is a catch-all category, originally designed to accommodate organisms intermediate between plants and animals. Protists are both diverse in type and myriad in number. They include all the single-celled organisms with a nucleus and distinct organelles, as well as some bizarre colonial life forms such as slime molds. Some are planktonic (floating) and photo-synthesize their nutritional needs; others are highly motile, feed on other organisms, and to a degree resemble animals. Collectively the Protista constitute a very large part of the ocean's living mass, lying at or near the base of the food chain. They are the soup kitchen of the ocean world, and produce at least half the planet's oxygen. These tiny cells—coccolithophores, dinoflagellates, foraminiferans, and diatoms, which between them dominate the plankton—also construct a huge variety of beautiful, intricate skeletons from the chemical salts of the sea.

Marine biologists have long debated where exactly to classify the algae, and they are often grouped with the protists. The familiar green, brown, and red seaweeds that festoon rock pools and wave in the shallow sea, as well as the giant kelp forests that shelter a rich and exotic fauna, are all types of algae. They are classified here as simple plants; that is, without an internal water transport system, true roots, stems, or leaves. Indeed, the only higher members of the plant kingdom that have returned fully to a life in the oceans are the seagrasses. Many other types of plants, including those of the uniquely adapted mangrove swamps, keep to the fringes and only flirt with seawater.

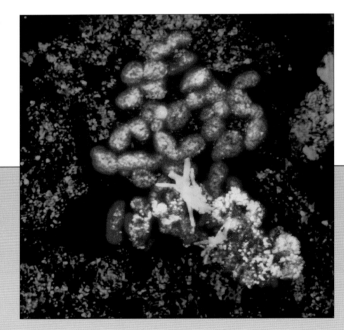

Marine microbes: adapting to extremes

These "rock-eating bacteria" (red cells, *right*) were found in 1995, sandwiched between layers of solid rock 1,000m (3,300ft) below the Earth's surface. Deprived of light and oxygen, they may be methanogens living on hydrogen in rock and dissolved carbon dioxide. They are called SLIMEs (subsurface lithoautotrophic microbial ecosystems) and may be members of the Archaea.

Imagine the most extreme and inhospitable environments on Earth and they will almost certainly be teeming with ultra-microscopic microbial life—mainly bacteria. Versatile and resilient, these single-celled organisms are, quite literally, everywhere. They are producers and consumers, with free-living, colonial, and symbiotic forms. Although extremely primitive, they outnumber every other type of organism on the planet.

These are the undertakers of the ocean world, as well as the waste collectors and sewage workers. At the first signs of death, a host of tiny microbes moves in and takes over. They live among the thriving plankton at the ocean surface, waiting for droppings or death; they hover in the ghostly region that is Middle Ocean to hitch a ride on falling carcasses; and they even lurk on the deepest seafloor. Their role in death is vital to life for, by decomposing the remains of dead organisms and animal waste products, they release essential nutrients back into the environment. Bacteria are skilled specialists, each with a different role in the breakdown of complex organic molecules and inorganic ions. A task force of numerous species works in close consort to effect almost complete recycling.

Even when the supply of free oxygen is used up, in stagnant waters enriched by decaying organisms and their waste, other types of bacteria take over. These are anaerobic microbes that operate the process of fermentation—much slower than aerobic decomposition, but effective nonetheless. Ever-increasing numbers of these tiny organisms have been found as exploration of the ocean floor has progressed and scientists have drilled ever further into the thick layers of sediment. Not only are there completely new species but they are present in unimaginable quantities, living in tiny pore spaces between sediment grains hundreds of meters below the seafloor. Here they must adapt to life without sunlight and oxygen, at extreme pressures and steadily increasing salinities within the compressed pore fluids.

These remarkable organisms have also adapted to extreme heat and cold. When black smokers were first discovered along mid-ocean ridges, spewing their noxious cocktail of toxic metals and scalding hot water, scientists were amazed to find a variety and abundance of deep-sea life around them. These oases of the deep are only tenable because of the countless chemosynthetic bacteria that have made them their home, often living in close symbiosis with the life they support, and harnessing hydrogen sulphide from the hot vent fluids. Bacteria and archaea have also been recovered from beneath several kilometers of Antarctic ice, and from cold methane vents on the seafloor. Nobody yet knows just what heat they can tolerate, the depth below the seafloor that they can withstand, nor the variety of energy sources they can exploit.

An intriguing implication of ongoing research into marine bacteria is that many other planets in the solar system and beyond may harbor similar ecosystems below their surfaces, adapted to whatever extreme conditions may prevail.

Members of the Kingdom Fungi are much less familiar in the oceans than on land. Widely dispersed, especially along the intertidal zone, the primary role of the marine fungi is to join the bacteria as decomposers of all dead organic matter. Eons ago, some species of fungi made a particularly successful alliance with algae. The strikingly resilient result of this symbiotic relationship was lichen—one of the first organisms ever to emerge from the safety of the seas and help colonize the continents over 450 million years ago. Once classified together with plants, fungi are now, thanks to molecular studies, thought to be more closely related to animals.

All animals are multicellular eukaryotes, generally motile, and wholly dependent on the primary producers for their food, but within this broad description are a number of deep divisions: between the sponges and the true metazoans (multicellular animals), between radially symmetrical animals such as sea anemones and the rest of the animal kingdom, and between alternative processes of embryonic development. Marine animals are found from near the base of the food chain to its very pinnacle, from simple sponges and jellyfish, worms and sea cucumbers, through an exotic assortment of lifeforms—oysters, starfish, crabs, shrimps, squids, and the dazzling variety of fishes that have colonized and colored every last corner of the oceans—to the powerful major predators: dolphins and Killer whales, marine crocodiles and sharks.

The unknown ocean Although the oceans offer 250 times more living space than the continents, and can therefore support uncountable numbers of organisms, less than a quarter of the 1.5 million known species on the planet are found within them. This is not surprising when we consider how little we actually know of our ocean world, how many of its habitats have never been seen nor even remotely explored. Merely sifting through the seafloor mud yields not only an incredible array of organisms, from tiny microbes to scavenging metazoans of the deep, but also many new species and unexpected forms. The work of marine biologists through the 21st century may more than double the number of known species. Some scientists suggest that there could be well over 20 million different species living in and beneath the oceans today.

We do know that ocean life is dramatically uneven in its distribution. Most productivity and diversity is concentrated either in the sunlit surface waters or along the busy shoreline and out across the shallow continental shelves. There are oases of profusion, such as coral reefs and mangrove swamps, kelp forests and plankton blooms, juxtaposed with regions of relative scarcity, for example beneath the ice-covered Arctic Ocean. But the real deserts of ocean space are many times more barren than the heart of the Sahara and a thousand times larger. These are the great voids of perpetual darkness and eternal cold that occupy the mid-water depths of every ocean.

Food is of paramount importance in the grand theater of life. Without the constant and efficient cycling of nutrients the fragile thread of existence would soon trail away.

Cycles, Webs, and Flows

PHOTOSYNTHESIS

CHLOROPHYLL

BIOMASS

AUTOTROPH

HETEROTROPH

PRIMARY PRODUCER

CONSUMER

HERBIVORE

CARNIVORE

DECOMPOSER

FOOD WEB

TROPHIC LEVEL

MARINE SNOW

DETRITUS

Life exists through the cycling of matter, the intake and metabolization of nutrients, and the transfer of energy. Every organism is affected by both the living and non-living elements of its surroundings. Temperature, salinity, water pressure, the availability of sunlight and of nutrient chemicals—these are some of the physical factors that influence ocean life, while the assemblage of organisms that occupy any one habitat has a profound influence on every member of the community. There is competition for living space, there is an urge to reproduce. But the endless quest for food is the priority of all living creatures.

Cycle of life The marvel that sustains life on Earth is a simple but all-important chemical reaction, developed at the dawn of time when cyanobacteria first appeared in the ocean soup. This is the process of photosynthesis, whereby energy from the Sun is absorbed by the green-colored pigment known as chlorophyll and used to convert carbon dioxide and water into a sugar called glucose.

Glucose itself is a useful source of energy, but is also essential in the construction of more complex carbohydrates. Oxygen is a byproduct of the reaction, released into the hydrosphere and atmosphere and taken up by the majority of other organisms.

On land it is the trees, shrubs, and grasses that constitute most of the biomass and carry out photosynthesis. In the oceans it is the phytoplankton—a diverse, floating population of single-celled organisms, which basks in energizing sunlight and is bathed in the inorganic nutrients present in all seawater. The phytoplankton are called "autotrophs," because they manufacture their own food. Together with blue-greens, algae, lichen, and marine plants, they allow the living marine world to function and survive. They are the vital base of nearly all food chains in the ocean, and therefore are many times more abundant than the organisms that feed on them—known as "heterotrophs." Only in extreme habitats without sunlight do chemosynthetic bacteria act alone as the primary producers. Different species of bacteria and archaea utilize hydrogen sulfide and other chemicals to construct the building blocks of life.

All primary producers are rapidly grazed by an ever-present army of hungry herbivores. In their turn the herbivores are eaten by carnivores. These are, respectively, the primary and secondary consumers within the ecosystem. Sooner or later all living organisms are either eaten or die naturally. Phytoplankton typically last only a matter of hours or days. The first consumers of this bounty, the microscopic creatures known as zooplankton, may live for a few weeks before they too are consumed or die. The lives of the larger fishes, turtles, dolphins, and whales may span anything from a year to several decades.

But if production, consumption, and death were the whole story, then life itself would soon become quite untenable. Its vital ingredients would pass right through the chain, ending up as a useless heap of remains, feces, and carcasses scattered over the seafloor. The ocean is bountiful indeed, but even its rich storehouse of nutrients would run out if it were not replenished. Fortunately, the decomposers—microscopic bacteria, archaea, and fungi—step in rapidly to clean up and recycle wherever death occurs.

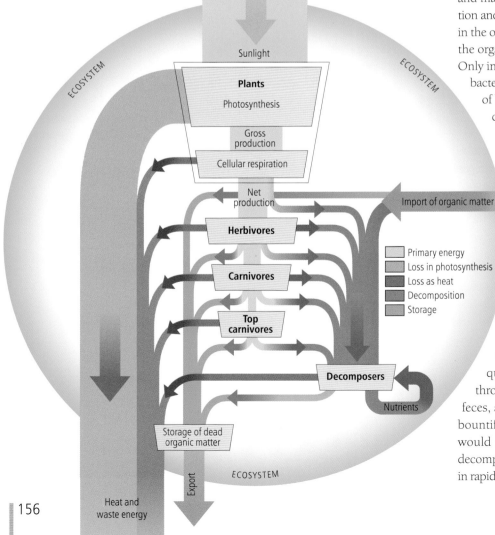

Sunlight

ECOSYSTEM

ECOSYSTEM

Plants

Photosynthesis

Gross production

Cellular respiration

Net production

Import of organic matter

Herbivores

Primary energy
Loss in photosynthesis
Loss as heat
Decomposition
Storage

Carnivores

Top carnivores

Decomposers

Nutrients

Storage of dead organic matter

ECOSYSTEM

Export

Heat and waste energy

Through the action of decomposition, the chemicals that have been so carefully built into the complex organic molecules of life are stripped away and returned to the ocean waters as dissolved inorganic substances. There they are once more available to be taken up by primary producers during photosynthesis. Chief among the nutrient elements essential to life are carbon, nitrogen, phosphorus, and silicon, as well as the hydrogen and oxygen in the water molecule itself. A host of other minor elements—iron, nickel, manganese, cobalt, and copper, for example—tend to be scavenged in minute quantities during the breakdown of organic matter, contributing to small but important processes within the wide spectrum of existence. And thus the cycle of life is completed.

The flow of energy through this cycle is by no means as efficient as the recycling of nutrients. Primary producers capture and store less than 1 percent of the available solar energy, although this is clearly adequate for their brief needs. Of this, only about 10 percent is passed from one trophic (feeding) level to the next. Most of the energy flow gained by eating is utilized for the processes of growth, movement, feeding, and reproduction, and is ultimately removed from the system as heat loss. The energy throughput, therefore, is not cyclic. This fact severely limits the number of feeding levels up from the secondary consumer. It also explains why biomass decreases so drastically from the base to the top of the food chain, and why the animals at the top, although larger in size, are far fewer in number.

Food chains and webs The concept of the food chain illustrates how, in the many different environments of our ocean world, organisms are linked by a steplike sequence of feeding relationships. For example, diatoms flourish in open surface waters and utilize energy from photosynthesis. Tiny crustaceans—copepods and others—harvest the diatoms, wafting them toward their open mouths with rhythmically flailing appendages as they, too, float amid the plankton. Small fishes, such as anchovies, feed on the copepods, and themselves fall prey to swift hunters, like tuna. This typical scenario is played out off the coast of Peru, where the plankton-rich waters support a huge anchovy population, which in turn feeds Yellowfin tuna and Pacific bonito, as well as pelicans and cormorants. Sharks and humans vie for the role of top consumer.

Closer to shore, from temperate to tropical regions, seagrass beds are a major source of primary production. However, apart from sea urchins, some fishes and birds, and the Green sea turtle, few of the many inhabitants graze directly. Most of the seagrass tissue is broken down into detritus and enters the food chain that way. In cold polar seas, in another variation on the theme, the 25m (80ft) Blue whale feeds directly on krill—small (approximately 5cm/2in long) shrimp-like crustaceans. Because energy is lost in each transfer to higher trophic levels, eating low down on the food chain in this way is highly energy-efficient for the whale, although it does require some 4 tons of krill per day to satisfy its enormous metabolic needs.

Phytoplankton and autotrophic bacteria are responsible for around 90 percent of primary productivity in the oceans. Seaweeds (*left*) and seagrasses also contribute to the process, especially in coastal habitats. More importantly, decaying plant and algal remains are a significant source of organic detritus for detrital food chains.

Where plankton productivity is high, the upper levels of the ocean become a whirl of activity. Bay anchovies (*below*) school in vast numbers in a reef cave off the Cayman Islands. Feeding mainly on zooplankton, these small fish are themselves a major food item on the reef.

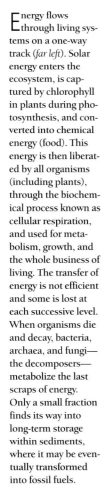

Energy flows through living systems on a one-way track (*far left*). Solar energy enters the ecosystem, is captured by chlorophyll in plants during photosynthesis, and converted into chemical energy (food). This energy is then liberated by all organisms (including plants), through the biochemical process known as cellular respiration, and used for metabolism, growth, and the whole business of living. The transfer of energy is not efficient and some is lost at each successive level. When organisms die and decay, bacteria, archaea, and fungi—the decomposers—metabolize the last scraps of energy. Only a small fraction finds its way into long-term storage within sediments, where it may be eventually transformed into fossil fuels.

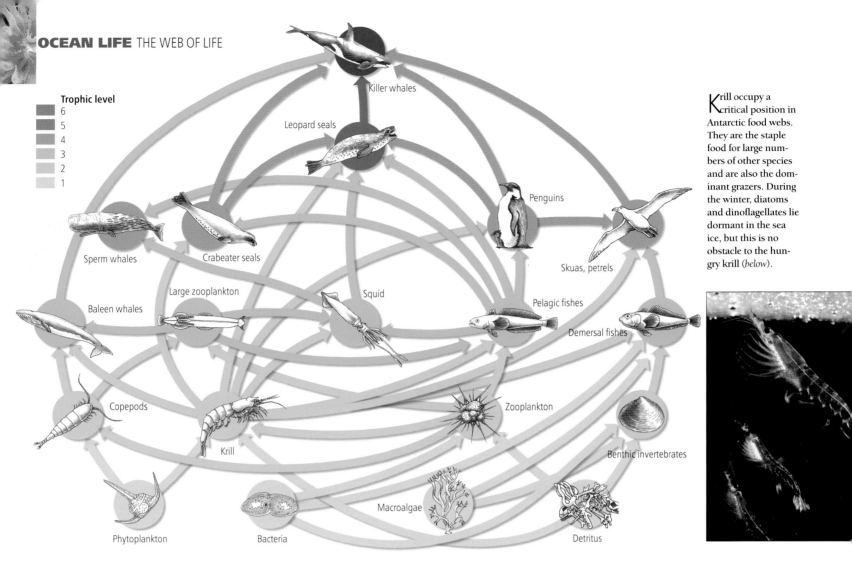

Trophic level

6
5
4
3
2
1

Killer whales

Leopard seals

Penguins

Sperm whales

Crabeater seals

Skuas, petrels

Large zooplankton

Squid

Pelagic fishes

Baleen whales

Demersal fishes

Copepods

Zooplankton

Krill

Benthic invertebrates

Phytoplankton

Macroalgae

Bacteria

Detritus

Krill occupy a critical position in Antarctic food webs. They are the staple food for large numbers of other species and are also the dominant grazers. During the winter, diatoms and dinoflagellates lie dormant in the sea ice, but this is no obstacle to the hungry krill (*below*).

The Antarctic food web is one of many interwoven feeding pathways in the oceans. Phytoplankton, autotrophic bacteria, and macroalgae (seaweeds) are the primary producers. Decomposed organic matter—detritus—is also an important food source, although not a primary product. But even at this level, and certainly at higher trophic (feeding) levels, the web is more complex than it appears. Bacteria also break down organic material and have an intricate relationship with the various nano- and pico- (ultramicroscopic) plankton in recycling nutrients. Much, too, is dependent on relative size—copepods miss this smallest plankton, while seabirds will predate penguin chicks but not adults.

The concept of food chains is a much simplified version of what actually occurs in the oceans. In any particular community or ecosystem, feeding patterns are usually far more dynamic, forming an interconnecting web of relationships. Consumers may prey on a wide range of animals at several different trophic levels, and species may change level at different stages of their lifecycle. A broad-based food web is ecologically more stable than a simple linear chain, for if one prey species disappears there are alternative choices on the menu. However, the more trophic levels that exist in a chain, the less efficient it is.

Of course, not all the food on offer at any level is consumed; much escapes or is rejected, and eventually sinks through the water column untouched or only partially consumed. Many types of phytoplankton and zooplankton co-exist and, if they evade predation, live out their undramatic lives, energized by sunlight and stirred by swirling currents. As they die and begin to fall through the water column, decomposition by microorganisms begins almost immediately. The recycling of nutrients stimulates more primary production, forming yet another loop in the web. Mid-water zooplankton and opportunistic fishes gobble a further slice as the leftovers pass through their domains. Food availability decreases with depth; the bottom-dwellers—a bizarre assemblage of fishes, sponges, sea cucumbers, the occasional Giant squid, and many other animals—are sustained by what few morsels remain. More bacteria, archaea, and fungi, some of which are still unknown to science, then set to work to decompose the final scraps.

Marine snow Everywhere in the oceans there is a constant, unremitting, downward drift of material from the surface toward the seafloor. Dead and decaying organic matter, feces, busily working bacteria, tiny protective skeletons and mucus "houses" from planktonic species, and inorganic detritus windblown or washed in from the continents all play a part. Many of the individual particles are miniscule and would take an interminable age to sink on their own, but together they clump into larger aggregates of material—sticky or fragile, porous or dense, elaborate or simple—all quietly falling as a kind of marine snow. In particular parts of the ocean and at certain times of the year when surface productivity is high, the falls are thick and fast, reaching to even the deepest parts in a matter of weeks, and blanketing the seafloor with a soft, gray carpet of organic-rich mud.

The fall of marine snow is often difficult to observe in nature, for it is actually highly dispersed. The marine snowflakes are even more tricky to sample, for they disaggregate at the slightest touch. Yet the study of this detritus, and the role of microorganisms, provides exciting insights into the complex webs of food and material recycling that occur, and it is a challenge in which many marine scientists are now engaged.

Chemical cycles The largest and most important cycle of nutrient elements on the planet is that of carbon. It is a cycle that spans land, sea, and air. The Earth's lifeforms are all carbon-based, so that the nature and cycling of carbon, both organic and inorganic, are crucial to the success and distribution of life across the world. Nearly all food webs start with the manufacture of organic molecules from carbon dioxide, for use either as food or as structural components, and the many links in the web of life ensure that this extraordinary and versatile element is recycled again and again.

In certain places, the fall of marine snow can be so thick and fast that even the combined efforts of mid-water and benthic bacteria are insufficient to extract and recycle all the organic carbon. This dead organic matter builds up within the sediment over geological time and is ultimately converted into fossil fuels—oil and gas at sea, coal on land. We know that the seafloor regions that exhibit intense organic carbon accumulation today are commonly located beneath zones of extremely high primary productivity at the surface. The same would have been true in the geological past, thus leading to the source rocks for marine hydrocarbon accumulation.

Even more carbon is locked up in the billions upon billions of tiny planktonic skeletons composed of calcium carbonate. With the passage of time, these become cemented into chalk and limestone rock, deeply buried beneath the seafloor. After many millions of years they may be uplifted into the chalk cliffs and limestone uplands that constitute so impressive a part of the continental landscape. Very slowly, weathering and erosion of the land will return even this carbon to the sea, so it may once more become incorporated into the building blocks of life.

The atmosphere, too, is closely involved, for an exchange of gases readily takes place across the air–sea interface, and the respiration of organisms on land and at sea pumps carbon dioxide back into the air and water, just as photosynthesis by plants and phytoplankton removes it. The volcanic release of carbon dioxide, episodic but hugely important, is another vital source of the gas.

The other significant nutrient elements (nitrogen, phosphorus, silicon) have their own complicated pathways and cycles on a similarly global scale. They interact with living beings and are essential to the very existence of life as we know it, but they also reside in the atmosphere and water, the sediments and rocks. The constant flow of these elements is important, just as it is for carbon, as is a general balance in the global system. The upsetting of the carbon balance in the geological past was one

of the prime reasons behind several of the mass extinction events the world has experienced. This fact partly explains why scientists are currently so concerned about the implications of human behavior on levels of atmospheric carbon dioxide. A marked increase in these levels can have a dramatic warming effect on global climate via the process known as the greenhouse effect. As we continue to burn fossil fuels, we are releasing carbon dioxide back into the atmosphere at a hitherto unprecedented rate. The short-term influence on climate and the longer-term effects on life in general are still unknown. Marine scientists are especially interested, therefore, in the ocean's part in the carbon cycle.

Atmosphere
700 Gt
3 yrs

Surface water
600 Gt
100,000 yrs

Living organisms
700 Gt
5 years

Decaying organisms
1100 Gt
20 years

LAND

Intermediate water
7000 Gt
100 yrs

OCEAN

Deep water
30,000 Gt
100,000 yrs

Fossil fuels & shales
12,000 Gt
1000 years

Sediments
30 million Gt
100,000,000 yrs

Particulates and dissolved carbon

Living organisms

Carbon is the element of life, the source of energy in fuels, and a key player in global climate. It cycles through both living and non-living systems. The simplified diagram (above) shows the principal carbon reservoirs in gigatonnes (1 billion metric tons) and the average residence time in each.

These tiny, beautiful foraminiferan skeletons are made of calcium carbonate. It is the slow accumulation of such shell debris on the seafloor that locks carbon into sedimentary rocks.

SEE ALSO

Phytoplankton are the foundation of the oceanic food web; the energy they capture supports the entire marine community. Their microscopic world is full of vitality, charm, and surprise.

The World of Phytoplankton

KEY TERMS

PLANKTON

PHYTOPLANKTON

DIATOM

COCCOLITHOPHORE

COCCOLITH

DINOFLAGELLATE

FLAGELLUM

CYANOBACTERIA

PRIMARY PRODUCTIVITY

UPWELLING

PLANKTON BLOOM

It is strange to think that the whole spectrum of ocean life depends upon an almost invisible world that we perceive only as different colors of the water. Crystal clear, vivid blue seas—the idyll of a tropical island paradise—are generally poor in nutrients and have a consequent dearth of phytoplankton. But the cooler, green seas of temperate latitudes contain countless billions of these tiny photosynthetic organisms, each with microscopic specks of chlorophyll enclosed in its single cell. When the phytoplankton is in bloom, patches of ocean may turn an iridescent green or may come to resemble blood-red stains or milky-white clouds as sunlight is scattered through the water by innumerable tiny, pigmented cells and their reflective skeletons.

Microscopic sculpture Plankton are the floaters and drifters of the ocean. Without independent movement, or at best as poor swimmers, both phytoplankton and zooplankton are carried along at the mercy of wind and currents, tide and storm. Most are transient members of the community, with very short lifetimes. Generally they are microscopic in size, although some of the zooplankton grow to several centimeters in length, while really gigantic drifters, such as the jellyfish *Cyanea,* may measure 2m (6.5ft) across. Some are so tiny that only the finest filters can isolate them from water samples.

The phytoplankton world can only be entered through powerful microscopes. Even so, its existence remained unknown to science until the early 19th century, some 200 years after the microscope was first invented. Ideally, it needs to be examined with the ultra-high magnification achieved by scanning electron microscopy, with whose aid ocean scientists can at last wonder at the myriad shapes and diverse lifestyles revealed.

The master sculptors of these incredible forms are nothing more than single-celled organisms, members of the Kingdom Protista, that work with chemical salts freely available in seawater. They create minute, intricate skeletons of pure silica glass—transparent spheres, some bristling with tiny, sharp spines; minute, sievelike disks and boxes; ovoid and pyramid-shaped sacks, tubes, and rods; bulbous forms like sets of miniature lifting weights. Other species cast their mineralized coats of brilliant white lime, sequined with multiple tiny platelets.

At times this plethora of exquisite forms resembles an exhibition of ocean art. Although we describe these magnificent sculptures as protective skeletons, protection is by no means the sole reason why the development of such shells, and the subsequent diversification of form and shape, has been such a successful evolutionary adaptation. Many phytoplanktonic organisms survive only for a few hours before they are consumed as food, despite their elaborate coats. Yet others manage equally well as soft-bodied cells, filaments, or mats. Perhaps even more significant than protection is the effectiveness of chambers, spikes, and other protrusions at improving flotation—maintaining that all-important position in the uppermost, sunlit waters.

Garden of Eden The phytoplankton "meadows" are the garden of the world—one of splendor and fertility. Each year over 6 thousand billion metric tons of phytoplankton grow worldwide wherever light penetrates, harnessing solar energy to synthesize their tiny bodies from the carbon and other nutrient chemicals in seawater.

The amount of solar radiation varies with latitude and season. The low-latitude tropical oceans receive plenty of sunshine yet plankton productivity is low. The intense surface heating creates a sharp thermocline (temperature gradient), preventing deep-

Gross primary productivity in the oceans shows as much variation as on land. The central regions of the open ocean, which have very low productivity, equate with continental deserts, while coral-reef ecosystems are as prolific as tropical rainforests. (Quantities are expressed in grams of carbon per square meter per year.)

Primary productivity

Quantity (gC/m²/yr)	Ocean ecosystem	Land ecosystem
750–500	Shallow inlets; kelp beds; coral reefs	Tropical rainforest; freshwater swamps
500–250	Coastal upwellings; deep estuaries	Intensive agricultural areas; lakes
250–50	Continental shelves	Temperate forests; grasslands; croplands
<50	Open ocean	Deserts, steppes

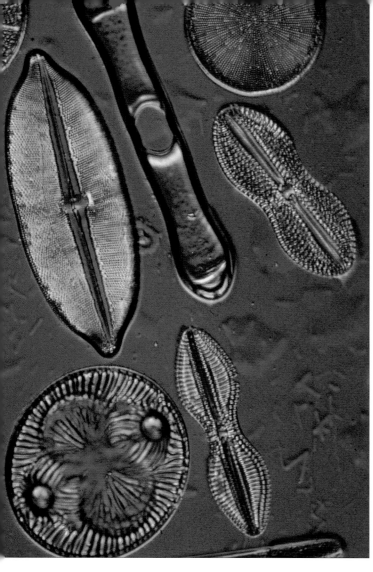

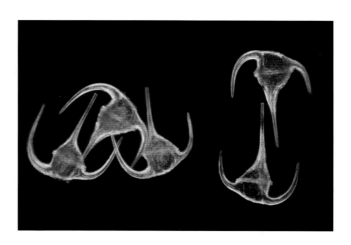

A tiny sample of more than 6,000 known species of diatoms (*left*), showing centric (circular) and pennate (long) types. Each forms its tiny glassy skeleton from pure silica, usually in two halves of different sizes—like a small box and lid. The shells are known as frustules.

water nutrients from reaching the sunlit upper levels. By contrast, the long days of summer in the well-mixed polar seas stir up a rich planktonic soup. Coastal areas, which receive dissolved nutrients from land, are also productive, while the richest spots of all are the regions of equatorial or coastal upwelling.

Diatoms are one of the major and most abundant groups of plankton, with over 6,000 different species. These miniature organisms inhabit variously ornamented silica shells, formed of two valves that fit together like a box and lid. They reproduce asexually every 12–24 hours, with one valve going to each daughter cell. A new valve is secreted within each original half, and so on, until the resulting cells eventually become too small for viable subdivision. Then they cast aside their glass cases and reproduce sexually, eventually growing to their former size and building a new home. Most diatoms store food reserves as little droplets of oils and fatty acids, which also help maintain buoyancy. Diatoms are the dominant phytoplankton in temperate waters, as well as along many shores and inlets.

Coccolithophores thrive in tropical seas and the open ocean. Their complex, spherical homes are constructed of tiny, patterned platelets (known as coccoliths) made out of calcium carbonate. They have two threadlike flagella, which they use primarily for staying afloat and so maintaining their position in the sunlight. Silicoflagellates, by contrast, are extremely abundant in cold polar waters. They have internal skeletons of clear glass, typically with silica spikes and one or two flagella, both of which help prevent them from sinking. Like diatoms and coccolithophores, they contain chlorophyll for photosynthesis as well as golden-brown pigments known as xanthophylls.

Most commonly seen in the tropics at night, some dinoflagellates glow with a ghostly, green-blue bioluminescence. They have a tough cellulose coating, and are among the more active swimmers, aided by their two long, whiplike flagella. As well as chlorophyll, they possess yellow-orange carotenoids, and are one of the groups whose rapid increases are responsible for the often dangerous phenomenon of "red tides." Some species lack the cell wall and flagella, and lead a symbiotic life within the bodies of certain jellyfish, corals, and mollusks. Their task is to supply food, while their host provides protection, carbon dioxide, and other nutrients. Other dinoflagellates have taken up a parasitic existence, living in the intestines of marine crustaceans.

Dinoflagellates (*above*) are a diverse group of microscopic organisms. They are important photosynthesizers within the phytoplankton. Like plants, they have cell walls composed of cellulose but, like animals, they can move using their thread-like flagella.

Colonies of filamentous cyanobacteria (*left*). Although these are usually blue-green in color, bright shades of pink and red also occur due to accessory pigments in the cells. Such microscopic organisms have been coloring the oceans since the dawn of existence. They are capable of converting nitrogen directly from the atmosphere into useable forms (such as ammonia) for protein manufacture, by a process known as nitrogen fixation.

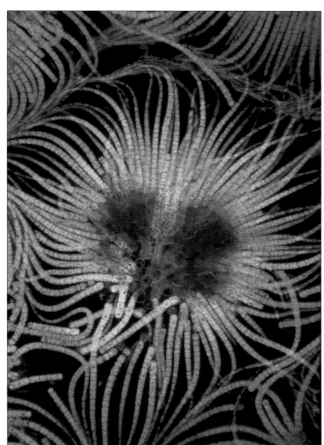

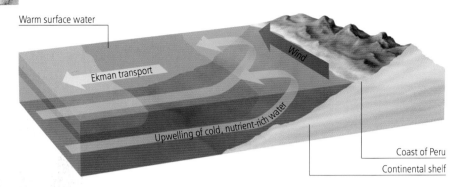

Warm surface water

Ekman transport

Wind

Upwelling of cold, nutrient-rich water

Coast of Peru
Continental shelf

The environmental conditions required to sustain a bloom are quite specific, but simple enough to fulfil. A plentiful supply of sunlight is, of course, the first essential. Clear, near-surface waters, free from the turbidity caused by storm stirring, are also needed, as are days that are long enough to allow photosynthesis, growth, and reproduction to take place. These days start in the spring and continue through the height of summer.

The second essential is a storehouse of the right nutrient elements—carbon, nitrogen, phosphorus, and silicon, together with certain trace metals, such as magnesium and iron for chlorophyll and other pigments. Such nutrients are not always sufficiently abundant at the surface to allow more than a minor background level of productivity, and their availability is a key constraint on fertility in the ocean. One of the principal supplies is from rivers, which constantly leach minerals from continental rocks and empty billions of metric tons of chemicals annually into the oceans. They also carry terrestrial organic detritus to the seas. The other main source of nutrients is from the bacterial recycling of dead and decaying organisms and their feces. But these materials have the unfortunate habit of sinking rapidly as part of the marine snow and taking vital nutrients with them away from the well-lit surface waters. Their return to the surface occurs most effectively in regions of upwelling, as well as by thorough mixing of the upper ocean layer during winter storms.

Finally, growth must be sufficiently rapid for the plankton to keep ahead of the fast-growing population of herbivores, which constantly graze the ocean pastures. In temperate and polar waters, the season which consistently brings together all these requirements is spring. Daylight hours are growing longer, winter storms have stirred up nutrients from the deep, and the

Coastal upwelling occurs on western coastlines where, due to the Coriolis effect, wind-driven shelf currents are deflected offshore—a process known as Ekman transport. Nutrient-rich waters from below rise to the surface, creating ideal conditions for plankton growth. These areas become rich feeding grounds for an abundance of fishes and other marine creatures.

Cyanobacteria occur as single cells, filaments, and chains. In shallow waters they are found carpeting the seafloor in the form of rather slippery, dense, green mats. Those forms that secrete calcium carbonate are the architects of the stony, cushion-like stromatolites. This highly successful group of photosynthesizing bacteria laid down the original template for life on Earth 3.5 billion years ago. Similarly ancient, yet formerly unnoticed in the phytoplankton because of their ultra-microscopic size, are the archaea and bacteria. More than half a million of their individual cells would fit onto a single pinhead. Yet we now know that they can account for a staggering 80 percent of primary production at the ocean surface.

Spring blooms When conditions are optimal for their rapid growth, phytoplankton can reproduce in such abundance that huge areas of ocean literally swarm with life. Such a phenomenon is commonly referred to as a plankton bloom and, as with spring flowering on land, it is most commonly a seasonal occurrence.

Map of global variations in primary productivity in the oceans. Each of the major ocean basins is like a bulls-eye of decreasing productivity towards the middle (light green). These central areas are regions of downwelling, where surface water sinks—the truly barren marine deserts. One equatorial and two high-latitude belts of greater productivity (mid-green) mark where different water masses meet, with the deeper, nutrient-rich waters being forced upward. Maximum productivity occurs in coastal regions where nutrients from both upwellings and rivers are plentiful (dark green).

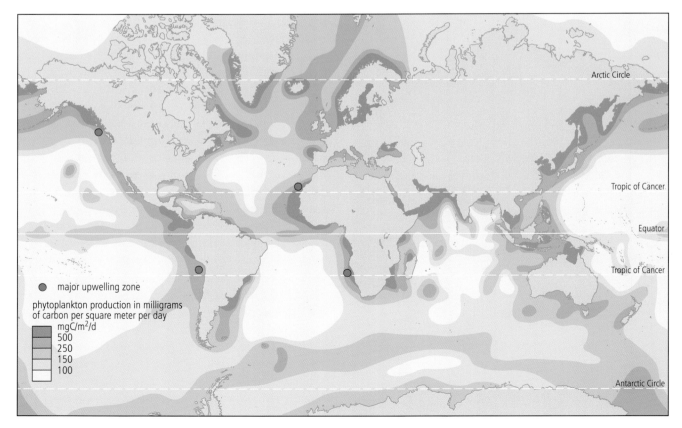

Arctic Circle

Tropic of Cancer

Equator

Tropic of Cancer

Antarctic Circle

● major upwelling zone

phytoplankton production in milligrams of carbon per square meter per day

mgC/m²/d
500
250
150
100

Marine research has recently elucidated a remarkable link between the microcosm and the macrocosm, as well as between plankton blooms today and the oceans of 80 million years ago. The common factor is a tiny member of the coccolithophore group, single-celled photosynthesizing organisms that live out their briefest of lives in the competitive world of ocean plankton. Its name is *Emiliania huxleyi*, and its near-spherical, armor-plated body reaches a maximum diameter of only 6 microns, which means that about 40,000 individuals would fit onto a pinhead.

Emiliania huxleyi has existed for around a quarter of a million years, and is now the dominant form among a multitude of different coccolithophore species everywhere from tropical to polar waters. Every individual is programmed to construct an outer shell of tiny, patterned disks, each one neatly overlapping the last, to form an impenetrable coat of delicate sequins. It uses the calcium carbonate dissolved in seawater for this remarkable feat, so that its shell glistens white in the sun. In the warmth of spring and summer, and where nutrients have been stirred up to the surface, these microscopic organisms multiply at a phenomenal rate. Concentrations can increase to over 100 million cells per cubic meter of water (76.5 million/cu yd), and spring blooms incorporating a staggering 25 billion billion individuals have been observed by satellite spreading across 250,000sq km (100,000sq mi) of ocean, an area roughly the size of Oregon. Each of these cells contains minute amounts of photosynthesizing pigment, so that the water changes color: first turquoise and then milky white as light is reflected from their surface.

Such profusion not only affects the color and composition of the sea but also influences gas exchange with the atmosphere. The blooms draw in vast quantities of carbon dioxide from the air above as well as from seawater, forming an essential link in the global carbon cycle that moderates Earth's climate. In exchange, the phytoplankton produce dimethyl sulfide, a volatile sulfur compound that oxidizes in air to yield sulfur dioxide, seeding local build-ups of cloud and contributing to acid rain.

Such activity forms a normal part of the annual cycle of growth and die-back in the plankton world. But if global warming continues, plankton blooms may spread still further. During the late Cretaceous 80–65 million years ago, for example, Earth's climate was much warmer than it is now, with warm, shallow seas spread across low-lying parts of the continents. Unimaginable productivity in the oceans over untold millennia led to the slow accumulation of layer upon layer of coccoliths on the seafloor—an endless snowstorm of calcite platelets derived from ancient relatives of *Emiliania huxleyi* that thrived in those far-gone days. The white cliffs of Dover and the banded chalk successions of northwest Europe and west-central America all now stand as testament to the warmer climates of a past age—and to the remarkable resilience of coccolithophores.

Professor Patrick Holligan,
Southampton Oceanography Centre, UK

Emiliania huxleyi: a microscopic ocean bloom

Satellite imagery captures a milky turquoise bloom of coccolithophores in Atlantic waters off Nova Scotia (*left*). At the other end of the scale, a scanning electron micrograph reveals that each cell (*below*) is covered with intricately patterned disks (coccoliths) of pure white calcite.

grazing zooplankton population has died back. At any one time plankton blooms normally consist of a single dominant species, resulting in a huge increase in biomass that lasts for just a week or two before dying back or giving way so that another member of the community can flourish in its turn. At times there can be as many as half a billion individuals in 1cu m (35cu ft) of seawater. Such blooms are very common along coasts and in estuaries, where they are generally small and localized events, covering little more than 1sq km (0.4sq mi). They also occur in freshwater lakes and rivers. In open waters, however, they are often much more extensive, covering hundreds of square kilometers. Truly gigantic mid-ocean blooms, up to 250,000sq km (100,000sq mi) in area, have been detected by satellite images.

Red tides Just as spring flowers create splashes of color across green fields on land, so the sheer abundance of phytoplankton blooms can paint the ocean surface. The beautiful greens and bluish-greens of diatoms and cyanobacteria are most common, while the brilliant white calcite platelets of coccolithophores create a milky turquoise color. These water colors, as well as a variety of reds, browns, and yellows, form the rainbow spectrum of the seas, with which people were already familiar long before they knew anything of the microscopic world that caused them.

Several species of plankton can produce brownish or yellowish seawater colors when in full bloom, as a cumulative result of the tiny amounts of pigment carried by each cell. Two particular species—the dinoflagellate *Noctiluca* and the ciliate *Mesodinium rubrum*—are known to contain symbiotic organisms with carotenoid pigments. In spring these plankton species can become so numerous that they create widespread patches, colored an alarming blood-red and known as red tides, which are common in many estuaries and coastal areas. They have been well-known in the Red Sea since biblical times and are often harmless. However, in other cases, dinoflagellate blooms can release highly toxic chemicals that have dramatic effects on marine life. One red tide off the Florida coast is believed to have killed about 100,000 metric tons of fish.

Both peaceful grazers and voracious predators are found in the miniature world of zooplankton, where a community of diverse animals, from single-celled protists to diaphanous jellyfish, compete for survival.

The Zooplankton

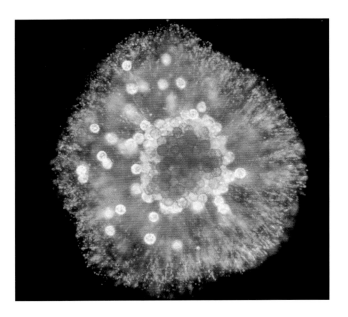

Radiolarians are single-celled zooplankton that have adapted to life at many different depths. This colonial form (*above right*) has a close symbiotic relationship with photosynthesizing bacteria.

Many marine animals and protists spend their larval stage as zooplankton. Some species, such as the pteropod —a form of snail— (*below*), remain among the plankton all their lives.

Were it possible to become a miniaturized traveller, then a journey through the surface waters of the open ocean would present itself as the most remarkable safari imaginable. Living off the colorful bounty of spring blooms and phytoplankton pastures, the tiny marine herbivores and the carnivores that prey on them inhabit a world of open warfare in which there is nowhere to hide. The pastures and the prey may be plentiful indeed, but an ever larger and better equipped predator is always close at hand; only the fittest, or the luckiest, will survive to maturity. The members of this community have therefore evolved a complexity and variety that belies their small size. Their hunting techniques, body armor, and assorted weaponry are used to devastating effect. Many species have developed near-transparent forms in an effort to avoid predation. But all is to little avail—the zooplankton and phytoplankton together feed the rest of the ocean.

Fast breeders One way of surviving, for the species if not for the individual, is to remain small and breed quickly. This reproductive strategy is employed by many creatures of the plankton community, and is well-known among several protozoan groups. These are all single-celled organisms that can build their tiny, sculptured shells, about the size of sand grains, at a phenomenal rate. Their numbers multiply to extraordinary proportions, as they are generally first at the feast when the phytoplankton blooms.

Both the elaborate, multichambered shells of foraminiferans, made from calcium carbonate, and the spiky glass spheres of radiolarians are speckled with tiny pores. Through these the organisms extend fine, radiating strands of their own cell material. Such glutinous and waving filopedia (literally, "feet of thread") easily ensnare other, even more minute creatures such as phytoplankton and bacteria, drawing the living material into the parent cell, where it is chemically broken down by powerful enzymes. Even organisms of comparable size can be caught and dismembered before being consumed.

Another group of helmet-shaped microprotozoans, called tintinnids, either secrete a hard outer cell coating or use a variety of foreign particles to further toughen their structure. These constructions are covered with hairlike threads (or cilia) that are used for capturing prey as well as for locomotion. A ring of wavy cilia surrounding the oral opening gently wafts water and phytoplankton toward a sticky end. These three smallest members of the zooplankton remain successful in the lottery of life only because they reproduce so quickly.

Eggs and larvae Sometime after the spring blooms have dispersed, the ocean world in summer can become a seething mass of floating eggs and growing larvae. Many free-swimming fishes of the open ocean also favor the technique of breeding in large numbers in their quest for survival. In their case, very large numbers of eggs, often tiny and transparent, are released into the plankton. Only a tiny fraction of the many millions dispersed ever develop into a larval stage and thence to adulthood.

In similar fashion, the young of a wide range of marine animals with quite different adult lifestyles pass through the surface waters as various types of larvae. Familiar bottom-dwellers such as polychaete worms and barnacles, sea urchins and brittle stars, crabs and lobsters, sea snails and bivalve mollusks all spend the first few weeks of life as ocean drifters in the plankton—a rite of passage into adulthood that must rank as one of the most arduous in the animal kingdom. The hunt for food may not be the

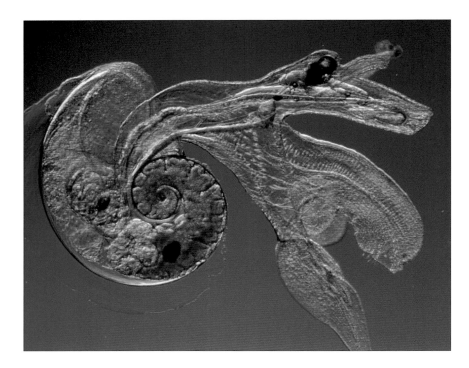

biggest test, but rather the gauntlet of hungry predators that must be evaded along the way. Most larvae do not survive. The few that do are swept by chance currents to the safety of some distant shore, where they can live a more secure adult life.

Miniature armies The chief marauders of phytoplankton are copepods—tiny, elongate crustaceans with broad, waving antennae—arriving at the spring feast only a short time behind the microprotozoans. They grow larger, up to 2mm (0.08in) in length, hunt more voraciously, and consume more rapidly than the microprotozoans. They have developed a highly efficient means of grazing: their minuscule legs beat furiously, setting up a steady flow of water that carries floating diatoms and other food items toward feeding cups spaced out along their limbs.

As they graze and reproduce, their numbers increase astronomically. Once again it is their tremendous fertility that leads to such abundance. After fertilization, a female copepod can produce over 100 eggs, and some species produce a new clutch of eggs every 4–5 days. New hatchlings take less than two weeks to reach maturity before they too can reproduce. Where food is plentiful, especially in the surface waters across broad continental shelves of upwelling areas (offshore Peru and northwest Europe, for example), their concentration can reach a staggering 100,000 individuals per cubic meter of water (almost 3,000 per cubic foot). At such times, they probably rank among the most numerous animals on the planet. Each one can consume as many as 120,000 diatoms per day.

In the open ocean, and especially the icy waters encircling Antarctica, high plankton productivity is related to the upwelling of nutrient-rich waters along the polar front that separates different water masses. The diatom blooms at these high latitudes generally begin in the early summer months, when pack ice begins to melt and the hours of sunlight increase. In close attendance are an ever-ready population of krill—open-water shrimps that grow rapidly to several centimeters in length. During the cold, dark winters, the krill population decreases and moves down the water column, feeding mainly on a sparse rain of organic detritus from above. But as the phytoplankton blooms, they migrate up to the surface to feed and breed. Single swarms in the Weddell Sea can be hundreds of square meters in area and as much as 5m (16.5ft) thick, and have been estimated at around 2 million metric tons in weight. Most of the 86 known species of krill are bioluminescent, so these unbelievably dense swarms produce an eerie light—an awe-inspiring sight for those who venture into the high-latitude southern seas.

Medusae In Greek mythology, Medusa was one of the three Gorgons—grotesque female monsters with snakes for hair who turned anyone unfortunate enough to see them into stone. In marine biology, a medusa is the free-floating stage of the cnidarian group known as jellyfishes. They are aptly named for their long, snakelike tentacles, which are coated with hundreds of stinging cells. These nemato-

cysts, as the stinging organelles within the cell are known, are tiny but lethal harpoons, designed to paralyze prey with a powerful toxin before drawing them into the large, bell-shaped gastrovascular cavity where digestion takes place. Small but most lethal of all are the box jellyfishes, some of which can kill a human in minutes. Also dangerous are the giants of the medusa world—the Portuguese man-of-war (a colonial jellyfish) and the Lion's mane jellyfish. The biggest of these animals grow very rapidly on a diet consisting mainly of copepods and small fishes. The largest recorded bell diameters are 2–3m (7–10ft) across, with tentacles extending 60–70m (200–230ft).

In another respect, the fearsome name is less apt, for these gently billowing clouds of flimsy, diaphanous material, with their fringe of trailing tentacles, are among the most remarkable and beautiful creatures in the open ocean. The tiny jellyfishes that feed mainly on copepods multiply rapidly if they can catch the peak blooming of these minute crustaceans before other predators arrive. Swarms of the lightly pigmented Moon jellyfish become so dense that they turn the waters pinkish white, like a wind-ruffled carpet of cherry blossom.

Jellyfish are the largest members of the zooplankton. One of the most common is the Moon jellyfish, *Aurelia* (*above*), measuring 15–45 cm (6–9 in) in diameter.

From underwater forests and floating meadows to brightly colored rock pools and seaweed-clad shores, the domain of the algae is one of the richest in the ocean realm.

Algal Forests

KEY TERMS

ALGAE
KELP
SEAGRASS
HOLDFAST
SUBSTRATE
WAVELENGTH (*light*)
VISIBLE SPECTRUM
PHOTIC ZONE
RHODOPHYTA
PHAEOPHYTA
CHLOROPHYTA
BLADE
STIPE

Some species of red algae, known as coralline algae (*right*), precipitate calcium carbonate out of seawater to form a solid coating, as do corals. They contribute to coral reef growth and may form as much as 50 percent of the reef bulk. They are especially important in cementing together fragmented pieces of reef material and, by forming a ridge on the reef crest, protect the leeward reef flat from wave action.

The world of algae is an ancient and mysterious one, both intriguing and profoundly beautiful. It ranges from microscopic cells and filaments in the open ocean to colorful seaweeds, and giant kelp forests in the cold, nutrient-rich waters off California, where individual growths can reach over 90m (300ft) in length. Surviving pounding waves, salty rock pools, and voracious herbivores, nearly 10,000 species of algae festoon coastlines everywhere from the Arctic islands to the tropics.

The algae that adorn the seaside world familiar to millions of vacationers harbor a wealth of animal life, just as gardens do on land. Delicate purple and pink filigree seaweeds and vivid, yellow-green "mermaid's hair" cling to the smoothed stone and crevices of low-tide rock pools, where they provide food and shelter for baby crabs, shrimps, and sea snails. Yet, unlike garden plants, the seaweeds possess no true leaves, stems, or roots; they do not flower, and they spread no seeds. Plantlike they may be, but many scientists are reluctant to grant seaweeds full membership of the plant kingdom, classing them instead as a unique group, transitional between unicellular protists and true plants.

Seagrasses are the only group of higher (flowering) plants that have fully and successfully recolonized the marine realm. They have both adapted to salt water and found an effective means of pollination and seed dispersal underwater. Eel grass is widely distributed in the north Atlantic and Pacific Oceans, even extending into the Arctic Circle. Turtle grass and manatee grass are the subtropical equivalent, forming dense underwater meadows throughout the Caribbean, and providing a unique habitat.

Producers and cleansers

Single-celled algae were among the very first photosynthesizing organisms that evolved in the hostile early-ocean world. Against all odds they survived and, together with the even more primitive cyanobacteria, slowly created a whole new atmosphere for planet Earth—the oxygen-rich air that all animals now breathe. Today there is still a wide range of tiny algal cells and filaments floating in the plankton, hidden between grains of sand on the seafloor, and even locked between the frozen crystals of sea ice. The ethereal, greenish tinge of the massive Antarctic ice cliffs and huge, floating bergs is due to the chlorophyll pigmentation in billions of microscopic algae and diatoms that became trapped as the seawater froze. Remarkably, when the ice melts, they are ready and waiting to spawn new life.

As part of the phytoplankton, unicellular algae join in the great primary food production of the seas and oceans. The larger seaweeds similarly make an important contribution at the base of many food chains and webs, especially those closer to shore. Here they may contribute up to 10 percent of the total primary production, daily providing fresh food for an astonishing variety of herbivores, which seek refuge as well as sustenance amid the dense mats and waving fronds. Just as important, because seaweed perishes as rapidly as it grows, are its dead and dying remains, which drift down through the water column to the host of detritus feeders lurking in the lower depths.

Rather surprisingly, seaweeds also act to keep coastal waters clean. The complex network of holdfasts that anchor seaweeds to the substrate traps fine particles in the surrounding water, and then creates a stabilizing influence on bottom sediments. It is only when major storms wrench up great swathes of seaweed forest and stir the sediment below that murky waters return and the cleansing process must begin anew.

carotenoid pigments to scoop up the shallow-penetrating yellow to red end of the spectrum. The different colors of algae that we actually see are due to the wavelengths reflected by them—in other words those that are not absorbed. One seaweed, *Codium*, absorbs all available light and therefore appears almost black!

Varied habitats There are some 4,000 species of red algae (Rhodophyta), most of which are marine. They are generally bottom-dwellers in deeper waters, especially in the tropics, where they can be found as far as 200m (660ft) below the surface. They tend to be small and even quite delicate, boasting spectacular frills of lacy purples, pinks, and reds. But some, known as coralline algae, have learned to secrete a thick coating of calcium carbonate around their cell walls, in order to withstand the buffeting of shallower seas. These play an important part, along with corals, bryozoans, and other invertebrates, in the construction of coral reefs. When times are tough or the salinity is high, they may take over as the dominant or even the only reef builder. Some small species are epiphytes, living on the outer surface of their neighbors. Others still have discarded their chlorophyll altogether and so have lost the ability to photosynthesize; instead, they live a parasitic existence on the coat-tails of larger reds.

The brown algae (Phaeophyta) are less diverse, with some 1,500 species, but are commonly present in very large numbers along shorelines everywhere. These are among the most familiar seaweeds—kelps, sea palms, oarweed, rockweed, wracks—that cover the rocks with a tough, slippery carpet. This treacherous surface is due to the presence of a gelatinous substance—mucilage—secreted by the cells of many types of seaweed. Brown algae are mainly marine and prefer colder waters. Some of the smaller forms are planktonic (free-floating) like their close relatives, the diatoms. One large floating seaweed is Sargassum gulfweed, whose clumps of multicelled brown algae are common in parts of the calm, central waters of the Atlantic.

The carpets of seaweed draping the shoreline, as here on a New Zealand beach (*left*), are as vital to the coastal ecosystem as the floating phytoplankton are to the open ocean. They not only offer bountiful food but also provide a safe haven for a host of creepers, crawlers, and small fishes. Exposed to the full force of the sea, intertidal plants need to be very tough and elastic to avoid being destroyed by wave energy.

Secret of the colors The first multicelled alga for which we have fossil evidence, *Grypania*, evolved near the dawn of the Proterozoic era around 2.2 billion years ago. The needs of this and other early forms that graced the prehistoric oceans were simple ones that have probably changed very little over the course of time. Most multicelled algae today still require sunlight, a plentiful supply of nutrients, and a suitable place to grow, which generally means a firm substrate. They flourish best where there is a healthy food web around them, including something higher up the food chain to keep the hordes of grazing invertebrates in check.

Sunlight is the first essential for all photosynthesizing organisms. Inevitably, this requirement involves living within the photic (lighted) zone, ideally in the upper few tens of meters where light penetration is at its most intense. But as solar energy enters the water, the different wavelengths that make up the visible spectrum are differentially absorbed. Even where the sea is perfectly clear, red and orange wavelengths are rapidly and completely absorbed within the upper few meters, 71 percent of red light being converted into heat in the first meter (3.3ft). Yellow, green, and then blue each penetrate a little further, but by 100 meters (330ft) only the dimmest blue light remains.

Intriguingly, algae have become adapted to harness different wavelengths and so maximize the depth range over which they can exist. They have achieved this feat by incorporating a suite of different accessory pigments in association with their chlorophyll. Red algae contain the pigment phycoerythrin, which is especially effective at absorbing green and blue light. Brown algae contain the golden-brown coloring agent fucoxanthin, which absorbs primarily green light. Green algae contain

Sargassum is a free-floating brown alga that forms large intertwined masses in the open ocean. It gives its name (from the Portuguese for "grapes," which its floats resemble) to the Sargasso Sea, where it provides a habitat for a range of unique organisms.

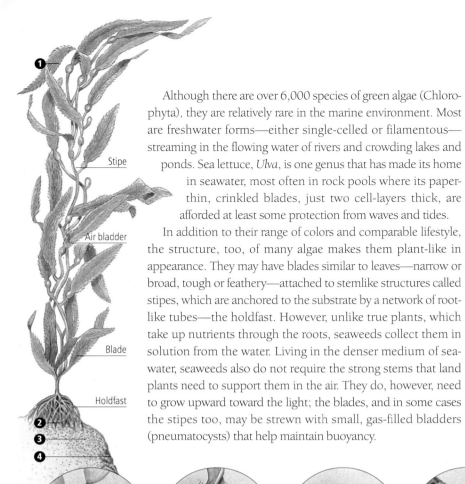

Stipe

Air bladder

Blade

Holdfast

1
2
3
4

1 **2** **3** **4**

Although there are over 6,000 species of green algae (Chlorophyta), they are relatively rare in the marine environment. Most are freshwater forms—either single-celled or filamentous—streaming in the flowing water of rivers and crowding lakes and ponds. Sea lettuce, *Ulva*, is one genus that has made its home in seawater, most often in rock pools where its paper-thin, crinkled blades, just two cell-layers thick, are afforded at least some protection from waves and tides.

In addition to their range of colors and comparable lifestyle, the structure, too, of many algae makes them plant-like in appearance. They may have blades similar to leaves—narrow or broad, tough or feathery—attached to stemlike structures called stipes, which are anchored to the substrate by a network of root-like tubes—the holdfast. However, unlike true plants, which take up nutrients through the roots, seaweeds collect them in solution from the water. Living in the denser medium of sea-water, seaweeds also do not require the strong stems that land plants need to support them in the air. They do, however, need to grow upward toward the light; the blades, and in some cases the stipes too, may be strewn with small, gas-filled bladders (pneumatocysts) that help maintain buoyancy.

Underwater forests Most impressive of all seaweeds are the 100 or so species of brown algae known as kelp. Those of the *Macrocystis* genus (Giant kelp) are the world's largest algae. Perennial plants, with a life span of three to seven years, they can grow 50cm (20in) in a single day and, in exceptional circumstances, reach a staggering length of over 90m (300ft). They favor cold waters of moderate depth, between 20m (66ft) and 90m (300ft) deep, moved by waves and currents but below the lowest tide and so away from the highest-energy coastal zone. It is only in these nutrient-rich environments that they can manage to sustain the phenomenal rates of photosynthesis and growth that have been observed.

From a tangled mass of thick, tubular holdfasts on the continental shelf, the kelps' tall, narrow stipes and waving blades grow in tight clusters, reaching to the very surface of the sea. Here the blades spread out, streaming in the sunlight and swift currents, bathed in the nutrients essential for survival. The surface canopy takes up the lion's share of the available solar energy, but thin shafts of light still filter into the shady depths. The Palm kelp, *Eisenia*, is a hardy understory species able to grow on even the minimal light that penetrates through. Its short, thick stipe is tough and flexible like rubber, bending in the strong undertow. Several species of red algae carpet the floor of this remarkable underwater forest.

Many other organisms have taken advantage of the plentiful food and shelter on offer. In the richest kelp communities off the coast of California, every square kilometer is home to an estimated 50,000 animals, as well as a host of other seaweeds. Giant sea

Kelp forest management: the ecological role of Sea otters

The Giant kelp forests of the Pacific coasts may be home to an amazing number and variety of animals but their survival depends on a simple food chain. Sea urchins that graze the kelp would, if left unchecked, completely devour the young algae and damage the all-important holdfasts of established plants. Although the buffeting of waves and constant patrol by several large predatory fishes prevent the urchins from damaging the upper portions of the kelp forest and the canopy itself, there are few predators that can manage the urchin population on the seafloor. Among those that can are Sea otters (*left*), which consume large quantities.

In the 18th and 19th centuries, Sea otters were hunted almost to extinction for their furs, and great swathes of kelp forest throughout the temperate northern Pacific Ocean vanished with them. Hunting was banned in 1911, and by the 1970s otter populations were recovering. Yet the delicate ecological balance that was thus eventually restored may once more be in danger. Marine biologists monitoring Sea otters in the Aleutian Island region noted a steady increase of up to 25 percent a year by the early 1990s, when the rise suddenly turned into a sharp decline. At first there seemed no obvious explanation; the otters, which had been tagged, showed no signs of mass emigration, they were not suffering from any apparent diseases, nor was there excessive pollution in the area. Meanwhile, as the otter population declined, the sea urchins multiplied; grazing intensity increased by more than 800 percent, and the kelp forests thinned.

Persistent observation has now revealed that the prime cause for the otters' decline is predation by Killer whales, which have shifted to hunting otters from their former diet of seals, sea lions, and the larger open-water fishes. This change in feeding preference may be linked to a decrease in fish stocks—as a result of a combination of overfishing and increased sea-surface temperatures—and thus a reduced pinniped population. Intelligent and efficient hunters, Killer whales are among the few creatures capable of snatching otters from their refuges among the giant kelp fronds. If feeding on otters alone, a single Killer whale would need to kill around 1,800 individuals a year in order to survive. Researchers in one of the study areas, Kuluk Bay, have estimated that the presence of just three or four Killer whales would be enough to explain the observed decline in that sector.

Kelp forests (*left*) which grow on continental shelves in cool temperate seas, are some of the most productive marine ecosystems. The Giant kelp, *Macrocystis*, can grow as tall as the tallest forest trees, forming a surface canopy. Smaller kelp species and several forms of red algae live in its shade. Every part of the Giant kelp is utilized by different communities of organisms (*far left*): polychaetes, isopods, and bryozoans live on the surface of the blades and stipes (**1**), sea urchins, brittle stars, polychaetes, and crustaceans within the holdfast (**2**), sea stars, urchins, understory algae, and benthic fishes on the horizontal substrate surfaces (**3**), and sponges, tunicates, bryozoans, and sea anemones on the vertical surfaces (**4**).

slugs, such as the Californian sea hare, munch their way through the uppermost fronds, growing to over 1m (3.3ft) in length and a staggering 15kg (33lb) in weight. On the seafloor, the kelp holdfasts harbor over 175 different species. Sea urchins and mollusks of all descriptions and sizes, kelp crabs and shrimps, brittle stars and anemones, all feast at will, grazing the living kelp or scavenging on plentiful remnants. Starfish, like the many-rayed Sunflower or smaller Bat star, move in with a rapacious appetite for urchins in particular. A veritable kaleidoscope of fishes—blennies, wrasses, eels, pipefish, rockfish—are all drawn into this intricate web of life; they dart and dive, twist and turn, a riot of commuters in a city rush hour.

Several larger marine animals—sea lions, Harbor seals, even Gray whales—are sometime visitors. But quite the most delightful are the Sea otters, with their plushy fur, playful faces, and engaging behavior. The only mammal, other than primates, to use a tool while foraging, Sea otters carry up stones from the sea bed on which to smash clams and other shellfish. They need to consume up to 25 percent of their body weight each day, mostly in order just to keep warm and keep diving. Fortunately, they are especially partial to sea urchins and so help to keep these voracious herbivores under control. Any decrease in the population of Sea otters can lead to a population explosion of urchins and a dramatic decline in the kelp forests.

Sunlight pierces the nutrient-rich surface waters above a submarine ridge. These Bigeye trevally gather in a tightly packed school of several thousand individuals, where there is a certain safety in numbers, and slowly circle in a resting pattern. Keen eyesight and finely tuned sensitivity to underwater movement allow the school to react in unison when a larger predator approaches, attempting to confuse the unwanted intruder. At night the school fragments to feed in the relative protection of the dark.

MARINE LIFESTYLES

CHAPTER 8

The myriad adaptations for survival in the ocean

Marine life exists in a world of evolution and change, of profusion and vitality, of physical challenges and extreme competition. From the vast expanse of surface water to the deepest, darkest abyss, every creature in the ocean must survive through adaptation to its environment. All must be able to breathe in water and see in the dark, to tolerate the extremes of temperature and intensity of pressure, to regulate body chemistry while immersed in saltwater, and to face the lottery of feeding as well as the dangers of predation.

The dynamic balance between prey and predator has led to an amazing range of strategies. The variety and ingenuity of movement is enormous; there are gently billowing floaters, streamlined speed merchants, an ungainly underworld of creepers and crawlers, even fish that can fly. There is also an astonishing assortment of weaponry, armor, and disguise. To make a living at sea further requires the ability to use and filter all manner of sensory information—visual and auditory signals, and the pressure waves from minute disturbances, as well as subtle electrical, magnetic, and chemical stimuli.

One of the principal driving forces in life is the need to reproduce. The cycle of life in the oceans is one of elaborate richness—a world of unbelievable fecundity and colorful promiscuity, of asexual reproduction and life-long partnerships, of the synchronous spawning of millions of coral polyps, and of animal migrations halfway across an ocean to mate in the place of their birth.

Surviving in the oceans involves perfect adaptation to the environment. Each marine species has developed its own ingenious way of coping with the unique physical conditions to which it is exposed.

The Marine Environment

KEY TERMS

METABOLISM

HOMEOSTASIS

ECTOTHERM

ENDOTHERM

OSMOSIS

OSMOREGULATION

Weddell seals (*right*), the most southerly living mammals besides visiting humans, must work hard to keep open small breathing holes in the packed sea ice. Their rotund shape and thick, blubber-lined bodies insulate them through the harsh Antarctic winters.

Mean temperature in different parts of the ocean lies within a relatively narrow range for both the open ocean and coastal seas (*below*). On land, temperatures reach greater extremes, with a recorded maximum of 58°C (136°F) and minimum of -88°C (-127°F).

Water is a simple yet unique substance. Its remarkable properties have allowed life as we know it to evolve in abundance and diversity across the face of the planet. For the first nine-tenths of the Earth's history, all living organisms were entirely confined to the marine realm and, even now, the numbers and variety of organisms that inhabit it is astonishing. The oceans provide vital nutrients, dissolved gases, and mineral salts. They are rich in oxygen, and their surface is bathed in an endless supply of sunlight. Water temperatures, though they vary from region to region, are far more constant than on land. But, despite these generally favorable conditions, marine life exists in a world of environmental change and biological evolution, of physical challenges and fierce competition.

For every organism, from the single-celled to complex giants, the ability to maintain an internal balance of cell structure and chemistry against the changeable outside world is the first condition of survival. This is the process known as homeostasis. The adaptations that have evolved to meet the challenges of the marine environment are stunning in their variety and ingenuity. But every species has its optimal range of conditions for survival. Pushed beyond these limits, the organism becomes stressed, normal metabolic functions cease, and reproduction becomes increasingly difficult or impossible. Eventually the tolerance level is exceeded and the organism can no longer survive.

Heat and light Sunlight at the sea's surface powers the process of photosynthesis that, directly or indirectly, provides energy for nearly all forms of ocean life. Unlike the atmosphere above, seawater rapidly absorbs light, so that only the uppermost layer is bathed in a plentiful supply of energy. Photosynthetic organisms have, therefore, developed many adaptations for maintaining a place in this photic zone.

The majority of heterotrophic organisms—those that live off organic matter—must similarly live where food is plentiful.

However, many organisms have adapted to a life entirely without sunlight. Beneath the twilight zone—where some vestige of light still penetrates—is a realm of eternal darkness. For some creatures this means long periods of starvation between meals that fall at random from above, and the heightening of senses other than vision in order to find food, avoid predation, and attract a mate. Others find a home near chemosynthetic bacteria, which utilize chemical sources of energy to grow.

Most of the solar energy absorbed by seawater is converted directly to heat. The oceans act as the world's thermal-energy storehouse, regulating and moderating climate as currents stir the waters. Although extremes of heat and cold are far less marked than on land, water temperature is still one of the most important regulators for the distribution of life in the oceans. Whereas mean temperature ranges from 0°–40°C (32°–104°F), 90 percent of the oceans are permanently below 5°C (41°F). Most marine animals are ectothermic (cold-blooded), obtaining their body heat from their surroundings, and are therefore directly affected by any temperature changes. They generally have narrow tolerance limits, and so are restricted to particular latitudinal belts or water depths. Eggs, larvae, and juveniles tend to be still more sensitive than adults of the species, so that the success of future generations is significantly affected by climatic conditions and by ocean currents, which may sweep the vulnerable young into waters too warm or too cold for their survival.

In contrast, marine mammals and birds are endothermic (warm-blooded); their metabolism generates heat internally, thereby allowing the organism to maintain constant body temperature despite external fluctuations. This ability typically requires greater quantities of food and good insulation. For all types of biota, temperature exerts a strong control on the rates of chemical reactions within body cells. Consequently, the organisms that live in tropical seas grow more rapidly, reproduce more frequently, and live generally shorter and more energetic lives than those of polar oceans.

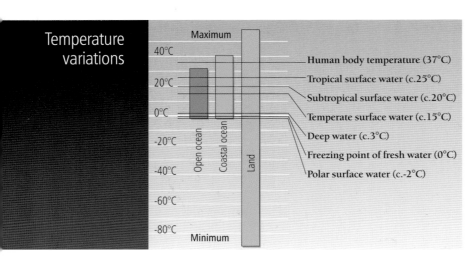

Temperature variations

Maximum

40°C

20°C

0°C

-20°C

-40°C

-60°C

-80°C

Minimum

Open ocean

Coastal ocean

Land

Human body temperature (37°C)

Tropical surface water (c.25°C)

Subtropical surface water (c.20°C)

Temperate surface water (c.15°C)

Deep water (c.3°C)

Freezing point of fresh water (0°C)

Polar surface water (c.-2°C)

Many species of whales spend the summer in high latitudes, where food supplies are plentiful, and then migrate to warmer seas as winter approaches. There are some mammals with extra-thick insulation, such as walruses and Weddell seals, that remain throughout the winter in some of the bleakest and harshest environments on the planet. Their metabolism slows down, and they must regularly break the ice to keep breathing holes open when the sea temperature drops below freezing.

Salt water tolerance

All organisms contain a large proportion of water, typically constituting over 50 percent of their body weight. The body fluids of marine creatures generally contain exactly the same dissolved salts as seawater, and in the same proportions, although the total concentration may be more or less than their surroundings; they have therefore had to evolve ways of maintaining homeostasis.

Cell membranes are semipermeable, allowing free passage of water and of certain other molecules. As a result, plant cells can absorb nutrient elements, while waste products from animal metabolism can be discharged. However, water from the more concentrated fluids found in most cells tends to diffuse through the cell walls toward fluids that are more dilute—such as ambient seawater. This process is known as osmosis and, if allowed to proceed unchecked, could lead to dehydration and death.

Such debilitating exchanges are prevented by a mechanism called osmoregulation, which adjusts the concentration of dissolved salts in the body. Marine fishes, for example, achieve balance by drinking much seawater, rarely urinating, and excreting excess salts through specialized cells in their gills. But most other creatures of the open ocean are unable to regulate their body fluids in this way and must tolerate salinity changes. Fortunately, the salinity of the open ocean is remarkably stable.

For organisms living in coastal areas—in bays, estuaries, and rock pools—salinity can change dramatically with every flood and ebb of the tide. Fiddler crabs (*Uca*) thrive in these environments, where their open-ocean relatives could not, because they have evolved a highly efficient system of osmoregulation.

Marine fishes, such as this wrasse, use their gills to excrete surplus salts from their bodies, as well as extracting oxygen from the water.

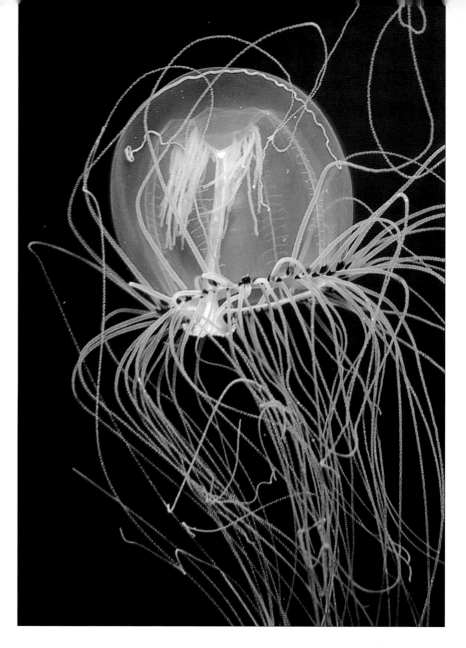

Coping with pressure

Undoubtedly, the most extreme physical condition anywhere in the ocean must be the hydrostatic pressure exerted at abyssal depths, caused by the weight of the overlying water column. For every 10m (33ft) of water, the pressure increases by one atmosphere—a unit equivalent to that exerted at the Earth's surface by the full weight of the atmosphere above. At a depth of 1km (0.6mi), the pressure is 100 atmospheres; at 10km (6.2mi), near the bottom of the deep oceanic trenches, it reaches a staggering 1,000 atmospheres. For most surface-dwelling species, these pressures would be catastrophic. Indeed, marine biologists of the 19th century generally believed that the deep sea must be devoid of life because the conditions were untenable.

We know now, of course, that a rich and diverse community of organisms populates the abyssal seafloor. How can this be? In fact, the explanation is really quite simple, for it is gases, not liquids, that are highly compressible. Deep-sea creatures, therefore, are composed mainly of water and contain no gases, and are apparently unaware of the oppressive hydrostatic pressures under which they live. Rather more specialized adaptations are required, however, for those fishes and whales that can live at the surface and then dive into the deep sea in search of food.

One of the best adaptations to living in the ocean is to become at one with that environment. Jellyfishes, for example, are composed of 95 percent water.

From burrowing and crawling to swimming and leaping, the variety and ingenuity of methods that marine animals have adopted to move about their watery world is a stunning tribute to evolution in progress.

MARINE LIFESTYLES

Movement Underwater

KEY TERMS

FLAGELLUM

JET PROPULSION

FRICTION DRAG

FORM DRAG

TURBULENCE DRAG

CAUDAL FIN

DORSAL FIN

PECTORAL FIN

VENTRAL FIN

SWIM BLADDER

TUBE-FOOT

Self-propelled motion is a fundamental ability in many organisms. From the beating flagellae of microscopic plankton keeping afloat near the ocean surface or the ungainly, almost comical, sideways scuttling of crabs to the playful perfection of dolphins surfing the bow wave of a ship, marine creatures have adopted a huge variety of styles, speeds, and methods of movement. Each species has its own particular need for the evolutionary developments that have taken place, but the basic requirements are the same—finding food, escaping predation, seeking a mate or a safe place to have young, or migrating to an area with more favorable conditions.

Several different styles of movement can be identified among marine organisms, and some will use two or more of these in the course of their daily routine. The dominant styles include floating, swimming, jet propulsion, creeping, crawling, and burrowing. A few species have also learned to hitch a ride on others, to catch the wind and sail across the sea's surface, or even to leap from the water and glide through the air. Yet other animals choose a sedentary existence, at least for most of their lives, dwelling in fixed refuges from which they extend tentacles or cilia in order to capture a passing meal.

The particular physical properties of water that most affect movement are density, viscosity, and buoyancy. Seawater is about 800 times denser than air, and nearly 100 times as viscous. Consequently, there is much more resistance to movement than on land, as anyone will know who has ever tried to wade through waist-deep water. However, with density comes much greater buoyancy, so that organisms need spend relatively little energy to stay afloat. As they move through the ocean environment, organisms seek to make wave motion, currents, and natural turbulence work to their advantage, not detriment.

Saving energy The vast majority of food is produced at or near the surface of the ocean by countless numbers of microscopic phytoplankton and bacteria. Many animals, especially the tiny zooplankton, have taken to a life of simply floating amid these bountiful pastures, contentedly grazing. They have little need to exert unnecessary energy in chasing after a meal when they are already fully immersed in a

Compared with air, seawater is a dense and viscous medium that offers considerable resistance, or drag, to any moving object. A rounded form with a smooth surface reduces friction, an elongate pencil-shape with minimal cross-sectional area reduces form drag, while a rounded front and tapered rear end reduce turbulence as the water passes along and around the body. Fast swimming fishes, such as tuna, mackerel, and sharks, combine these forms in a streamlined, torpedo shape that is a direct evolutionary adaptation to cope with movement underwater.

nutritious soup. These floating animals may be at the mercy of tides and currents but then so is their food—and life is generally short. Delicate skeletons and tiny tentacles help keep them near the surface, as do the bulbous floats of the larger jellyfish.

Although life among the plankton might seem easy, passively yielding to the elements, there is in fact a remarkable range of movement. Some tiny plankton only 1–2mm (0.04–0.08in) in length actually travel long distances each day. Every evening, species that live well below the photic zone, at depths of several hundred meters, swim up to the surface to feed in the relative safety of darkness. At dawn they sink back down in an effort to avoid predation—a double journey equivalent to a person swimming 700km (400mi) a day. Some animals, such as copepods, may only move slowly and clumsily through the surface waters, but are able to maintain a rhythmic beat of their tiny appendages, which sets up a stream of water and food toward the mouth region. Even jellyfishes use a weak form of jet propulsion, squeezing water from their floating sacs to effect motion.

Swimming for survival Most of the larger marine animals, including all fishes and mammals, have adopted some form of active swimming as their mode of life. Given the number of species and the sheer abundance of fishes that exist, it is clear that they have had remarkable success in adapting this form of movement to their needs. It has taken millions of years of evolutionary trial and error to overcome the chief deterrent to motion through a dense medium such as water—that of drag resistance. Swimming efficiency has been achieved by minimizing the three types of drag, created by friction, turbulence, and body form. To reduce surface friction, the body must be smooth and rounded. In addition, the scales of most fishes are covered with slime to lubricate their passage through water. To reduce form drag, the cross-sectional area of the body should be minimal—a pencil shape would be ideal. To reduce the turbulent drag created as water flows around the moving shape, a rounded front end and tapered rear are required. The combined shape, taking into account all three types of drag, is the streamlined torpedo form of a tuna, the fastest-swimming of all fishes.

Speed is only one of three important aspects of swimming ability. Tuna, swordfish, and mackerel all specialize in fast, steady cruising. However, although they encounter a lot of prey in their travels, their attacks are only successful 10–15 percent of the time. Other fishes, such as barracuda, maximize rapid acceleration over sustained speed. The freshwater pike, which lurks in the shadows until its quarry is within striking distance and then attacks with great rapidity, achieves a remarkable 70–80 percent

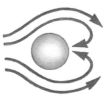
Minimal friction drag

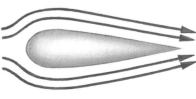
Minimal form drag

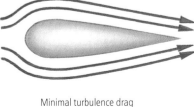
Minimal turbulence drag

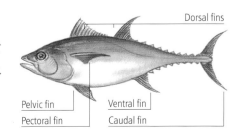
Dorsal fins

Pelvic fin

Pectoral fin

Ventral fin

Caudal fin

success rate. The third specialization is maneuverability, best demonstrated by the butterfly fishes. These have disk-shaped bodies that permit abrupt changes in direction. Many fishes are generalists, being at least partly proficient in all three modes of movement. Bass and trout, for example, have an intermediate success rate of 40–50 percent.

Almost all fishes swim by undulation. Strong W-shaped muscles along the side of the body progressively contract and relax in sequence, from head to tail and from side to side, creating a traveling horizontal wave. The body is thrown into a series of curves that press sideways and back against the water, producing a forward thrust. The narrow, elongate forms of eels, ribbonfishes, and sea snakes allow easy undulation along their full length. In contrast, the more stubby and inflexible bodies of armor-plated trunkfish and cowfish render them clumsy and slow-moving; they use only the swish of their short caudal (tail) fins to propel them through the water. Most other fishes combine elements of both methods, coordinating powerful strokes of the caudal fins with subtle body undulations. In flatfishes and marine mammals, the wave travels in a vertical rather than a horizontal plane.

Fins play a crucial and versatile role in fish mobility. The vertically oriented dorsal and ventral fins control rolling and yawing (sideways instability) while pitching (up-and-down movement) is controlled by the laterally paired pectoral and pelvic fins. While the shape of the caudal (tail) fin relates directly to speed—lunate (crescent-moon shaped) for fast cruising, broad and flat for acceleration—the style and arrangement of the other fins are most important for maneuverability. Puffer fishes scull with tiny, oscillating pectoral fins, and the equally slow-moving butterfly fishes undulate

their broad dorsal and ventral fins, twisting and turning with great precision amid the magnificent confusion of the coral reef.

Most remarkable of all are the strangely shaped seahorses, which move by vibrating their delicate and nearly transparent dorsal fin as many as 70 times a second. This minute wave action ripples through the fin from top to bottom with each vibration. A pair of pectoral fins behind the head maintain vertical poise and allow the smallest of turns, while also helping determine whether the motion is up or down, forward or backward. In all cases it is incredibly slow, for seahorses can take a full minute to travel just 20cm (8in).

Dolphins are the swiftest and most agile of all marine mammals. Their stunningly graceful leaps clear out of the water may appear effortlessly playful but, in reality, require remarkable speed and power.

1 Butterfly fish
2 Mackerel
3 Barracuda
4 Sea bass
(not to scale)

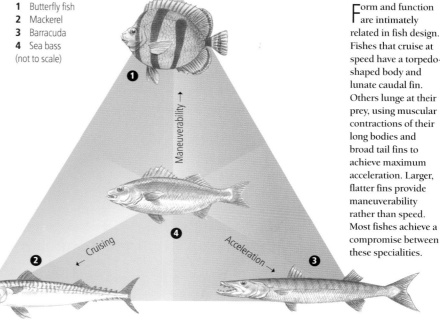

Form and function are intimately related in fish design. Fishes that cruise at speed have a torpedo-shaped body and lunate caudal fin. Others lunge at their prey, using muscular contractions of their long bodies and broad tail fins to achieve maximum acceleration. Larger, flatter fins provide maneuverability rather than speed. Most fishes achieve a compromise between these specialities.

A flight of Cownose rays in the shallow shelf seas surrounding the Galápagos Islands. These magnificent creatures are perfectly adapted to life just above the seafloor, where their flattened form, keen senses, and graceful movement make them the top predator of a rich benthic food web. All skates and rays, with the exception of sawfishes and guitarfishes, have a unique swimming style, powered by movement of their greatly enlarged pectoral fins. These are also highly effective at uncovering mollusks and crustaceans hiding in the soft substrate.

Buoyancy and flight

Keeping afloat is not a great problem in the world of swimmers, although most fishes do have swim bladders to help them offset the density of their bodies and so maintain neutral buoyancy with minimal effort. These small, gas-filled chambers contain specialized networks of blood vessels that can add or remove gases such as oxygen and carbon dioxide. The ability to remain indefinitely at a constant depth without moving or expending energy is especially important for the slower-moving rock and reef fishes that seek food and safety in the shallow underwater world, or for those that hunt and scavenge in kelp forests. Active swimmers, such as mackerel, skipjacks, and sharks, do not have swim bladders, because they need to change depth more rapidly than they could regulate the gas content. These fishes must swim forever or they will sink. Bottom-dwellers and deep-sea fishes have no need for swim bladders and, indeed, the deeper they live the more difficult it would be to cope with a high-pressure gas cylinder as an internal organ.

The giant Manta ray must be one of the most beautiful dancers on earth. Its greatly enlarged pectoral fins, which are attached to the head, rhythmically beat the water, just as a bird moves its wings in flight. These great muscular flaps lend the rays a grace and delicacy hardly credible for creatures that can weigh as much as 1.5 metric tons. Even more remarkable is their ability to leap clear out of the water. But skates and rays are primarily adapted to life on the bottom, gliding over the sandy seafloor to feed mainly on crustaceans and mollusks.

Several other fishes can leap from the water, and do so as a much more routine part of their lifestyle. For the flying fishes, leaping is probably the most important evolutionary adaptation that has enabled them to survive through millions of years of predation by equally swift hunters. Launching into the air requires a fish to swim upward at high speed; then, as it begins to leave the surface, it thrusts strongly from side to side with its enlarged caudal fin, and extends its broad, winglike pectoral fins for complete lift-off. Astonishingly, these fishes can leap as high as 10m (33ft), soaring through the air at speeds of 60km/hr (37 mph) and traveling 250m (800ft) in a single flight. The fast, whirring action of their caudal fins as they return to the sea allows them to take off again and make several long flights in a row—even to turn a near right angle in mid-flight. Such a unique ability would seem to offer no hope whatsoever for even the fastest swimming predator—but not so! Some will follow a flying fish's shadow, waiting for it to splash back down, while others, notably the dorado, will actually leap out of the water themselves and catch their prey in midair.

Marine mammals surface to breathe—breaching whales spout water and gulp in air, dolphins race across the sea's surface in a series of graceful arcs, seals and walruses break open breathing holes in the sea ice during winter. For air-breathing mammals and sea birds, this facility is just as essential to survival as keeping submerged is for fish. Sea otters wrap themselves in fronds of floating seaweed between dives, seals clamber about clumsily onshore, and penguins huddle together on ice floes. But all are more efficient and streamlined in the water than on land. In reality, we still do not fully understand the reasons why dolphins, sea lions, and even the mighty Humpback whales sometimes leap clean out of the water—for joy or play, in anger or frustration, for show or distraction.

Creepers and crawlers For every creature that swims or floats, there are many more that have opted for a slower or more sedentary existence on or around the seafloor. These include not only the photosynthesizing algae of the photic zone and the microscopic bacteria and fungi that clean and recycle waste, but also a host of larger animals of all types and sizes. Their simple principle in life is that food will eventually fall to the seafloor or wash up along the shoreline, and so it is simply a matter of waiting. There is little need to expend energy in moving fast. While scavengers such as worms, shellfish, anemones, and sea slugs are content to wait for their meal to arrive, it is not long before other, slightly larger predators join the feast, slowly moving in for the kill.

Living in the sediment, generally soft sand and mud substrates, has several advantages—protection from predators, from damage by waves or currents, and from desiccation in the intertidal zone. However, these animals must be able to move in a much denser and more difficult medium than water. Polychaete worms have tapered heads and long, tubular bodies which they move by the alternate contraction of longitudinal and circular muscles. These create a bulge that travels down the body, pushing against the sediment. Many bivalve shellfish spend at least part of their lives buried in the substrate. One of the most effective and rapid burrowers is the razor clam, whose lightweight, smooth, and elongate shell can be easily dragged down into the soft sand by its muscular foot.

Other bivalves, including oysters, clams, scallops, and mussels, as well as a variety of sea snails (gastropods), extend their single, muscular foot below and in front of their shell. By alternate expansions and contractions, they creep very slowly over the seafloor. Most sea slugs use a similar method of movement

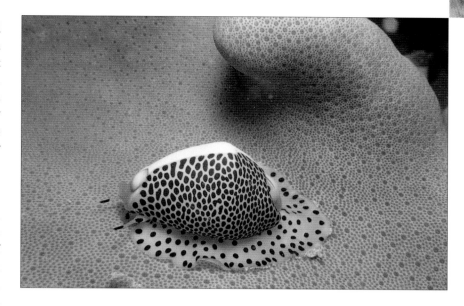

as they munch their way through an array of coral polyps or travel along the trailing tentacles of large jellyfishes. In one of the most amazing feats of coordinated movement by a simple animal, the starfish slowly inches forward on as many as 40,000 individual tube-feet. Each one must be carefully extended in the desired direction; water is then forced through the tubes from tiny bulbs located just above the tube-feet. Starfishes use hydrostatic pressure to move over a grainy substrate at about 15cm (6in) a minute. Over smoother surfaces, they prefer to move by using minute suction cups at the end of each foot.

In hot pursuit come an army of crustacean scavengers—lobsters, crabs, and shrimps—picking up any animal or vegetable detritus that remains. They use four of their five pairs of jointed legs to walk about on the ocean floor—an awkward form of movement like that of heavily armored knights. When the creatures are threatened, this normal, cautious stalking can be replaced by very rapid backward swimming as the animal frantically flips its tail beneath the rest of its body.

About 60 different species of batfishes, living mostly in tropical waters, have adopted a similar ungainly walk in order to stalk both mollusk and crustacean scavengers of the seafloor. They have evolved jointed and heavily built pectoral fins, which extend downward and are used rather as a person might use elbows to creep forward on the floor.

This beautiful, speckled ovulid—a type of snail also known as a "false cowrie"—emerges at night, to creep slowly across the rough surface of corals and sponges on its large single foot, grazing on a thin coating of algae. At dawn it will crawl into any available crack or crevice, hiding out during the perilous daylight hours.

Maintaining their alignment in the column by touch, spiny lobsters march across tropical shallows in the Caribbean. Like many crabs and other crustacean relatives, lobsters have traded efficiency of movement for body armor and hunting equipment.

SEE ALSO

EVOLUTION & EXTINCTION
March through Time   142–47

THE WEB OF LIFE
The Zooplankton   164–65

COMPLEX COMMUNITIES
Wetlands   200–01
The Open Ocean   208–09

Bombarded with sensory information of all kinds—visual, chemical, pressure, electrical, and other stimuli—each marine animal is able to filter the signals it needs for survival.

Heightened Senses

uli that permeate their watery habitat. In the simplest animals, individual cells respond to any sustained stimulus, whereas in more complex creatures specific cells have become specialized for receiving and interpreting particular sensory inputs.

KEY TERMS

COMPOUND EYE

TWILIGHT ZONE

BIOLUMINESCENCE

PHOTOPHORE

LATERAL LINE

ULTRASONIC

INFRASONIC

ECHOLOCATION

MELON

The compound eye (*right*) evolved half a billion years ago and is still present in insects and crustaceans alike. The multiple facets are well-honed receptors of even very low levels of light, and are sensitive to tiny movements over a wide field of vision.

With excellent eyesight and the most highly developed nervous system of any mollusk, octopuses make formidable predators. They move with slow deliberation across the seafloor and then pounce with unexpected rapidity. Their large eyes are capable of seeing color and shape, while their complex brains display good memory and simple problem-solving ability. They also possess a very keen sense of touch coupled with great delicacy of movement, used both to extract prey from tiny crevices and to demonstrate the utmost care in tending their eggs.

The paramount sense among terrestrial animals is sight. The most effective hunters as well as the most elusive prey are often those with the sharpest vision—but not so underwater. Certainly in the uppermost 10 percent of the ocean, in the clear blue seas and emerald green waters that we know best, the ability to see well is vital for nearly all the larger resident marine animals. Even in the deepest and darkest reaches of the ocean, where ghostly bioluminescence creates a minimal and patchy glow, many animals still have a general perception of light, even though they cannot see well. But nowhere underwater is sight the only important sense, for animals have generally developed a heightened awareness of several different sources of sensory information. Sound is especially important for some, whereas for others it may be pressure or heat differences, smell, touch, taste—or even gravity, magnetic, or electrical fields.

Senses are crucial to all aspects of undersea survival—hiding or seeking, competing for food, territory, or mates. Every animal receives a constant flow of information from its environment and from other lifeforms, and every species has found some way of decoding this data stream to its advantage. Different organisms are able to decipher more or less of the great variety of stim-

Seeing beneath the sea Animals of the underwater world have experimented with a wide range of adaptations for vision. The primitive flatworm has an organ analogous to an eye, with tiny pigment spots that allow it to distinguish light and dark. As there are two receptors, it can also determine the direction of the light. The Giant clam, *Tridacna gigas*, has a whole series of such organs scattered along its protruding mantle (the fleshy rim of its body). These, too, give a simple sense of light and dark, perhaps warning the mollusk to clamp shut when the shadow of a potential predator approaches.

Over 500 million years ago, the now extinct trilobites dominated the seafloor. This highly successful group of arthropods

Compound eye

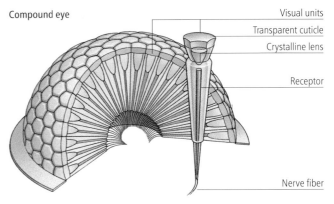

Visual units

Transparent cuticle

Crystalline lens

Receptor

Nerve fiber

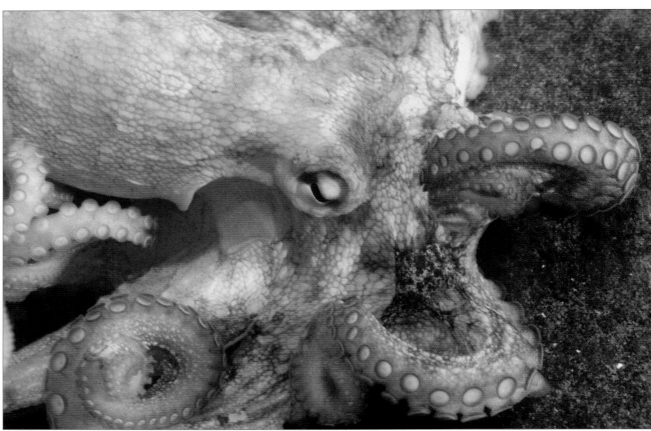

As night falls, the ocean becomes a mysterious realm, disturbed not only by the ghostly flicker of moonlight across the surface but by an ethereal, blue-green glow from billions of tiny planktonic organisms. These creatures produce their own light by a process known as bioluminescence, either in specialized organs called photophores, or by harboring bioluminescent bacteria within their tissues. Light is produced when the protein luciferin combines with oxygen in the presence of an enzyme and adenosine triphosphate (ATP). The chemical energy locked up in the ATP molecule is converted to light energy with almost 100 percent efficiency; the light produced may be quite bright but not warm. Photophores may be turned on or off at will, while the pockets of luminescing bacteria can be controlled by varying the supply of oxygen to that particular area.

Most single-celled dinoflagellates live a dual life, photosynthesizing in the sunlit surface waters by day and transmitting their own cryptic bioluminescent message by night. Many types of krill are also bioluminescent, flashing signals that apparently attract individuals into huge swarms, perhaps to facilitate mating.

But it is the deep sea that provides the most spectacular display, for an estimated 75–80 percent of all species are bioluminescent. Photophores vary in size, shape, color, and location on the body, so different species and even different sexes have their own unique patterns and coded signals. The viperfish, for instance, has rows of lights along its body, with others inside its mouth and on the tip of a long fin spine. These are used to lure prey closer, in the same way that the anglerfish uses a spiny fishing rod baited with light that dangles in front of its awesomely equipped mouth.

Deep-sea squid often sparkle with light in order to signal their presence and their species, and can further release bioluminescent fluid as a defense strategy, clouding the water around them with a sudden burst of confusing light. The scarlet-colored opossum shrimp has adopted precisely the same trick. In addition, there are over 200 species of lanternfishes in the twilight zone, each like a miniature constellation of stars, using their remarkable bioluminescent ability to illuminate their world and to attract prey, or perhaps in some mating ritual that is not yet understood.

Professor Josef Gitelson,
Krasnoyarsk University, Russia

Professor Peter Herring,
Southampton Oceanography Centre, UK

Bio-luminescence: lanterns in the dark

Bioluminescent bacteria form a symbiotic colony in an organ below the eye of this flashlight fish. The emission of light is continuous, but it can be blocked by raising a black shutter mechanism. By blinking, the fish can lure prey, confuse predators, and communicate with its own species. The vibration-sensing lateral line is also illuminated.

had evolved a much more advanced system than their contemporary rivals for viewing that early ocean world—the compound eye. Beautifully preserved Cambrian-period fossils from the Burgess Shale formation in Canada reveal a gridlike pattern in which the eye is divided into many hundreds of different facets. Exactly the same structure is found today, not only in that extraordinary "living fossil," the Horseshoe crab, which closely resembles its trilobite relatives, but also in many other crustaceans, as well as in most insects on land.

Each facet of a compound eye is a lens for a cone-shaped visual unit, containing light-sensitive pigments. Deep-sea crustaceans living in the near darkness of the twilight zone have an abundance of such pigments, which serve to enhance even minimal amounts of light. Compound eyes are best suited for detecting movement, with each unit pointing in a slightly different direction so that even very subtle events can be perceived over a wide area. The lobster's eyes, mounted on the tips of movable stalks, are still better able to scan the whole vicinity.

However, the compound eye does not provide great clarity of vision. Instead, the integration of different signals by a primitive brain produces a blurred, mosaic-like image. Most other marine creatures have a single, spherical lens in each of two eyes, which gives them sharp vision underwater. The strong curvature needed to better focus light in a medium almost as dense as the lens material itself also creates near-sightedness. Such myopia, however, appears to present little handicap when long-distance vision is anyway impaired by low light levels and cloudy water.

Most fishes have eyes set on opposite sides of the head, using them independently for monocular vision. As they weave their heads and bodies from side to side in swimming, each eye can scan a wide range—around 190°—with reasonable overlap to improve perception in front of them. Some fishes can swivel their eyes forward and so achieve limited binocular vision. Having no muscles to alter the shape of the lenses, they must move the entire lens forward to view near objects and back to see distant objects, in exactly the same way that we focus a camera.

Among the 25,000 or more fish species so far recorded there is a wonderful variety of adaptations to match the range of lifestyles and environments. Flatfishes that spend most of the time lying on the bottom have both eyes on the same side of the head. Others, like the tropical freshwater "four-eyed" fish *Anableps*, which lies at the surface, have eyes adapted to see above and below the water simultaneously. Some that live in the almost complete and permanent darkness of the abyssal region have no eyes at all, while others retain vestigial eyes that can sense light only dimly. More commonly, species have evolved to make use of the minute amounts of light available, either through very large eyes and pupils or by reflecting the light that does enter many times over until it can be seen. A similar adaptation is seen in sharks, which have spectacularly reflective eyes caused by an iridescent layer of carefully positioned mirrors behind the retina. This feature dramatically enhances the predator's vision as it glides silently through murky surface waters.

Most remarkable and mysterious of all in those little-explored abyssal regions is the fairground glow of light and color that sparkles intermittently, and often in splendid isolation. Where once we thought there was nothing but pitch-black desolation, we now know that as much as 80 percent of the total deep-sea population—from bacteria and worms to sea fans and fishes—actually generate their own bioluminescence.

Sharks appeared early in the fish lineage and, together with skates, rays, and chimaeras, have a flexible skeleton composed of cartilage rather than bone. Evolving for more than 350 million years, they are perfectly adapted to their environment and role—mostly as top predators, like this Whitetip reef shark patrolling off Hawaii. They have sharp vision and an amazing sensitivity to the small amounts of electrical energy given off by all living things. Even more acute is their sense of smell, with over two-thirds of the brain involved in processing olfactory information.

Sense of smell Where vision is imperfect or sight fails completely, organisms must rely on other senses for survival. Chemical stimuli are readily transmitted through an aqueous medium and so form an important source of information for many. Sea anemones gradually expand their bodies, reach out their waving tentacles, and grasp at any suitable food particles that come within their chemical range. A great many other creatures, from primitive protozoans and colonial sponges to barrel-shaped salps and sluggish sea cucumbers as well as carnivorous mollusks and crustaceans and all types of fish, have well-honed chemoreceptor cells.

While many mollusks are non-specific seafloor scavengers, the *Murex* snail definitely prefers a live oyster or clam to a dead one. A small chemical fingerprint exuded by the live shellfish is enough to attract a *Murex* from a considerable distance. Once found there is no contest, for the predatory snail attaches itself firmly, bores a hole into the helpless oyster, and rasps out the soft, still-living tissue. Atlantic lobsters use a powerful language of chemical pheromones to attract mates, signal aggressive intent, or demand territorial rights.

Most fishes, too, have a well-developed sense of smell, some much more so than others. The moray eel is a common reef resident, skulking about in caves and crevices by day and hunting by night, that locates and tracks its prey largely by scent. The Nassau grouper is one of many similar species that move very little, hovering nearly motionless, apparently sniffing the water for the least whiff of potential food. Its double nasal opening gives it a highly efficient olfactory sense, and one that is essential for obtaining wide-range information about its environment.

There are generally two olfactory receptors in fishes, located on either side of the upper portion of the snout in nasal pits that are lined with delicate folds of sensitive tissue and covered by a protective roof. One or two holes in the roof allow water, with its invisible chemical messages, to flow in and pass across the sensory tissues. The normal motion of continuous swimmers like most sharks forces water through the nasal pits, whereas less active fishes encourage circulation by a pumping action linked to breathing through their gills, or by the beating of hairlike cilia inside the nostrils, as in eels.

Sixth sense Sharks famously have an uncanny sense for the least trace of blood within a wide circle of their location. They have excellent vision up to about 15m (50ft), but even more remarkable olfaction, which allows them to detect food or blood in the water at levels of only a few parts per hundred million. But, together with the other cartilaginous fishes—skates, rays, and chimaeras—they are also able to sense electricity through special organs located over the top and sides of the head region. These sensors are known as ampullae of Lorenzini, after the Italian marine biologist Stefano Lorenzini, who first discovered them in 1678. They are small sacs filled with a jellylike substance that coats folds of sensitive tissue, and are connected to the surface via narrow ducts opening as tiny pores. Quite amazingly, these ampullae can detect as little as one one-millionth of a volt of electricity. Since all animals involuntarily generate minute electrical charges as they live and move, the ability to detect and interpret such charges is a powerful weapon indeed.

A noisy silence Far from being silent, the sea is filled with a veritable cacophony of sounds. Animals of all types contribute clicks, wheezes, whines, rumbles, scrapes, and bumps. The songs of whales may echo for unbelievable distances across the oceans, utilizing a natural communication channel 1,000m (3,300ft) beneath the surface. The roar of surf, the rustle of sand, raindrops spattering on the ocean surface, or the nibbling sounds of fishes grazing on algae are all heard with amazing clarity in the underwater world.

The sheer noise of the oceans is all the more remarkable in that fishes, mollusks, and crustaceans have no vocal chords to help them emit sound. Instead, they bring different, ingenious adaptations into play. Fish of the grunt family grind together their upper and lower pharyngeal teeth, located deep in their throats, to emit a sound not unlike that of feeding pigs. Croakers and drummers are also aptly named, vibrating strong muscles attached to the swim-bladder walls. Rather like a guitar's soundbox, the bladder allows resonance and amplification, while the rate and nature of vibration result in noises of varied pitch, from deep thumps to higher clicks. The toadfish, in addition to grunting, has mastered the art of boat whistles—long, low-frequency bursts of sound used typically in the mating season.

A number of crustaceans emit intentional sounds. Pistol shrimps, though less than 5cm (2 in) long, can produce a sudden, sharp retort by snapping together two parts of their outsize "pistol" claw. The shrimps use this technique very effectively in hunting, to stun small creatures into unconsciousness, as well as in defending their own burrows. Spiny lobsters possess a stridulatory (noise-making) organ near the base of their large antennae, which they rattle to frighten away predators. Fiddler crabs use their claws to rap out warning signals, and leg movements to produce a lower-frequency honking that is audible, and supposedly attractive, to females over 10m (33ft) away.

All these sounds, as well as the displacement effect on water caused by animal movements, are transmitted by compressive wave motion far more effectively underwater than in air. As a result, sound travels twice as fast in water, and about four times as far. The audible or sonic range for humans is between 20 and 20,000Hz (hertz , or cycles per second). Lower frequencies are

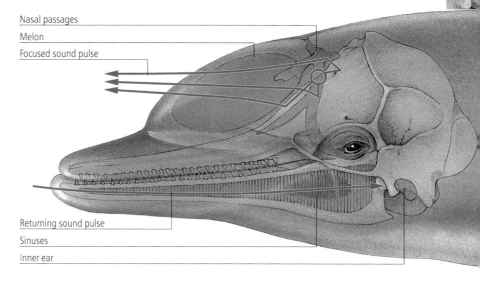

known as infrasonic, while higher frequencies are referred to as ultrasonic. Marine animals have a variety of auditory and pressure sensors for picking up these signals; lobsters have receiving antennae, crabs are sensitive to ground vibrations, while fishes have both an inner ear and a lateral line.

Even without an external ear to capture and amplify signals, the inner ear in some fishes can detect sound vibrations up to several kilometers away, while just below the surface of the skin is a system of mucus-filled canals containing flexible nerve endings that stretches the length of the fish—the lateral line. Infrasonic vibrations or pressure differences cause the fluid in the canals to move and distort the nerve endings. Fishes are therefore very aware of the least movement underwater, which is one reason why a whole school of many hundreds of fishes is able to twist and turn in exact unison, whether or not the sea is crystal-clear or dark and murky.

Songs of the sea Sound is undoubtedly the principal information-retrieval system for the larger marine mammals, such as whales and dolphins. Dolphins emit high-frequency sounds, from audible clicks to ultrasonic vibrations of over 100,000Hz, and then listen for the echoes. From these they are able to build up an astonishingly accurate image of their underwater domain—the distance and direction of objects large and small, their size, shape, texture, and density, species of fish as well as their every movement. The clicks are produced from the dolphin's large forehead region, probably by recycled compressed air, and the echoes are received over a broader area around the sides of the head and lower jaw that includes the inner ears. A gigantic auditory lobe in the brain helps decode this complex array of signals.

There is undoubtedly an element of higher communication in the range and variety of sounds emitted by dolphins, as well as in the hauntingly beautiful songs of the Humpback whales as they migrate to and from their polar feeding grounds. It remains an intriguing and important challenge for scientists to learn the language of these other large-brained, intelligent residents of Planet Earth, and even to wonder at the possibility of inter-species communication.

Nasal passages
Melon
Focused sound pulse
Returning sound pulse
Sinuses
Inner ear

Toothed whales, such as dolphins, are able to to find objects by means of echolocation. Sounds are produced in the nasal passages, then focused and sent out by the melon, a fatty organ in the forehead that acts as an acoustic lens. Returning sound waves are channeled through oil-filled sinuses in the lower jaw to the inner ear, which is insulated from extraneous resonances within the skull.

When alarmed, the spiny lobster (*left*) can produce an abrupt rattling sound by scraping the end of its antennae against the hard plates under its eyes.

Life in the oceans is a dynamic balance between prey and predator. Agility, keenly developed senses, and well-designed arms and armor are all essential parts of an animal's survival strategy.

MARINE LIFESTYLES

Making a Living

KEY TERMS

PREDATION

SYMBIOSIS

FOOD CHAIN

GRAZER

RADULA

FILTER FEEDER

BALEEN

SUBSTRATE

POLYP

NEMATOCYST

CRYPTIC COLORATION

DISRUPTIVE COLORATION

COUNTERSHADING

MUTUALISM

COMMENSALISM

PARASITISM

Many of the animal world's weird and wonderful adaptations for life in the sea are ultimately strategies for feeding. Any creature unable to find an adequate and constant supply of food will die slowly of starvation. This is the harsh reality of survival, in which all animals tread the finest of lines between life and death. It is small wonder, therefore, that the evolutionary eons have thrown up such an astonishing range of offensive and defensive weaponry and disguise. From the top predators to tiny zooplankton, each animal has evolved to suit its own ecological niche, together with an appropriate feeding strategy—grazing, hunting, or simply sitting and waiting for lunch. Next comes the refinement of tools and techniques, of weaponry and camouflage. The campaign is played out by all in deadly earnest.

Peaceful grazers Eating low on the food chain is the most energy-efficient means of nourishment for animal life, and the preferred option for countless marine organisms. From the ubiquitous sea snail to the rare Blue whale, the principal determinant is locating enough to eat. Each species then has its own special harvesting method.

Many types of marine gastropods (sea snails) have chosen to graze the algae that grow in such profusion around the ocean fringe, coating rocks along the shore, as well as those strewn out across the continental shelf. Limpets and cowries, like all herbivorous gastropods, possess a tooth-studded, ribbonlike device known as a radula to scrape away at their tough, leathery meal. Conchs use their radulas to eat the rather softer seagrasses, their single foot clinging by powerful suction to the swaying stems and leaves.

Sea slugs (or nudibranchs) commonly sport bright colors and dazzling patterns. They have abandoned the protective shell of their gastropod relatives and feed on sponges, corals, anemones, jellyfish, and other prey, with their soft bodies fully exposed. For protection, they recycle the stinging cells of their digested prey, using them "second-hand" for defense against predators.

They are joined at the feast by a great many other creatures. Green sea turtles are omnivores with a healthy appetite for algae and mangroves, while the Marine iguana—the world's only marine lizard—grazes on dense mats of seaweed festooning the rocks of the Galápagos Islands' shorelines. Manatees and dugongs, the large, gray marine mammals with wrinkled skin and whiskers that figure in many a seafarer's tale, are partial to tropical mangroves and seagrasses; manatees can eat as much as 30kg (66lb) per day. A host of fishes have also become specialists in vegetarian diet. Surgeonfishes have small, flat teeth with sharp cutting edges, not unlike our own, which they use to scrape algae from rocks and corals. Parrotfishes actually feed on the symbiotic algae within coral polyps, with beaklike mouthparts and strong teeth that are capable of crushing the coral to extract the algae.

The fishes of the open ocean are mostly carnivores, but some have adapted well to filter-feeding on plankton. Anchovies and other small fishes that graze these plentiful pastures generally live in large schools, in order to minimize the chances of individual predation. But the schools make easy pickings for hungry predators. Only the really large fishes are relatively safe because of their great size. The Whale shark, growing to over 15m (49ft) in length, is the biggest of all. It has between 3,000 and 5,000 very small teeth, which it uses to strain food from large gulps of plankton-rich water. The Manta ray, with its 8-meter (26ft) wingspan, uses its giant fins to fan thousands of liters of water and food into its gaping mouth; spines in its throat are able to hold prey until it is ready to swallow. The Megamouth shark, only discovered in 1976, is a shy, deep-ocean creature that ascends to mid-water depths at night to feed. Its broad mouth stretches over 1m (3.3ft) across and has 200 rows of tiny teeth, almost identical to those found as fossils from rocks some 26 million years old. Clearly a life of plankton grazing can be highly successful in evolutionary terms.

The great baleen whales are perhaps the ultimate in perfection when it comes to harvesting a meal from the ocean plankton. The Blue whale, for example, the largest creature on the planet, eats an average of 4 metric tons of plankton every day. Mostly it eats krill, a small, shrimplike member of the zooplankton, and small fishes, but simply by opening its enormous mouth it swallows everything in its path. In place of teeth, a Blue whale has great plates of baleen composed of keratin (a tough protein found in mammalian nails and hair) forming a tight mesh. When the huge mouth is full of food and water, it closes and expels the water through the baleen plates, straining out the krill and other food, which it then swallows. Humpback whales often feed by ascending slowly from below a rich patch of plankton, blowing a ring of bubbles. The ring acts as a bubble net, trapping the krill as a concentrated broth near the surface.

Slow and sedentary Grazing the ocean pastures may seem like an easy option, but other creatures make even less effort and settle down for most or all their life in just one place. The sedentary lifestyle is an excellent way to conserve energy; its practitioners do not have to move but simply let food and water waft toward them, perhaps helping it along by flailing cilia or waving tentacles. The chief problem is to find a suitable location where food will be naturally plentiful.

Different animals select different substrates in or on which to make their homes. Polychaete worms burrow into a soft or sandy seafloor, extending their delicate tentacles to intercept what food may pass. One genus, the beautiful feather duster worm, lives in stiff tubes made out of sand grains bound with mucus, from which its feathery arms protrude, and to which it retreats for safety and rest, or when the buffeting waves and currents become too strong. Razor clams—unlike some other bivalves, such as the scallop, that can move over long distances—also sit tight. They are well adapted for burrowing into

soft sediment and filter-feeding, at the mercy of the tides and currents. Others prefer a hard, rocky bottom: mussels attach themselves to rocks by a strong, rubbery, threadlike material known as a byssus; oysters produce a tough cement. Even the Giant clam, which grows to over 1m (3.3ft) and weighs over 200kg (450lb), is an unaggressive filter-feeder common to reef communities in the Indo–Pacific region. In fact, because of its enormous size it must supplement its diet by hosting photosynthetic symbionts in its colorful mantle tissues. However, larger animals can also get trapped when these huge mollusks clamp shut their heavy shells in self-defense.

Much smaller bivalves, such as members of the *Teredo* and *Bankia* genera, evolved the ability to bore into timber and so became the bane of early ship-owners. Once housed in a secure wooden home, these tiny shipworms extend two miniscule siphons in search of food. They harbor enzyme-producing, symbiotic bacteria capable of breaking down cellulose, and so have developed the ability to eat, as well as live in, wood. Their relatives belonging to the *Pholas* genus are even able to bore into the softer rocks, such as limestone.

Living in a moving home has distinct advantages over remaining in one place, for the potential for a periodically renewed food supply as well as for wider dispersion of the larval stage is greatly increased. Barnacles, the small, hardy crustaceans whose conical encrusting shells are so common on rocky shores, have perfected this technique, not only by attaching themselves to ship's hulls but also to marine animals including crabs, turtles, fishes, and whales.

These decorative, feathery feeding fronds belong to a colony of tubeworms (polychaetes) that have settled down to an entirely sedentary existence. Different species use a variety of construction materials for their tubes— protein, calcium carbonate, pebbles, or sand held together by mucus. The delicate crowns of fronds extend to trap and filter plankton, mainly during the night. They are extremely sensitive to light, and to pressure changes in the water around them, and quickly withdraw into their tubes when disturbed.

The sponges, among the morphologically simplest of all animals, settled into an entirely sedentary existence over 600 million years ago. These large colonies of semi-independent organisms have specialized collar cells with hairlike cilia that beat in unison, wafting a weak current of water and food particles into the main body structure. As only limited firm substrate may be available, the chief problem facing sponges soon became one of competition for living space from corals, bryozoans, tunicates, and sea lilies. Remarkably, some species have learned to release chemical traces into the water that kill or inhibit the growth of their principal competitors.

Despite the sponges' efforts, however, corals in particular flourish in warm tropical waters. The individual coral polyps are tiny, soft-bodied creatures that live in and over the little cups of calcium carbonate they have secreted. Sometimes brightly colored, they display an ingenious range of tricks to enhance their filter-feeding ability. Their tentacles are lined with nematocysts (stinging cells) and may have harpoon tips for firing poison; they may take the form of long, fine threads to wrap around prey, and sometimes they are coated with a sticky substance, which further prevents escape. Polyps are typically covered in minute cilia that beat rhythmically, directing a flow of water and foodstuff over the polyp's mucus-lined interior.

Blue shark with its nose in the dinner bowl—in this case a dense school of Northern anchovies off California. These sharks are accomplished predators, patrolling both tropical and temperate seas. They are especially adept at diving for octopuses and squid, and will dive repeatedly to depths of 600m (2000 ft), often at night when there is less difference in temperature between surface and deep waters.

Hunting instinct

For animals that hunt for a living, locating and trapping a meal is a very serious business. They work hard and constantly in this endeavor, aided by a quite astonishing diversity of hunting equipment—teeth that come in all shapes, sizes, and numbers, rapierlike spikes, swords, and tusks, stun guns and electric shocks, suckers and stings, poisons that immobilize and kill.

The cruising fishes and toothed mammals are among the most voracious predators in the ocean, capable of inflicting serious wounds on even the largest prey and of devouring large quantities of smaller prey in moments. There are some 350 species of sharks, ranging in size from 15cm to 15m (6in to 50ft) long, of which about 35 are potentially dangerous to humans. Their several rows of large, sharp, triangular teeth, coupled with immensely powerful jaws, keen senses, and swift movement, cast them among the top marine predators. They tear off large chunks of flesh with consummate ease and swallow them whole. Their often indiscriminate feeding behavior explains some of the more unusual items that have been found in their stomachs and intestines, including half a bulldog, parts of a reindeer, tin cans, leather jackets, and a bottle of Madeira wine!

For sharks, Killer whales, tuna, barracudas, marlins, and their ilk, it is speed and teeth that count the most. Some species

Poisonous perfection

The art of poisoning seems even more prevalent at sea than on land. Five invertebrate phyla and a large number of vertebrates—fishes and sea snakes—contain species that regularly use venom, either as a defensive mechanism or for immobilizing prey. Many sedentary animals rely on batteries of stinging cells or nematocysts for both defense and attack. Some species advertise this fact with bright colors—the Red moss sponge and Fire sponge, Strawberry anemone and Hell's Fire sea anemone, for example—whereas most of the 2,700 varieties of hydroid display their armory of feathered tentacles more discreetly. Fire coral, Stinging seaweed, and Stinging hydroid are some of their common names, and the infamous Portuguese man-of-war is also a hydroid rather than a true jellyfish. This softly floating killing machine trails its deadly stinging tentacles up to 20m (65ft) below the surface, each loaded with hundreds of poisonous barbs which, when touched, shoot into the victim.

True jellyfish are free-swimming scyphozoans, which more often appear to float with the currents than swim of their own volition. They tend towards colorless or translucent pale blues, pinks, and milky white, and all trail tentacles with rows of stinging cells. The box jellyfishes and sea wasps of northern Australia and the north Indian Ocean are undoubtedly some of the most dangerous and venomous of sea creatures. People have been known to die within 30 seconds of being stung as a result of circulatory or respiratory failure.

Cone snails, common throughout the Indian and Pacific oceans, are among the more sophisticated and deadly poisoners. Some have a long, brightly colored proboscis that unwary fish apparently mistake for an edible worm. When a fish moves in to bite, the snail fires its poisonous radula into the soft tissues of the fish's mouth. The venom inflicts paralysis and causes rapid death, even of large animals, after which the snail can begin a leisurely meal. To date, at least 20 swimmers in South Pacific waters are known to have been killed by this tiny creature. Slow and deadly is the motto of many marine animals,

The huge gape and formidable dentition of the viperfish (*left*) show it to be a highly specialized predator. The upper front teeth have sharp ridges near the tip and are used for stabbing. The longest teeth are at the front of the lower jaw. Normally they lie outside the upper jaw, but when they impale prey their curvature pushes the prey toward the mouth.

regularly discard and grow new sets of teeth—as frequently as every week in Nurse sharks. Others have rows of teeth in reserve. A special instrument has been designed to measure the actual force of a shark bite, which can exceed 100 metric tons per sq cm (650 tons per sq in) for some of the larger species. But it is the frenzied feeding pattern involving a large number of sharks together and large quantities of food that is most awesome to witness. Equally impressive, and with even larger teeth, is the Saltwater crocodile, especially those of mangrove swamps and estuaries throughout southern India, southeast Asia, and northern Australia. These alarming beasts can grow up to 6m (20ft) in length, lie deathly still like a floating log, and then pounce with incredible speed and deadly results.

Many types of fishes also prefer to lie in wait, suitably hidden or camouflaged, and then lunge forward with great alacrity when prey strays within range. Groupers can grow very large and have enormous jaws. When they lunge and quickly snap open their mouth, the unsuspecting victim is literally sucked inside—almost as though they were being inhaled. Gulper eels are also lungers, with enormous jaws for swallowing fishes larger than themselves. Garden eels dig tunnels up to 1m (3.3ft) deep into a soft substrate and lie in wait. They live in tightknit colonies of 200–300 individuals, their thin, tubular bodies weaving about like strange garden plants.

The ingenuity of hunting techniques is almost endless. Many reef-feeding fishes have long snouts for prying into cracks and crevices. Starfishes often use strong suckers to lever open shellfish, while the Crown-of-thorns starfish forces its stomach through its mouth, secreting digestive juices that dissolve coral polyps into a "soup" which it then ingests. Sea spiders feed on the juices of cnidarians and other soft-bodied invertebrates, which they extract with a long, sucking proboscis. The Bat ray swims lazily over the seafloor spurting water from its gills to stir up the sediment and reveal any hiding animals; it then settles and eats, sucking the helpless victims into its mouth. Electric and torpedo rays are among the 250 species of fishes known to possess specialized organs capable of delivering painful electric shocks, stunning or even killing their prey. The shocks can be as mild as 8 volts or as powerful as 220, and can be repeated several times in quick succession before a recharge is necessary.

A group of hungry starfish ambush some sea urchins, quite undeterred by the vicious-looking spines. Starfishes are efficient predators, feeding on a wide range of other invertebrates. They are ubiquitous and voracious, can "smell" the chemical signal of the presence of suitable prey, and move quickly toward it on hundreds of tiny tube-feet. Those starfishes that prey on bivalves use their tube-feet to prize them open.

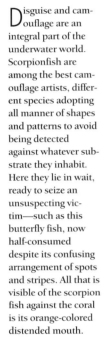

including a wide variety of worms, sea urchins, and starfishes. These small creatures may not be lethal to humans, perhaps, but are very effective at dealing with their choice of prey.

The fishes are one of the oldest and certainly the most diverse group of vertebrates in the world, so it comes as no surprise that numerous species have evolved mechanisms for producing and utilizing venom. The dogfish injects a mild poison into the wounds it has already inflicted, whereas lionfishes, scorpionfishes, stonefishes, and pufferfishes all carry a much more lethal poison in their dorsal or lateral spines. Among the most dangerous of all is the superbly camouflaged stonefish, which can kill small prey in seconds and even an unwary human swimmer within a couple of hours, and the pufferfish, whose toxin is 200,000 times more poisonous than the deadly plant extract used to tip poison arrows. The lionfish, although a relative of the stonefish and similarly venomous, deters would-be predators with its bold colors and elaborate decorations.

Stingrays are another group of fish with poisonous spines in the tail region, which they use principally for defense, preferring, on the whole, to flee rather than fight. The numerous species include some of the largest rays such as the Manta ray. However, they are mostly shy bottom-dwellers in shallow waters from temperate to tropical regions, typically burrowing into sand or mud and quietly feeding on worms, mollusks, and crustaceans.

Although seldom larger than 60cm (2ft) in body size, the octopus is a formidable predator. It has eight long and very strong arms with powerful suckers that may extend over 10m (33ft) across. It has keen eyesight, an ability to change its color in response to its background or mood, and a highly developed nervous system including a well-developed brain. Octopuses use a deadly mix of strategies—slow stalking and camouflage, pouncing with amazing speed, enveloping the prey with powerful arms and suckers, biting and injecting searing venom that paralyzes it. Once the prey stops struggling, the octopus crushes

Disguise and cam-ouflage are an integral part of the underwater world. Scorpionfish are among the best camouflage artists, different species adopting all manner of shapes and patterns to avoid being detected against whatever substrate they inhabit. Here they lie in wait, ready to seize an unsuspecting victim—such as this butterfly fish, now half-consumed despite its confusing arrangement of spots and stripes. All that is visible of the scorpion fish against the coral is its orange-colored distended mouth.

its shell or carapace with beaklike jaws and tears the flesh into bite-sized morsels. The Australian Blue-ringed octopus has a body only 3cm (1in) long but produces such powerful venom that it probably kills more people every year than any shark.

Armor and camouflage Sophisticated weaponry and hunting techniques are one way to make a living and ensure survival, but quite other strategies are needed to defend, to hide, or to escape. Marine animals of all types and sizes display a bewildering array of such methods. Those with poison glands—sponges, corals, sea anemones, and nudibranchs—often advertise the fact as a warning to potential predators; the Poison-fang blenny, for example, is a small Pacific fish with bright aposematic (warning) coloration, avoided by most larger predators. Other creatures shut up shop, packing away their vulnerable soft parts, hide away in burrows or crevices, or escape the danger area with great rapidity. Sea squirts release a stream of water as a warning or deterrent, while many cephalopods, including squids, release a dense, inky substance to temporarily blind and confuse predators. The Tiger cowrie shoots out mild sulphuric acid when disturbed, and several types of sea slug release noxious-smelling chemicals, or mucous coats. The primitive hagfish exudes an extremely effective milky white fluid, which renders its body too slippery to catch.

Sluggish, fleshy sea cucumbers (a unique, mainly soft-bodied group of echinoderms) may appear extremely vulnerable as they creep along the seafloor. In fact they are capable defenders, often with unpleasant odors and poisonous skins, an ability to expel water and contract in size, and the unusual practice of releasing a sticky tangle of tubules from the anus. These long, spaghetti-like strands inconvenience, entangle, and temporarily immobilize any animal that comes too close. An even more extreme line of defense is to eviscerate completely, spewing out guts and other organs all over the hapless intruder, and then make an escape. Quite remarkably, the eviscerated organs regenerate within a few weeks. Other echinoderms—starfishes and especially brittle stars—are able to lose one or more arms, and even part of the central disk, and then quickly grow replacements.

Body armor is one of the most common defenses employed by legions of marine organisms, from the microscopic plates and spines of unicellular plankton to the massive calcite shell of the Giant clam and the leathery scales of the Saltwater crocodile. Sponges, corals, bryozoans, and sea lilies all have exoskeletons, constructed and shared by many individuals. Crustaceans—including crabs, lobsters, shrimps, and barnacles—also have tough outer coats, often jointed; they are awkward for movement, but effective against many predators. Spiny protection, as exemplified by sea urchins, is popular too. Oysters, mussels, scallops, clams, sea snails, and a host of other mollusks are best known for their shells, tough and gnarled, ribbed and grooved, or patterned and colorful. The adoption of hard parts by organisms some 550 million years ago led, of course, to their better preservation as fossils from that time onward. It was also instrumental in the explosion of form and shape that followed, and in the remarkable evolutionary longevity of many of these groups.

Although the oceans may sometimes seem a harsh world in which neither predator nor prey can ever rest from battle, there are other times and places in which life appears to flow in tranquil harmony. Many organisms have achieved at least a degree of stability by developing close relationships with one another in a partnership known as symbiosis. Three types of symbiotic relationship are distinguished: mutualism, which benefits both organisms; commensalism, which benefits one more but does no harm to the other; and parasitism, in which the host may eventually be destroyed by the parasite.

Sometimes the symbionts are inseparable, behaving as one organism and being unable to survive alone. Such is the case for coral polyps and the symbiotic dinoflagellates, known as zooanthellae, that live within their tissues. Other reef animals, such as giant clams and related mollusks or the beautiful Blue dragon nudibranch (sea slug), also support large numbers of symbiotic zooanthellae. Microbial organisms of all kinds—bioluminescent bacteria and chemosynthetic bacteria in vent communities, for example—often form inseparable relationships with a variety of different plants and animals.

In many cases, however, the symbionts can live apart if separated, although their chances of survival become greatly diminished as a result. The bright-orange Pacific clownfish, with its broad white stripes and protective mucous coat, has developed an ingenious way of living within the stinging tentacles of an anemone. By so doing, it receives protection from would-be predators wary of the toxic stings, while in return it defends the anemone from other fishes that are able to eat the tentacles. Another species of anemone attaches itself to a mollusk shell occupied by a hermit crab, in a position just above and behind the crab's mouth. It feeds on the scraps that escape, and in return offers the crab both protection and camouflage. Eventually the anemone overgrows and absorbs the mollusk shell, but remains as a permanent home growing at the same rate as the crab.

Over 40 species of marine cleanerfishes, cleaner shrimps, and one cleaner crab are now known, performing the important task of removing dead skin, parasites, and pests, as well as leftover food, from all parts of their customers' bodies. These tiny animals dart in and out of the delicate gills, mouths, and even throats of larger fishes with complete impunity. The groupers, sharks, and others who visit never harm the cleaners and leave when they are spotless and satisfied. The Banded coral shrimp specializes in dentistry for moray eels, picking over the moray's needle-sharp teeth with its delicate claws.

The secret of symbiotic success is mutualism, where both parties benefit equally. Species that practice commensalism, such as the remoras that attach themselves to sharks, rays, or turtles via a powerful vacuum sucker, are quietly tolerated but can become a nuisance. Marine parasites—the flukes, worms, and lampreys, for example—risk losing their unwilling host entirely if their demands become too extreme!

Animal partnerships: the secret of symbiosis

The tiny Spotted cleaner shrimp works with delicacy and impunity among the lethal tentacles of a Pink-tipped sea anemone. Both benefit from this symbiotic mutualism.

Camouflage is always intriguing in its ingenuity. Flatfish and cuttlefish are masters of disguise, able to change color and pattern to match almost any backdrop. Species that live mainly in one habitat often have cryptic coloration—the poisonous stonefish closely matches the rocky seafloor, for example, and the mottled scorpionfish, a coral reef. Many reef fishes display disruptive coloration, usually one of vertical lines that make their overall outline difficult to detect. Open-ocean fishes such as tuna and mackerel employ countershading; their darker-colored topsides mimic the water coloring from a birdseye view, while their pale, whitish-colored bellies are difficult to distinguish from the sky when looking upward. Many planktonic invertebrates, such as jellyfishes and pteropods, as well as little-known creatures of the deep sea, adopt transparency as a disguise.

Camouflage by mimicry is often the most remarkable of all. Pipefish and seadragons look almost indistinguishable from the seaweeds they live among, as does the Tassled wobbegong shark that hunts them, with its marbled skin and branched, fleshy mouth lobes. The Sargassum fish mimics the floating *Sargassum* weed of the central Atlantic gyre that is its home. It even has white body patches to match the casts of tubeworms on the seaweed. The shrimpfish is long and thin with a black stripe along its length, ideal for hiding among sea-urchin spines. Certain mimics have learned to disguise themselves as other fishes, and so derive an element of surprise or protection where needed. Others, like the decorator crab, attach a variety of seaweeds and other organisms to their bodies to better blend in with the ever-risky seafloor across which they must scurry in search of a meal.

Marine organisms, perhaps even more than their counterparts on land, have evolved an elaborate richness of reproductive techniques, in a bid to win the mating game and ensure survival.

Sexual Encounters

The sheer scale of reproduction in the world's oceans is staggering. In an average liter (2.1 pints) of seawater from the ocean surface, there is a teeming variety of life—over half a million unicellular diatoms and microscopic plants supporting thousands of zooplankton. During the spring bloom, each cell divides and reproduces so rapidly that it can yield more than one billion progeny in a month.

This fecundity of asexual reproduction in single-celled organisms is almost matched by the remarkable egg-laying ability of animals higher up the evolutionary scale. The Green turtle, exhausted after her long crawl out of water across the sand, still lays about 100 eggs at a time; the Blue crab produces several thousand, while Horseshoe crabs each lay 20,000; their forebears have probably been doing the same each year in spring for 500 million years, with clearly proven success. But such abundance is as nothing compared with that of some fishes. The average mackerel will lay 100,000 eggs at a time, a hake as many as a million, a haddock up to 3 million, and a cod from 3 to 9 million. Some marine snails such as the Purple sea hare produce 20 million eggs, and an oyster will lay an astonishing 500 million eggs in a single year! For many species the mortality rate of their offspring is well over 99 percent, and it has to be so in order that the delicate balance and precise foundation of life pyramids in the ocean remain intact.

Strategies for survival Those creatures that are the most fecund simply abandon their eggs to fend for themselves, trusting that from such great numbers at least a few will hatch and grow to maturity. When, for example, the individual polyps of a large coral reef spawn together in synchronized, multi-species mass spawnings, the simultaneous release of billions of eggs and sperm clouds the water with a fantastic pinkish

Coral spawning is one of the truly great spectacles in nature, timed exactly by the season, the Moon, and the water temperature. Bundles of eggs, like tiny beads, and even smaller sperm rise to the mouth of each individual coral polyp and are then released in perfect synchronicity across the entire reef.

haze. Although predators gorge until their appetites are sated, enough eggs still remain for some to survive and hatch.

Less prolific species find other ways to ensure survival. Their eggs may be larger, with a yolk or with hardened shells; they may lay their eggs in cracks and crevices, bury them in sand or mud, or attach them to plants or rocks. Alternatively, the parent may even nurse the growing embryos to the point where they emerge as hatchlings. In one way or another, the egg is the foundation of all modes of sexual reproduction.

On hatching from the egg, the young of many marine creatures bear little resemblance to their parents. These larval forms move, feed, and live quite differently, commonly in the plankton and often in very large numbers. They, too, are an essential part of the ocean's rich food supply, and most are consumed by predators. Those that survive may pass through several different stages and forms before becoming adult. This is notable among crustaceans, where some decapods (prawns, shrimps, lobsters, and crabs) have as many as 18 juvenile stages, and most have more than one.

The larval form is mainly about dispersal and colonization of a new underwater terrain or patch of ocean. Adult sea squirts are small, translucent or leathery, baglike animals, that live by filter-feeding, permanently attached to the substrate. They are hermaphrodite, releasing sperm into the water but often retaining the fertilized eggs within the body. When hatched, their larvae resemble tiny tadpoles, actively swimming, first toward the light and then, a day or so later, toward the seafloor and a new home. The Edible blue mussel produces up to 12 million eggs at a time—good odds, one would wager, that at least some of the free-swimming larvae might survive their short spell in the plankton, drift to a new area, and start a fresh colony.

Energy and rivalry in the mating game permeate most marine sex, as graphically exemplified by these Broadclub cuttlefish. The female (left) is laying eggs, closely protected by her mate (center). A rival male (right) is being warned off by the first male, in an aggressive display of waving tentacles. In common with other cephalopods, cuttlefish can produce a startling range of color changes. As well as for camouflage, they use color and pattern to communicate, and can simultaneously signal sexual interest to a female and warning messages to a rival.

Sex: the pros and cons

In the beginning there was no sex in the world, and so it remained for over 1.5 billion years. Single-celled organisms reproduced asexually, simply splitting into two equal parts as they grew larger. Such a strategy has several advantages. It can take place at any time and can yield many progeny within a very short period. There is no need to find a mate, to release eggs, or to ensure their fertilization. The principal disadvantage is that no genetic variety is introduced as it is by the mixing of contributions from two parents. Without sex, all offspring are clones, genetically identical to their parents, and evolutionary variation is restricted to rare chance mutations. In addition, the ability to disperse and so colonize other parts of the planet can be severely restricted.

The advent of sexual reproduction about 2 billion years ago represented, therefore, a major evolutionary breakthrough. In this process, male and female sex cells (sperm and egg) fuse to form a cell with chromosomes from both parents—the zygote, or embryo, which then develops into a new individual with characteristics inherited from both parents. Each sex cell, or gamete, is slightly different, having received a random combination of chromosomes during its formation. Add to this the factor of mutation, and the potential exists for rich genetic diversity, and thus evolutionary change. Such diversity is truly a necessity for survival, as the environment is far from static, and a species that cannot adapt to change soon becomes extinct.

In fact many simpler life forms still practise asexual reproduction, especially by simple cell-division in single-celled organisms. Some sedentary bottom-dwellers such as corals, sea whips, sea fans, and anemones reproduce by budding, a process by which a small new organism grows as part of the adult, and then either detaches as a free-living individual or continues to expand the colony. In some planktonic crustaceans such as ostracods and branchiopods, the females have evolved a means of reproducing that does not require the input of male sperm; this behavior is known as parthenogenesis.

Evolutionary advantage often seems to be gained by keeping both options open, with alternate phases of sexual and asexual reproduction. Tiny diatoms, with their beautiful pill-box skeletons of transparent silica, reproduce very rapidly by repeatedly dividing in half until the individual cells become too small. At that point they release male and female gametes, which fuse and develop into completely new organisms. Corals and sponges can reproduce by asexual budding, fragmentation, and rejuvenation, but periodically also release gametes. These join to form a zygote, followed by a larval planktonic stage (called a planula larva in corals), thereby ensuring dispersal. Sessile branching hydrozoan colonies bud asexually to release small, free-floating medusae (jellyfish stage). It is the medusa that then releases eggs and sperm into the water column, the embryo developing into a planula before settling down as part of a new polyp colony. In true jellyfishes (scyphozoans), the medusa stage is the dominant phase of their life cycle.

Another very common development among some animals is hermaphroditism, in which a single organism can act as both male and female, either at the same time or sequentially. Slipper limpets (*Crepidula*), for example, tend to live a more or less sessile existence in large groups in which several individuals are stacked on top of one another. This arrangement allows the penis of the uppermost limpet to reach the female opening of the individual below, and so on through the stack. New young limpets are always male, but may then grow into either male or female forms depending on the sex ratio of the group. This information is imparted by the release of pheromones into the surrounding water. The Giant clam develops in a similar fashion, although the older members of the group (aged between 6 and 8 years) become both male and female simultaneously, and so release both eggs and sperm.

In the darkness of the abyss, the male anglerfish will avidly follow the least scent of a female, although she can be as much as 25 times his size. Once he has caught up he fearlessly grips her and plunges in his specialized teeth, slowly changing and fusing to become a permanent parasite—a strategy unique in the vertebrate world. Nourished by the female's blood in the manner of a fetus, his organ systems atrophy and he becomes little more than a sperm bank, available to fertilize her eggs as soon as they are released.

Female Horseshoe crabs drag smaller males up a New Jersey beach, in an age-old mating ritual (*right*). These shy creatures are seldom seen, as they live at considerable depth on the seafloor off southeast Asia and northeast North America. But every spring, on three successive nights when the Moon is full, they slowly emerge from the sea in the hundreds and thousands. Near the high-tide line they dig shallow pits with the edge of their shells, into which the females shed their eggs and the males their sperm. The young will emerge at the next full moon, when high spring tides stir the sands once more.

Manatees are marine mammals belonging to the order Sirenia and possibly the source of mermaid myths from long ago. This group of adults are seen caring for a juvenile.

Attracting a partner

To gain the advantages of sexual reproduction, animals have developed many techniques for attracting a mate. The males of many species become highly colored during the mating season, some with spectacular rainbow displays for courting. Those species that use color for camouflage or escape, such as lobsters, squid, and many types of reef fishes, are especially adept at signalling their readiness to mate. Bioluminescent displays are used in a similar way, particularly in the twilight zone and below, where bright colors become progressively redundant.

In dark or murky waters and for benthic or sessile organisms, chemical messages are often more important than color signals. Oysters do not move to mate, but when the male oyster releases sperm into the water the female responds to the scent by laying eggs inside her shell. She then takes in the sperm-laden water to fertilize the several million eggs she is protecting. In certain species of crabs and lobsters, the female releases a distinctive pheromone when she is ready to molt. This attracts the male, who seizes and mounts the pre-molt female, so that mating can occur when she first sheds her shell. In a seemingly chivalrous manner he continues to protect his now defenseless mate until she has grown her new shell.

In the rituals of reproduction it is sometimes difficult to distinguish between a display of ferocity and a mating dance. Marine males fight for the right to mate. They establish territories or build nests that they defend earnestly by ritual displays or fierce battle, chasing away lesser rivals but yielding to those who are stronger. Tiny worms engage in violent territorial battles over a few millimeters of mating tube, while mighty tusked walruses become badly scarred in defense of their hard-won patch of beach. Many fishes and cephalopods engage in ritualistic displays until the smaller combatant gracefully withdraws.

In contrast, the male cod performs a most beautiful and carefully choreographed dance before his chosen female, in the cold and eerie light of subsurface waters. If she responds to his overtures, he guides her gently toward the surface and mounts her.

She spawns, while he releases copious sperm to fertilize the many millions of her eggs now floating in these warmer, lighter waters. The eggs hatch in about 10 days, and the juvenile fish remain to feed and grow for a further 2–3 months before they return to their favored mid-water hunting grounds.

Some of the most beautiful mating scenes are enacted by cetaceans (whales and dolphins). Male dolphins perform before females in a dozen different ways, brushing and caressing their bodies to stimulate interest, displaying their own skill and agility by elaborate swimming maneuvers and spectacular leaps. These amorous acts are accompanied by much lively vocalization.

Parental care

The most elaborate courtship rituals are often followed by the release of a single egg or just a few, but even so yield optimal chances for fertilization, followed by parental care for the growing young. All marine mammals, as well as sharks, are viviparous in that the eggs are fertilized within the female body; the embryo grows and the young are born only when fully developed. Cetaceans give birth to one (or, rarely, two) offspring, and then live and travel in pods that include both adults and young of both sexes. Whale mothers, who can be aggressively protective of their young, feed them on a diet of very rich milk from their mammary glands. This practice results in rapid growth in relative safety; baby Blue whales double their weight in the first week and then gain as much as 90kg (200lb) a day. Other mammals such as Sea otters and pinnipeds may care for their young in the same way for several years, with much caressing as well as vocal exchange. Young otter pups cry when hungry, frightened, or lost, and the mother responds with a gentle cooing.

While mammals have learned that caring for their young is the best way of ensuring their survival after birth, a number of other species go to considerable lengths to protect the eggs before they hatch. Female sharks have a modified portion of the oviduct that acts as a uterus and so allows internal protection of the young. They are born fully developed and, although they remain close to mother for a short while, they quickly become accomplished hunters in their own right.

Other fishes are fully oviparous, with the eggs being fertilized externally, and yet still contrive to protect their clutch from predators. Many bottom-dwelling fishes build and guard egg nests. The jawfish lays her eggs in the male's mouth where he incubates them until they hatch, only temporarily transferring them to a protective burrow while he feeds. Male seahorses and pipefishes have special brood pouches for the eggs, as do some starfishes and brittlestars. The female octopus establishes a safe home on the seafloor for her many strands of eggs. Here she waits for five or six weeks, guarding against would-be egg snatchers, cleaning and oxygenating her brood, but failing to feed herself. By the time the tiny octopus babies burst from their egg cases, she is near to starvation and soon dies. Sadly, her efforts seem largely in vain, for hungry fishes are hovering nearby and of the 200,000 or so hatchlings only one or two will ever reach maturity and reproduce in their turn. For the ocean world, however, this still represents parental success.

Mating migrations

Large-scale movements of animals in the oceans generally reflect either seasonal change in the availability of food or else breeding patterns. Some 500 million years ago, the ancient relatives of Horseshoe crabs were migrating to the coast to mate and lay their eggs, and their descendants continue to do so to this day. Many other crabs and lobsters have adopted similar strategies, a remarkable programmed behavior that provides their offspring with a better chance of survival. Saltwater crocodiles, like all marine reptiles, must lay their eggs on land, and can be very particular about which beach, shore, or swamp they select. Seals, sea lions, walruses, and many species of seabird commonly migrate to the same shore at the same time year after year, even though their arrival in large numbers means space is severely limited.

Larger open-ocean fishes and benthic crustaceans tend to migrate toward shallow waters for breeding, especially favoring kelp forests, seagrass beds, and mangroves. These tangled environments provide ideal nurseries for the larvae, offering both protection and nourishment until the young are ready to swim or crawl back to the more favored adult habitats.

Some species have captured the public imagination, as well as arousing scientific curiosity, by their exceptionally long migrations and their astonishing ability to find the exact beach, headwater, or sea in which they were born. Turtles, salmon, and eels, as well as many cetaceans, are among the animals with these credentials. Not all turtles migrate long distances, but the Green sea turtle will swim halfway across the South Atlantic Ocean, from the manatee grasses along the coast of Brazil to the tiny target of Ascension Island some 2,330km (1,450mi) away. The turtles mate at sea and come onto the beach to lay about 100 golfball-sized eggs in shallow, sandy nests. This behavior is repeated four or five times at 12-day intervals, after which the adults return to their feeding grounds. About two months later the baby turtles hatch and dig their way out. The tiny hatchlings crawl quickly to the water, feeding on plankton as they drift in the currents towards Brazil. Those that survive will eventually return to breed on the same mid-ocean island, and so on through the generations. There are several hypotheses—the Sun, magnetic fields, ocean currents—as to how the turtles manage to navigate so unerringly, and just why they choose a beach so very far away. These questions are not yet resolved.

The great whales are the supreme lords of the oceans. With their keen senses and high intelligence, their cooperative behavior and long-distance communication, they travel the vast underwater world as if it were their home town. Gray whales migrate from the Bering Sea to Baja California to mate, and Humpback whales turn up each spring off Hawaii, performing feats of navigation that humans can only achieve using sophisticated satellite communication. After several months of calving, mating, playing, and singing, they all disappear within a few days, reappearing some weeks later off the Alaskan coast, to feed on the bounty of the northern seas.

Tide pools are just one of the many complex communities in which the great diversity of ocean creatures play out their lives. At first glance, the pools may appear to be peaceful, miniature oases, protected from the alternating rigors of wave shock and dessication. Even so, their inhabitants must be very adaptable. Tide-pool organisms—such as these colorful starfish and sea anemones—must be able to adjust to sudden changes in temperature, salinity, pH, and the oxygen content of seawater, all of which can fluctuate rapidly and frequently. For example, a heavy burst of rain can reduce the salinity of a rock pool nearly to that of fresh water.

COMPLEX COMMUNITIES

CHAPTER 9

Marine ecosystems and the secrets of their diversity

Part of the charm of any voyage of discovery through the ocean world lies in witnessing the varied habitats in which creatures have made their homes, and in unraveling the often extraordinary and always complex relationships between the different species. In comparison with land, the oceans offer a relatively stable environment, while presenting a quite different set of physical conditions to challenge life's evolutionary adaptability. Very subtle changes in temperature and salinity, in the provision of light and of nutrients, in water depths, current speed, and wave intensity, can combine to create extreme differences in habitat.

Coral reefs, kelp forests, and temperate wetlands, for example, all provide safe nurseries for fishes and oysters while also presenting a smorgasbord of delicacies for aerial predators. These bustling marine metropolises are full of vitality and sexual opportunity, but are also plagued with danger. It may be easy to eat or find a mate in them, but it is equally easy to end up as someone else's lunch.

It is little wonder, therefore, that colonization has pushed some creatures to the extreme boundaries of the marine world—to arctic sea-ice and the lonely shores of the high southern seas, to jagged, wave-battered cliffs, to the narrow intertidal zone on shorelines, and to the mysteriously illuminated depths of an abyssal realm we are only just beginning to explore. In such difficult or hidden environments, many marine creatures have carved their own specific niches in relative safety.

Although the marine environment is more stable and uniform than that on land, the distribution of life within it is far from even, with many contrasting habitats and unique ecosystems.

COMPLEX
COMMUNITIES

Marine Habitats

KEY TERMS

COMMUNITY

POPULATION

ECOLOGY

ECOSYSTEM

NICHE

PELAGIC

BENTHIC

NERITIC

NEKTONIC

LITTORAL ZONE

SUBLITTORAL ZONE

BATHYAL ZONE

ABYSSAL ZONE

HADAL ZONE

EPIPELAGIC ZONE

EUPHOTIC ZONE

DISPHOTIC ZONE

The stunning range of organisms that inhabit the oceans, and the interactions that link individual marine species with one another and with their environment, are the realm of marine ecologists. The ancient Greeks were among the first to conceive of this scientific discipline, and even today our word "ecology" derives from the Greek *oikos*, meaning "home."

The study of home environments in the marine world unveils a rich spectrum of many integrated communities of varied life-forms. The different populations are linked together and controlled by several biological factors—predation, competition, symbiosis, parasitism, and disease. Ecologists refer to the homes themselves as habitats, and to the interactions that join the living creatures to each other and to their physical–chemical environment as ecosystems. In this respect as in so many others, the marine world turns out to offer great diversity. Coral reefs, mangrove forests, and kelp forests are among the richest ecosystems on Earth, while the barren reaches of the open ocean, inky-black abyssal plains, and the rocky shorelines of polar seas must count among the toughest and most extreme places to survive.

Angelfish (*above right*) are nektonic (free-swimming) animals. Here they patrol a reef crowded with benthic (bottom-dwelling) organisms such as soft corals and sea fans.

Biozones Because the abiotic (non-biological) features of any given ecosystem are so important in determining the community of organisms that live in it, ocean space is generally subdivided according to its physical attributes. In the first instance, a clear distinction can be made between the pelagic province, which includes the whole of the water column and its host of nektonic (swimming) and planktonic (drifting) organisms, and the benthic province, or ocean floor, which supports a largely separate biota of its own. Divisions of the bottom extend from the littoral (shoreline) zone, to the deepest seabed of all, the hadal zone of the deep-sea trenches. A fundamental division is also made between the neritic zone of shallow water that overlies the continental shelves, and the oceanic zone of the open sea. The pelagic province can be further subdivided on the basis of depth—key to the availability of light in the ocean, and hence to the nature and distribution of living organisms.

The broad climatic belts that clothe our continents in arctic tundra, temperate grasslands, and tropical rainforest find a parallel series of divisions in the oceans. Great rafts of floating ice and perpetual winter darkness make the polar seas a harsh environment, forcing species seeking to avoid their fiercest rigors to make some of the most remarkable seasonal migrations on the planet. In comparison, the temperate zones are mild in climate but are still markedly seasonal, promoting a strong ocean circulation pattern, the overturn of water masses, and a steady recycling of nutrients. The result is a productive bonanza each spring and summer in the form of an explosive series of planktonic blooms that support many different colonies and ecosystems and can quite literally be said to feed the world. As for the tropical rainforest, its marine equivalent is an equatorial zone of permanently warm seas girdling the Earth. Although the productivity of these tropical waters is distinctly patchy and less than that of the temperate seas, it nonetheless supports some of the most diverse and rich of all marine habitats.

The ocean can be divided into two major provinces—the water column (pelagic province) and the seafloor (benthic province). Each of these provinces is further subdivided into zones as a function of both water depth and light penetration.

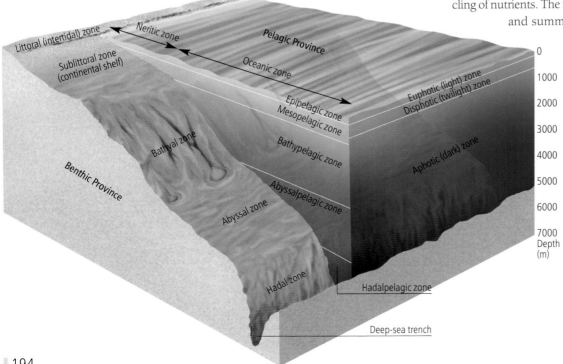

Littoral (intertidal) zone

Neritic zone

Pelagic Province

Sublittoral zone (continental shelf)

Oceanic zone

Epipelagic zone

Mesopelagic zone

Euphotic (light) zone

Disphotic (twilight) zone

Bathyal zone

Bathypelagic zone

Aphotic (dark) zone

Benthic Province

Abyssalpelagic zone

Abyssal zone

Hadal zone

Hadalpelagic zone

Deep-sea trench

0
1000
2000
3000
4000
5000
6000
7000
Depth
(m)

Finding a niche Among them, the oceans offer an almost endless range of habitats, some of which are notably more welcoming to life than others; to take an obvious example, the rich bounty of the temperate shelves has evident advantages over the near-starvation diet offered on the deep-sea floor. The prime marine habitats are, however, usually crowded; for instance, rich green seas that change color with each new plankton bloom usually support very large numbers of organisms, along with a whole series of complex food webs. Such popularity can conceal dangers. While these lush marine pastures may offer rich pickings and the prospects of an easy meal, they are also often home to many different varieties of predators, each provided with that special adaptation needed to penetrate even the most secure of hiding places or the most impenetrable armor.

In the oceans as on land, food availability often stimulates competition between species, each one eager to maximize its own share of the feast. In such circumstances, those species that are less well adapted to the particular demands of any given environment will be forced to look elsewhere for their sustenance if they are not to succumb. So Darwinian pressures of the survival of the fittest force certain organisms to colonize even the harshest habitats, where evolutionary processes of adaptation may eventually permit them to flourish. In time, even the grimmest marine desert may turn out to offer a satisfactory, if specialized, home.

Whatever its chosen habitat, each organism must also find its own particular niche if it is to survive. In an ecological context,

an organism's niche is intimately linked to its functional role and its behavior; for example, barnacles stick to wave-pounded rocks that few other creatures could survive, filtering food from the foaming surf, while sea cucumbers crawl quietly across the muddy seafloor hidden from the light, scavenging and recycling the organic waste deposited by other creatures. The more unique the niche and the less it overlaps with those of other organisms, the more successful the organism will be.

Cosmopolitan jet-setters The world's oceans provide opportunities not just to the inhabitants of their waters but also to those of the skies above. Of more than 9,000 bird species worldwide, around 250 are especially adapted for life in and around the ocean. Some have their own restricted niche in a particular marine habitat—for example, sandpipers and oystercatchers are specialist shorebirds, albatrosses and storm petrels favor windswept open-ocean waters, while most species of penguin swim the shelf seas around Antarctica in search of a fishy meal. Others are much more cosmopolitan in their choice of a home, and have become immensely successful almost everywhere, from the tropics to the polar regions. These are the gulls, and some of their relatives such as terns, skuas, and jaegers. They are noisy, aggressive, gregarious birds, efficient predators and scavengers, and garbage cleaners par excellence of the multifarious debris left by humans anywhere. In some areas they play such an important role in cleaning beaches and waterways that they are protected by law. But their success has also seen them reach nuisance levels at times.

Although the climatic regions in the oceans (*below*) are more muted than those of the continents, they are a key control on the distribution of certain marine habitats. Coral reefs and mangrove forests are only found in the shallow waters of the tropics and subtropics. Salt marshes line the shores from temperate to subpolar regions, while the polar seas offer their own harsh ecosystems. The distribution of hydrothermal vents reflects the limits of our discoveries, rather than the influence of climate.

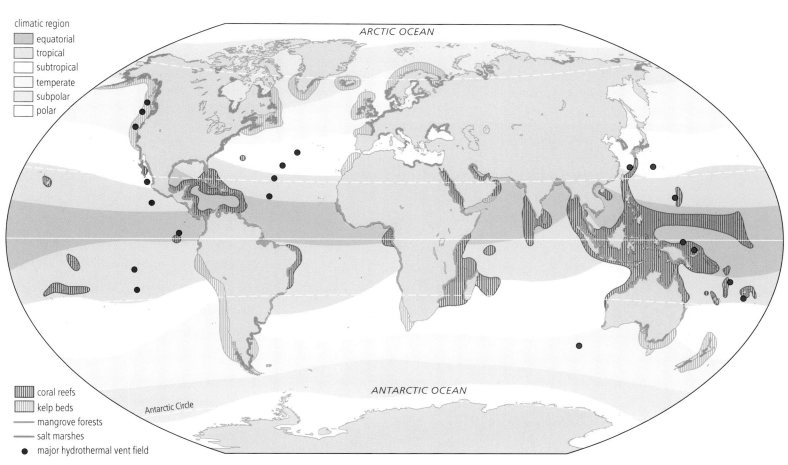

climatic region
- equatorial
- tropical
- subtropical
- temperate
- subpolar
- polar

ARCTIC OCEAN

ANTARCTIC OCEAN

Antarctic Circle

- coral reefs
- kelp beds
- mangrove forests
- salt marshes
- major hydrothermal vent field

High cliffs and fallen boulders, polished pebbles and soft sands may make attractive playgrounds for humans but they present their resident biota with some of the most challenging of all marine habitats.

COMPLEX
COMMUNITIES

Shoreline Communities

KEY TERMS

SPLASH ZONE

INTERTIDAL ZONE

COMPETITION

BEACH

MEIOFAUNA

SUBTIDAL ZONE

Along that narrow border between land and sea, there exist sparse but resilient communities made up of the hardiest and most adaptable organisms, as well as other, opportunistic animals that simply visit to feed or breed. Human populations are drawn to the edge of the sea for its scenery and convenience, and many other land animals and plants enjoy the temperate climate and fertile coastal plains, but for marine life it is a different story. For those marine organisms that make this their home, the shoreline presents one of the most extreme tests of survival of any habitat on Earth.

It is the constant change of a tidal regime, alternately submerged by seawater and then exposed to the ravages of weather and extremes of heat and cold, that offers the greatest challenge of all. Along most coasts the shoreline community experiences this dramatic alteration twice daily and even then the amount of rise and fall varies with the phases of the Moon. Winds are ever-present, and the full force of storms can be unleashed dramatically and without warning.

Breeding colonies of Cape gannets along the South African shoreline seem crowded places to attempt family life. These large, powerful birds spend practically the whole of their lives on the open sea, but crowd together, tens of thousands strong, to mate, nest, and rear their young. These, in their turn, come back to their natal colony to breed.

The splash zone Rocky coasts are often marked by clear color bands that stretch horizontally across the rocks and along the face of cliffs, formed as one dominant species or community gives way to another better adapted to life further away from the open sea. Each band, or zone, boasts a different, characteristic suite of organisms which lend it their color. The patterns of zonation vary, depending on the

slope, the tidal range, and the exposure to wave action. Above the highest high tides, the splash zone receives water only from the surf of breaking waves and spray whipped off the sea surface by onshore winds. Where winds are strong and a mighty ocean swell rolls in with waves breaking up to 50m (165ft) and more in height, then the splash zone can be quite extensive.

But very few organisms can tolerate the splash zone's relative absence of seawater, coupled with the intense, drying heat of summer days and the extreme low temperatures of winter nights. Gray and orange lichens—a partnership between photo-synthesizing algae and moisture-retaining fungi—are the most tenacious residents, along with tarlike patches of cyanobacteria. Lower in the splash zone, the filamentous green algae known as sea hair drape the rocks from spring to autumn, dying back again in winter. Above the high-tide mark, acorn barnacles of the *Chthamalus* genus dominate the available space. They have a greater tolerance for prolonged desiccation than their close relatives, *Balanus*, which are so abundant in the intertidal zone.

Only a small number of other animal species are fully adapted to life on the edge. The bivalve *Littorina* (Rough periwinkle)—a voracious grazer of algae and lichen—is typical, while isopod crustaceans such as sandhoppers and sea slaters swarm in the thousands to scavenge the least trace of organic detritus. Along the rocky shores of the Galápagos Islands they are joined by brightly colored Sally Lightfoot crabs, and by the world's only marine lizard, the Galápagos iguana. These large lizards (up to 1m/3.3ft in length) tend to be either wholly black or else mottled black and red in order to better absorb heat from the sun as they bask on the rocks above high tide. Basking is essential for these cold-blooded animals if they are to forage for food in the cold water of the Humboldt Current that swirls past the islands. To reach the densest mats of seaweed means swimming into the subtidal zone and feeding entirely below water, so that they typically swallow small pebbles, which helps them sink more easily. In common with other marine reptiles and sea birds that consume large quantities of saltwater while feeding, they have specialized tear and nasal glands for salt extraction and discharge.

The most common marine animals above the high-tide mark along the coast are only visitors—sea birds that need to rest, nest, mate, and care for their young, as well as reptiles and many marine mammals. As populations of successful species grow larger and visits from land animals become more predatory, the competition for suitable breeding sites and the pressures of predation become ever more challenging. Safety in numbers, particularly for the survival of vulnerable hatchlings, is one of the principal reasons why breeding colonies of sea birds often contain such staggering numbers of individuals. One of the remote islands of the Falklands archipelago, for example, plays host to about half a million Black-browed albatrosses each spring. The cacophony of courtship rituals and the stench of guano accompany one of nature's most spectacular displays—and this from a species noted for its solo flights across the Antarctic ocean. Even larger colonies are found among penguins—an estimated 2 million Chinstrap penguins nest on the bare, volcanically warmed slopes of Zavodovski Island in the South Sandwich group.

Exposed and vulnerable As the tide recedes, so the true intertidal community begins to face the same problems as that of the splash zone—how to cope with desiccation, temperature extremes, soaking by freshwater whenever it rains, gas exchange, and feeding. The fine filaments of green algae are covered by a gelatinous secretion that retards water loss. Mussels, barnacles, and periwinkles resist drying out by retaining water within their shells, and some animals—like the army of small fishes that invade daily to feed on the more permanent residents—simply swim out to sea with the tide.

For the less mobile creatures, however, the first challenge is to stay in one place. The force of the waves and a powerful undertow of currents have encouraged animals to evolve dorsally flattened bodies and strong adhesive mechanisms. Barnacles, which can number as many as 9,000 individuals per square meter (835 per sq ft) of rock, and common oysters attach themselves permanently to the rocks with durable calcite cement—an extension of their own shell material. Mussels moor themselves with strong, elastic byssal threads, made from proteins they secrete from glands in their foot, while chitons and limpets

hang on with suction power, and common periwinkles crawl into narrow crevices or hide in seaweed with starfishes and urchins. The seaweeds have very effective holdfasts anchoring them to the rocks, and are also able to bend gracefully with the waves, rather than trying to resist their movement altogether.

Many of these creatures, as well as hermit crabs, sea anemones, and small fishes, crowd in among the seaweed and crevices of rock pools as the tide falls. Yet seeking refuge in this way does not always turn out to be the best solution. Rockpool refugees risk becoming trapped as the blistering sun evaporates the pool to little more than a salt crust. Similarly, leaving oneself stranded within the feeding zone of a predatory dog whelk may turn out to be a fatal option.

Prey–predator relationships are not the only resource issues facing intertidal communities; there is also intense competition for living space. Though predation by carnivorous snails and starfishes limits the downward spread of barnacles, competition from mussels can be a major factor, too, while competition for space and light can limit the growth of different seaweeds as severely as critical tide levels.

Large clumps of Edible, or Blue, mussels cling to the wave-pounded rocky shoreline of Cornwall, England. Their tough, slightly flattened, blue-black shells are firmly anchored to the rocks by means of strong byssal threads, secreted from the bivalve's muscular foot. At low tide, their two valves close together tightly to avoid drying out. As the tide advances, they open up once more, filtering tiny food particles from the incoming waters.

Splash zone

Upper–middle shore

Rocky beach
1   Green lichen
2   Pale orange lichen
3   Orange lichen
4   Black lichen
5   Channel wrack
6   Bladder wrack
7   Carragheen
8   Serrated wrack
9   Oarweed (a kelp)
10 Dulse (a red seaweed)
11 Wing kelp
12 Sugar kelp
13 Sea slaters
14 Barnacles
15 Common periwinkles
16 Flat periwinkles
17 Edible mussels
18 Limpet
19 Beadlet anemone
20 Shore crab
21 Dog whelk
22 Starfish
23 Cushion star
24 Sea urchin

Low shore

25 Sponge
26 Sea squirts
27 Lobster
28 Goby
29 Sea bass
30 Sea scorpion
31 Rock pipit   Upper subtidal

A temperate rocky shoreline is carpeted with life in quite distinct zones. Even the uppermost, or splash zone, with its mosaic of lichens, displays horizontal color bands. Seaward, these give way to patches of slippery cyanobacteria and brown algae (wracks), grazed by limpets and periwinkle snails; a broad zone crowded with rough barnacles, followed by mussels and varicolored seaweeds; and then the tough, leathery kelps of the subtidal zone.

Shifting sands Sandy beaches are especially common along temperate to tropical shorelines. The sand may be golden if it originates from quartz, black from volcanic rock, or white from coral, while its consistency ranges from coarse-grained or pebbly to the finest grain size that prolonged erosion can produce. But the wave-swept sandy beach is an unwelcoming habitat; even the smallest grains can abrade the tissues of smaller organisms, and the surface is constantly shifting. Very few plants can survive here apart from a few species of seagrasses, so the upper reaches above the high-tide mark usually appear quite desertlike, occupied only by the lifeless flotsam and jetsam washed up by storm waves. Beyond, there may be wind-blown dunes patchily colonized by a hardy terrestrial flora.

The main scavengers visible in the area above the tideline are sand hoppers and ghost crabs, together with shorebirds and terrestrial insects, but a host of tiny organisms, called meiofauna, remain concealed in the spaces between the sediment particles.

Lower subtidal

Meiofauna range in size from the equivalent of fine to coarse sand grains (60 micrometers to 0.5 millimeters, or 0.002–0.02in) and include various protozoans, nematodes, cnidarians, flatworms, and polychaete worms. This hidden larder is eagerly sought after by great armies of sand bubbler crabs, which sieve through hectares of beach for microscopic organisms, discharging the unused waste as perfectly round balls of sand. Less fussy scavengers are the small, pale-colored ghost crabs, well camouflaged on the shell and coral sands of the tropical shores they prefer, which seem to appear from nowhere at the least scent of food—anything from rotting seaweed to turtle hatchlings.

A typical beach will harbor a great deal more life lower down toward the water line, although most of the permanent inhabitants are burrowers. At high tide they emerge to search for food and to find mates. Some are creepers and crawlers; others simply extend flimsy specialized appendages to filter the water for food and oxygen. Tiny mole crabs and coquina clams move up and down the beach with the tides; they use the waves to wash them forward, then hurriedly burrow into the sand and extend their appendages to filter-feed from the retreating water. Plankton is rich and varied in coastal waters—diatoms and dinoflagellates, coccolithophores and copepods, algal blooms and red tides. But the incoming tide also brings dangers: Blue crabs and Green crabs, small killifish and silversides, skates, rays, and a variety of other opportunistic carnivorous fishes from the subtidal region, all experts at foraging in the shallow, murky waters for their own particular choice of snack.

Airborne attack At low tide, the burrowers return to their underground refuges and wait, while the subtidal visitors withdraw, only to be replaced immediately by another set of eager predators—the shorebirds or waders. These birds are beautifully adapted for the task at hand, with long legs and thin, sharp bills, shaped for digging in the sand, for carefully extracting crustaceans and lugworms, for slicing through obstinate adductor muscles, for prying limpets off rocks, or for crushing crabs or cockles. Oystercatchers tend to work over the surface mollusks, redshanks dig a little deeper for clams and tellins, while curlews use their 20cm (8in) bills like forceps to reach for deeply buried lugworms and crustaceans. Turnstones are heavy-set birds with slightly upturned bills, which they use as miniature crowbars to turn over stones and other beach debris in search of food. Surfbirds and sanderlings are small but fearless, working at the very edge of the breaking waves—surfbirds take off vertically moments before the wave crashes into the sand, while sanderlings scurry about like clockwork toys across the smooth, wet surface.

Even those organisms that retreat to subtidal waters are not wholly safe from airborne attack. The coastal zone is rich in plankton, making it an ideal place for many types of fish to breed and set up their nurseries—an irresistible attraction for a wide variety of fishing birds, including divers, skimmers, the ubiquitous gulls, and the graceful, skillful terns. Common coastal birds such as gannets, cormorants, darters, frigatebirds, and boobies all range out to sea wherever pickings are richest.

Sandy beaches (*top inset*) appear deserted places at low tide. Most tracks and trails are washed away by the ebb tide, so that the only traces of life are numerous casts of lugworms. Worms and bivalves burrow into the beach to avoid detection and desiccation, becoming active only when the tide returns. Crustaceans such as sand hoppers and crabs can be found scavenging amid the debris on the strand-line.

Thick beds of mussels (*bottom inset*) are typical of many temperate rocky shores. Although still firmly attached to the rocks, many of the shells are empty, their succulent contents removed by seabirds and other predators. Even where there are no large rocks, mussels can usually gain a foothold, with individuals clinging to each other and smothering smaller organisms. However, once the colony is established, a variety of other species, such as barnacles, move in.

Sandy beach
1 Sand eel
2 Brittle star
3 Masked crab
4 Sand mason (a polychaete worm)
5 Sea potato (an urchin)
6 Thin tellin (a bivalve)
7 Sand gaper (a bivalve)
8 Lugworm

Mussel bed
1 Dog whelk
2 Stalked barnacle
3 Common mussel
4 Rough winkles
5 Barnacles

Where rivers meet the sea at the edge of the continent,
life abounds in a maze of tidal creeks and mangrove forests,
and across the naked gray expanse of mudflats.

COMPLEX
COMMUNITIES

Wetlands

KEY TERMS

ESTUARY

WETLANDS

SALTMARSH

MANGROVE

SALT WEDGE

SALINITY

ANAEROBIC

INFAUNA

The boundary between land and sea plays host to an intricate variety of shapes and contours. For every jagged headland there is a narrow inlet, for every stretch of sandy shore, a gentle embayment. Streams and rivers carry freshwater runoff, sometimes laden with sediment, dissolved chemicals, and organic detritus, into the headwaters of the inlets and embayments. These areas are known as estuaries, and the unique environments that spread out along their margins are wetlands. Although less exposed to storm and wave action than adjacent rocky or sandy coasts, they, too, are regions of constant change, presenting challenging conditions for life.

Wetlands boast fewer species than nearby habitats both landward and seaward; yet, with ample nutrients in a protected, semi-enclosed setting, they are highly productive, containing very large numbers of individuals. Interestingly, many of the animal species that inhabit wetlands originally came from the sea and have therefore evolved to regulate against excess freshwater, whereas most of the plants colonized from the land and have had to develop ways of ridding themselves of salt build-up.

Wetlands can develop around a great variety of estuaries and inlets. These range from the fjords of Norway and the Arctic to the drowned valleys of mighty tropical rivers such as the Congo or Amazon, and from large, complex embayments such as Chesapeake Bay, Long Island Sound, or the Severn Estuary to

the more modest rias of northwest Europe, into which river input is now negligible. The multiple tidal inlets of the Ganges Delta shelter extensive wetlands, as do the crocodile-infested mangroves of Florida's Everglades. Most have formed over the last 10,000 years of post-glacial rise in sea levels, as a result of the marine flooding of rivers and coastal plains, the building and breaching of barrier islands, or the inundation of land that has sunk due to tectonic adjustments, such as San Francisco Bay.

Saltmarsh maze In temperate and subarctic regions where the land falls almost imperceptibly toward the sea, great swathes of marsh grass clothe the margins, creating magical realms broken only by a maze of tiny channels

The tidal creeks and shallow lagoons, broad mudflats and expansive saltmarshes of a temperate estuarine environment (*below*) are home to a remarkable number of marine organisms. These varied habitats provide breeding sites, nurseries for the young of many different species, and feeding grounds in abundance.

Mudflat
1 Sandwich tern
2 Sanderling
3 Turnstone
4 Ringed plover
5 Mullet
6 Salmon

7 Seagrass
8 Brittle star
9 Razor shell
10 Green algae
11 Peacock worm
12 Fish fry
13 Lugworm
14 Cockle
15 Sea lettuce
16 Ribbon worm
17 Prawn
18 Shore crab
19 Wedge clam

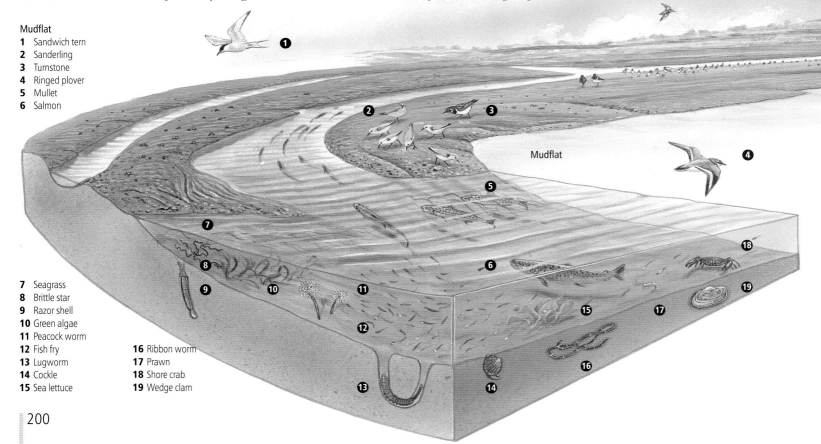

Mudflat

time, trapping more and more fine sediment so that the salt-marshes expand toward the sea.

The marsh grasses form the basis of another complex and distinct community, providing both food and cover. They are home to myriad hidden creatures—marsh periwinkles clinging to the underside of leaves and feeding on their algal coated surface, marsh snails that may number more than 1,000 individuals per square meter (100 per sq ft) without being seen, hermit crabs and fiddler crabs that crawl into their borrowed homes or carefully excavated burrows to avoid predation and the tide.

When the water recedes, hundreds of thousands of fiddlers emerge from the mud and scurry across the surface in search of food. They are omnivores and scavengers, cleaning the detritus, nibbling on algae, and catching small animals. The vast, subsurface network of burrows they create is essential for releasing trapped nutrients and aerating the dark, sulfide-smelling muds.

The rich pickings are also tempting for many kinds of land animals. Insects and spiders set up permanent residence, while snakes, rodents, Swamp rabbits, raccoons, mink, and otters are common visitors. Even Grizzly bears, fearless hunters of Alaska and remote northern Canada, will saunter across the low-tide mudflats to dig for succulent clams and worms.

Wetlands everywhere host an aerial concert. The complex melody of the skylark soars aloft, while Marsh hawks circle and hover, in search of insects and the many rodent visitors. Marsh wrens, Clapper rails, Red-winged blackbirds, and seaside sparrows all nest in the tall grasses. Marsh ducks glide silently through the tidal creeks, while the mudflats beyond ring to a veritable cacophony of beating wings and urgent cries.

through which the tides rise and fall twice daily. These intertidal lands are carefully stratified according to subtle differences in their height above low tide. Closest to shore, where only high spring tides can reach, the flora is varied and brightly colored, with succulent samphire, soft sea meadow grass, sea asters, and thrift. Fringing this zone is a thick carpet of short, fine grasses—salt meadow hay, spike grass, and cordgrass. Deeper into the tidal realm, however, only taller cordgrasses (*Spartina*) can survive. They have an extensive network of horizontal stems that grow within the mud, acting as anchors and stabilizers and, with

The fecundity of the saltmarsh ecosystem is vividly captured by this spring seascape of thrift (or Sea pink) carpeting an extensive saltmarsh on the shores of the Outer Hebrides, off the Scottish coast (*left*). The flora is more varied in these higher parts of the saltmarsh, passing seaward into monospecific cordgrasses. But everywhere the flora masks a hidden, busy world of intense activity subject only to the ebb and flow of the tide.

Saltmarsh
1 Cordgrass
2 Glasswort
3 Sea lavender
4 Sea purslane
5 Green algae
6 Seagrass
7 Brown rat
8 Oystercatcher
9 Natterjack toad
10 Dragonfly
11 Freshwater eel
12 Goby

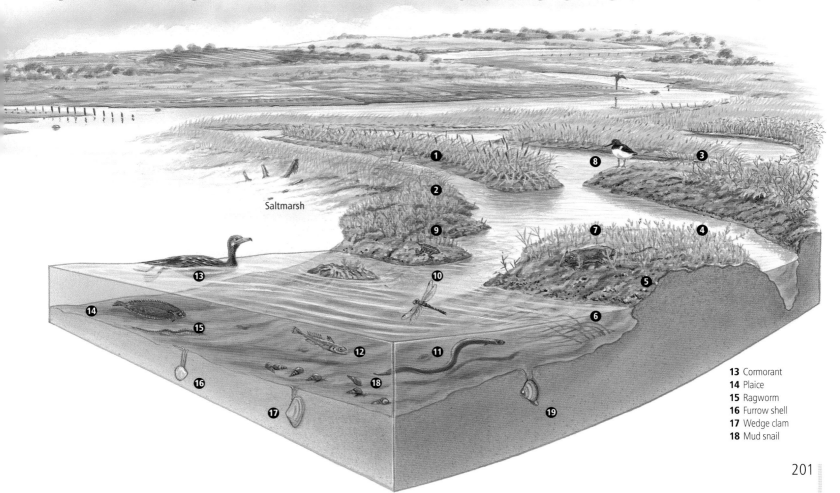

Saltmarsh

13 Cormorant
14 Plaice
15 Ragworm
16 Furrow shell
17 Wedge clam
18 Mud snail

201

The long legs and curved bill of the avocet are well adapted to wetland life. It obtains its food by stamping around on the mud bottom, thus stirring up countless small crustaceans, which it captures with rapid sweeping movements of its beak. It also nibbles on seagrasses and tiny mollusks.

The wetland environment

The biological productivity of wetlands is largely controlled by the mixing patterns between seawater and freshwater. Estuaries with a rapid outflow tend to develop a distinct saltwater wedge, with the outflowing river water lying at an angle above the incoming seawater. The salt wedge moves upstream and downstream with the rise and fall of the tide. Where the river outflow is low, however, there is more time for exchange between the water masses, and the estuary becomes well-mixed. Many estuaries fall somewhere between these extremes and display partial mixing. Distinct stratification can occur when a sill or other barrier restricts water flow and exchange across the wetland area. These physical differences affect both temperature and nutrient mixing.

Oysters are among the dominant mollusks that have succeeded well in adapting to the low or variable salinities of these environments, being able to close up completely in order to retain water when exposed. Once the free-swimming veliger larvae—those that have developed a velum, or swimming organ—have found a solid surface onto which they can cement their tiny, nascent shells, the colony is underway. Typically the larvae will settle on suitable large boulders or rocks swept clean of mud by the ebb and flow of strong currents somewhere in the network of tidal channels.

As the colony grows, new larvae attach themselves to the shells of previous generations; a large oyster reef may comprise hundreds of thousands of individuals, their shells fused into a prominent bank oriented at right angles to the tidal flow. The currents, rich in a variety of plankton, bring food to the oysters, and also carry away their waste products and ensure that the surface is kept free of sediment. Oyster reefs soon become busy habitats, providing shelter and a substrate for algae, sponges, bryozoans, hydrozoans, barnacles, sea urchins, and mollusks. Where seaweed takes hold, it can provide further food as well as protection for other species.

Estuaries are also safe nurseries for the young of many different fishes—striped bass, shad, croaker, and bluefish, for example. Young salmon pass through on their way from the river headwaters where they were born to the open sea that will be their home as adults, taking advantage of the abundant food to

The Everglades: Florida's mangrove maze

Mangrove forests, or mangals, are the tropical equivalent of saltmarshes, a remarkable in-between world where the ocean spills across the fringes of the continents. They thrive where wave action is tempered and muddy sediments accumulate, covering large tracts of coast in Malaysia, Indonesia, India, and the South American–Caribbean region. The mangals of southern Florida are especially dense, rich, and subtle. These are the Everglades—a national park and World Heritage Site that seeks to preserve a unique, fragile habitat and a range of endangered species.

Mangrove trees are uniquely adapted to an intertidal life, alternately bathed in salt water and exposed to air, with their roots swamped in soft, oxygen-starved mud. Their tangled prop (aboveground) roots are shallow and widely spread; these help anchor the plants, stabilize the substrate, and allow gas exchange directly with the air. They are also very good at keeping salt out of their inner tissues, while any that does get through is efficiently expelled onto the surface of leaves, trunks, and roots. Only three main species flourish in the Everglades, compared with up to 40 in the more diverse mangals of the Indo-Pacific region, and upward of 1,000 tree species in a single hectare (400 per acre) of tropical forest. The Red mangrove is the most pioneering, creeping closest to the water, followed by the Black and then White mangroves respectively.

The lack of floral diversity in the Everglades is in contrast to the plethora of animals that find food and shelter in the mangal maze. Mangrove roots are rapidly encrusted by coon oysters, the favorite delicacy of visiting raccoons, as well as by barnacles, oysters, mussels, anemones, sponges, tunicates, hydroids, and bryozoans. Periwinkle snails graze on the thin algal coatings that cover the roots, while Mangrove crabs feed on the leaves. Largest of the herbivores are the gentle manatees. As the tide flows in, schools of small fishes arrive.

Such rich food supplies attract many predators. Alligators, turtles, and snakes compete with land-based animals such as raccoons and otters, Black bears, and the sleek, elusive Florida panther. But it is the 350 species of birds for which the Everglades are best known. Herons and egrets, spoonbills and ibises, cranes and flamingoes number among the more magnificent waders, each with its own well-honed fishing technique. Large colonies of dip-netting pelicans are regular visitors, while the beautiful anhinga, or Snake bird, is an excellent swimmer that stalks and then impales fishes on its straight, sharp-pointed bill.

The delicate ecology of the Everglades is still at risk. While natural disasters such as hurricanes are unavoidable, there is serious competition from invasive alien species—including humans.

In the calm low-energy waters along the margins of sheltered inlets and estuaries, only the finest mud settles out of suspension. Large expanses of low-lying mudflats build up, becoming colonized by salt-marsh plants on their landward border but still subject to current erosion along the tidal channels and creeks. Although dark and fetid, and almost completely lacking in oxygen below the top few centimeters, they are home to myriad burrowing fauna—worms, mollusks, crustaceans, and others— as well as a whole world of micro-organisms.

grow and to adapt to saltwater. Juvenile freshwater eels, known as elvers, live and feed for a while at the mouths of estuaries around North Atlantic coasts, recovering from their long and remarkable migrations from the Sargasso Sea breeding grounds. The males then remain in the brackish waters while the females continue upstream, adapting to life in freshwater conditions.

Life in the mud The secret of wetlands productivity lies in the complementary mix of nutrients supplied by freshwater runoff (organic carbon, nitrogen, silica, iron) and by seawater (phosphorus, sulfur, and minor nutrient elements), as well as in the ability to retain and recycle these chemical substances. The fine-grained inorganic materials that are deposited, especially clay particles, readily absorb from the surrounding water any excess nutrients and organic matter that are not immediately claimed by plants, algae, bacteria, and plankton, and so act as a reservoir for their subsequent release. The mud rapidly becomes dark-colored and organic-rich, but also anaerobic (oxygen-depleted) below a surface layer of only a few centimeters, as there are high numbers of bacteria present, and the very fine sediment particles restrict water interchange.

Provided that organisms can adapt to both the variable salinity and low oxygen content of the water trapped within the sediment, then these organic-rich muds are an excellent source of food for all types of bacteria, as well as a prolific meiofauna and a varied infauna (burrowing animals) of mollusks and crustaceans. These in turn support the scavenging community—crabs, lobsters, and shrimps—while the nutrient-charged waters make filter feeding almost a life of luxury.

Each square meter of mud can support considerable numbers of burrowers of a wide variety—lugworms, bamboo worms, fringed worms, parchment worms, nematodes, and many more. Innkeeper worms are fat, pinkish, sausagelike monsters up to 30cm (12in) long that filter-feed by spreading a mucus net across one end of their U-shaped burrows and pumping a constant stream of water to supply food and oxygen, as well as to remove waste. These residents of California mudflats derive their name from the hospitality that they afford to several other species, including scale worms, minute pea crabs, soft-bodied clams, and tiny goby fishes. Their relationship is typical of symbiotic commensalism—the ghost shrimp, for example, takes in a very similar suite of boarders.

Like jewels in a sapphire sea, multifaceted and brightly sparkling, coral reefs are a treasure trove of marine life, quite unrivalled for their diversity, beauty, and scientific interest.

Coral Reefs

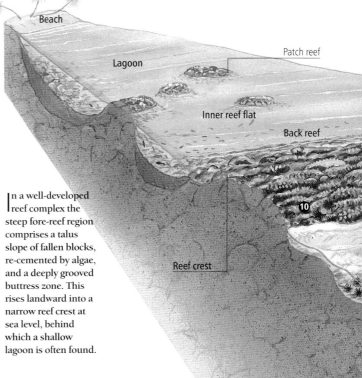

KEY TERMS

CORAL REEF

FRINGING REEF

BARRIER REEF

ATOLL

PATCH REEF

FORE REEF

REEF CREST

BACK REEF

BUTTRESS ZONE

TALUS SLOPE

ALGAE

POLYP

ZOOXANTHELLAE

PHOTOSYNTHESIS

SYMBIOSIS

HERBIVORE

CORALLIVORE

Set in the clearest of blue oceans, coral reefs' variety and profusion of life are unparalleled. These complex and intriguing ecosystems demonstrate graphically the energy of evolution and the ingenious ability of organisms to adapt and survive. They provide countless habitats and endless opportunities for life to experiment, and have done so for more than half a billion years. On at least four occasions in the past, coral reefs have all but vanished, only to be replaced each time with a vista still more intricate than the last. Their splendor today embraces an immensely long and rich heritage. Yet this fantastic richness presents a paradox, for the clear tropical waters they prefer are low in nutrients, with little of the plankton production that usually supports marine food webs.

In general, coral reefs are sparsely scattered across the oceans, for the animals that form them need specific conditions in order to flourish. They require water temperatures between 18° and 35°C (65–95°F), a solid substrate on which to build, and normal marine salinity. Clear water, free of smothering sediments, and shallow depths (less than about 60m/200ft) that allow sunlight to penetrate are also crucial. To satisfy these needs, coral reefs are distributed in a broad swathe between latitudes 25°N and 25°S across the central and western Pacific and Indian oceans, but are far more restricted in the Atlantic, whose waters are more turbid, and subject to much greater input of freshwater from large tropical rivers. Not all corals, however, build extensive reefs. Some occur as solitary individuals, while others build only small colonies. Many of these can tolerate a wider range of conditions, including temperate and even sub-arctic seas.

This magnificent fringing reef mantles the coast of Palau, Micronesia. The fore-reef baffles energy from incoming waves, while small patch reefs grow in the sheltered lagoon on the island side.

In a well-developed reef complex the steep fore-reef region comprises a talus slope of fallen blocks, re-cemented by algae, and a deeply grooved buttress zone. This rises landward into a narrow reef crest at sea level, behind which a shallow lagoon is often found.

Reef structure Ever since Darwin first classified reefs in 1842 on the basis of their structure and occurrence, we have followed his broad divisions: fringing reefs, barrier reefs, and atolls. Both fringing reefs and barrier reefs grow close to land, either island or continent, with a barrier reef being separated from the landmass by a broader, deeper stretch of sea. Atolls are generally smaller, somewhat circular and discontinuous rings of coral around a shallow central lagoon. They rise out of mid-ocean deep water, on top of submerged islands, typically of volcanic origin. In addition, there can be small table reefs in the open ocean with no central lagoon, and even smaller patch reefs within lagoons.

The three main types of reef share a common organization— a fore reef, reef crest, and back reef. On the seaward side, the fore reef rises abruptly from the ocean depths, often becoming precipitous or overhanging where the coral's outward growth has been especially rapid. The lower part of this steep slope is characterized by a chaotic talus of fallen fragments and broken blocks, while the upper, buttress zone is shaped into protruding spurs and deep grooves by the energy of incoming waves. The reef crest is where the reef reaches the surface—a generally narrow and relatively smooth zone constantly battered by the highest wave energy of all. Behind this, the reef flat extends shoreward, often giving rise to a broad, shallow lagoon, smaller pools and channels between individual patch reefs, and a seagrass bed leading up to a narrow beach of glistening coral sands.

Dynamic balance The different reef zones support a range of different coral species and other reef-building organisms, depending on the strength of the waves, the availibility of light, and the amount of sediment suspended in the water. Heavy branching corals such as Elkhorn, which can survive greater wave stress, grow higher up the fore reef. Massive, dome-shaped brain corals and sturdy, columnar pillar corals grow at intermediate depths, while the broader-spreading elephant ear and lettuce-leaf corals enjoy quieter, deeper waters

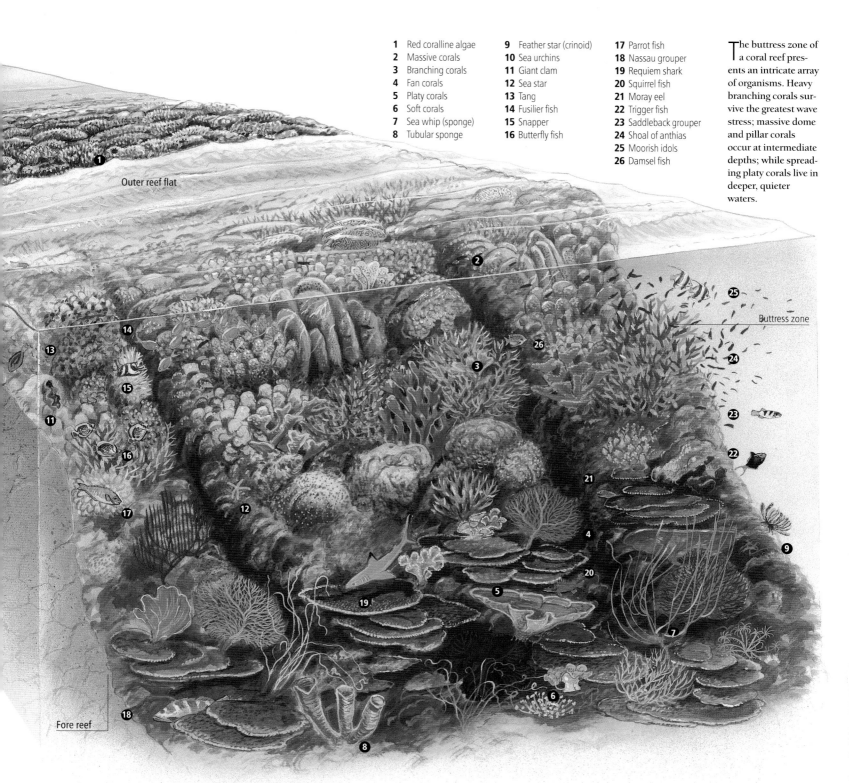

1 Red coralline algae
2 Massive corals
3 Branching corals
4 Fan corals
5 Platy corals
6 Soft corals
7 Sea whip (sponge)
8 Tubular sponge

9 Feather star (crinoid)
10 Sea urchins
11 Giant clam
12 Sea star
13 Tang
14 Fusilier fish
15 Snapper
16 Butterfly fish

17 Parrot fish
18 Nassau grouper
19 Requiem shark
20 Squirrel fish
21 Moray eel
22 Trigger fish
23 Saddleback grouper
24 Shoal of anthias
25 Moorish idols
26 Damsel fish

The buttress zone of a coral reef presents an intricate array of organisms. Heavy branching corals survive the greatest wave stress; massive dome and pillar corals occur at intermediate depths; while spreading platy corals live in deeper, quieter waters.

Outer reef flat

Buttress zone

Fore reef

below. A variety of bryozoans, sponges, soft corals, crinoids, sedentary polychaetes, and coralline algae all contribute to the construction of these towering, intricate, underwater cities.

The reef crest itself is the least favored place to live, alternately battered by waves and exposed to the tropical sun. It comprises a jumble of coral fragments thrown up by the waves and water-worn boulders, re-cemented to the reef by calcareous red algae that are particularly well adapted to life on the edge. Although protected from the full force of the waves by the algal ridge, the shallow waters of the reef flat and lagoonal area can become too hot for many species. Those that live there are the smaller forms

such as staghorn coral, finger, flower, star, and lace corals—a less diverse assemblage than on the fore reef, but with a delicacy all the more enchanting.

The whole process of reef formation involves a dynamic balance between constructive growth by this multitude of corals and associated lime-secreting organisms, and an occupying army of other animals bent on destruction. Moreover, there can be intense competition between the various types of coral. Erect branching corals are fast growing and soon spread out to shade and inhibit the growth of their neighbors, while the slower-growing massive forms slowly eat them!

Coral reefs are more than just coral; algae and mollusks contribute strongly to the carbonate reef structure, while other prominent groups are the soft corals, gorgonians (sea fans), echinoderms, sponges, polychaete worms, and a veritable kaleidoscope of fish species.

The protruding mouths of this pair of Masked butterfly fish (*right*), are well adapted for nibbling at coral polyps.

Coral polyps open, flower-like, on the branches of this soft coral (*bottom*). Each tubular polyp body has a ring of tentacles surrounding a central mouth. Symbiotic dinoflagellates (zooxanthellae) live within the tissues of the polyps (*below*), providing carbohydrates from photosynthesis.

Many types of reef fish are corallivores and directly consume the coral polyps, while other, herbivorous species such as parrot and surgeon fishes destroy the polyps in their search for algae. On the other hand, if the algae were not kept in check by grazing fishes and urchins, the more vigorous forms could soon overgrow the corals. Wherever the hard coral surface is left exposed, away from the stinging cells of living polyps, it is quickly attacked by borers—clams, gastropods, or sponges. Worst of all is the arrival of a platoon of the formidable Crown-of-thorns starfish, whose devastation of the living coral can sometimes be irreparable. This destructive activity soon breaks up parts of the reef structure into fragments, which may then be cemented together by coralline algae, to support new colonization and regeneration. But if erosion by organisms is quickly followed by aggressive wave or current action, then coral sand, silt, and even mud are eventually produced.

Secret of success Not only are corals the principal reef-building organisms, they are also the dominant primary producers at the base of these richly complex food webs. The secret of their success lies in a remarkable partnership. The tiny individual coral polyps are symbiotic hosts to single-celled dinoflagellates known as zooxanthellae, which line the inner wall of the polyp tissue. Each measures about 0.01mm (0.0004in) across, so that there are as many as 1 million cells in just 1sq cm (0.15sq in) of coral. These microscopic organisms use the sunlight that penetrates the transparent tissues of the polyp to photosynthesize carbohydrates. The zooxanthellae receive a safe, well-lit home, further protected by waving rows of stinging cells, while the coral animals can grow and flourish with an ever-ready supply of internal nourishment.

Tentacles
Mouth
Skeleton
Digestive cavity
Epidermis
Zooxanthellae

Corals are cnidarians, a group of invertebrate animals that also includes sea fans, gorgonians, hydroids, and jellyfishes. Colonies of corals begin when a planula larva settles down and attaches itself to a solid surface, commonly the broken-off remains of earlier coral or coralline algal constructions. Once attached, the larva transforms into a miniature coral polyp, with a tubular body and a ring of tentacles around a single mouth opening. In addition to its internal supply of photosynthetic energy, it can capture and feed on any zooplankton that drift within reach; these supply other nutrients. Each polyp is also able to extract calcium and carbonate from seawater and to secrete this as a hard lime cup around its body.

The life and growth of different coral species are fascinating. Most reproduction takes place asexually by a simple budding process, or sometimes by delicate branches breaking off and reattaching elsewhere. A thin layer of living tissue stretches over the limestone skeleton of the entire colony, sometimes colored pink, red, green, yellow, brown, or purple. The colony grows upward and outward, assuming the shape of its species—large cabbage structures, delicate fronds, or interwoven platelets—but modified by the force of waves and currents. At times, the whole reef erupts into an orgy of sexual reproduction as coral polyps of all different species release eggs and sperm into the water column in a simultaneous display of fecundity.

Coral city Corals and their symbiotic zooxanthellae are not the only primary producers on the reef. Several different species, such as Giant clams and other mollusks, also host vast numbers of zooxanthellae, while some of the most magnificent sponges that form a flourishing part of the reef have evolved a similar relationship with photosynthesizing cyanobacteria. Algal mats, strands and filaments, seaweeds, and seagrasses all play a part in feeding the bustling, hungry coral city.

Compared with productive, green, temperate waters, the plankton community above a coral reef is limited, but it is still able to support myriad filter feeders and whole battalions of tiny fishes—blennies and gobies, for example—that simply lie facing

Stretching 2,000km (1,250mi) along the northeastern coast of Australia, the largest complex of coral reefs and islands in the world covers an area greater than Britain and half the size of Texas. It has nearly 3,000 individual reefs and about 900 separate islands, providing homes for an astonishing diversity of life. There are about 400 different types of coral, 4,000 mollusks, and thousands more sponges, worms, crustaceans, and echinoderms. Fourteen species of seagrass grow throughout the area and provide important grazing for dugongs, and there are nesting grounds of world significance for the endangered Green sea turtle. Many other lesser-known creatures live out their lives barely noticed in the everyday commotion of this underwater megacity, while millions more inhabit an unseen microscopic world without which the greater fabric would not exist.

Some 1,500 fish species feed and breed in and around the reef. Their feeding habits range from the generalist to the highly specialized. There are plant-eating surgeonfishes with razor-sharp tails to warn off attack from behind, scavenging sweetlips and emperors that help keep the reef and its waters clean, cleaner wrasse that feed on the parasites and dead tissue of larger fishes, parrotfishes that scrape algae from the coral surface, and the ubiquitous butterfly fishes, which use their long snouts to extract small crustaceans from the maze of branching corals as well as to graze directly on the coral polyps. Barracudas, groupers, moray eels, and numerous species of sharks are among the most common predators.

One of the outstanding characteristics of reef fishes is their bright coloration. This serves several different functions: species and mate recognition, visual communication, confusion of predators, and warning that the species is toxic or unpalatable. Poison is the most widely used defense against predation. The spines of many bony fishes —rabbitfishes, porcupinefishes, and stingrays, for example—are venomous, as are those of the extremely potent, well-camouflaged stonefishes. Other species, like the pufferfishes and toadfishes, concentrate a lethal toxin within their bodies, while all members of the cnidarians (including corals and their relatives) have stinging cells in their tentacles.

The Great Barrier Reef: marine cornucopia

One of the most successful and widespread forms of coral, with more than 350 species, is the *Acropora* genus of staghorn corals, shown here on the Great Barrier Reef. Its success is partly due to its lightweight skeleton, which grows quickly and so outpaces and displaces its neighbors. It can occur as plates, tables, columns, or bushes. The main problems facing all corals on the reef, including the staghorn, are "bleaching" due to stressed environmental conditions, and periodic outbreaks of coral-eating Crown-of-thorns starfish.

Such an enormous diversity and range of adaptations reflects a fully mature ecosystem. The Great Barrier Reef is a remarkably ancient feature. It first appeared over 18 million years ago off the northernmost coast of the Australian continent, spreading gradually southward over the next 16 million years. Even the rise and fall of sea level through the past million years of ice age did not destroy the reef, but simply pushed its growth to and fro across the shelf.

By 6,000 years ago sea level had stabilized, and the reef began to develop into its present form. Growth rings on individual corals suggest that some living colonies may be up to 1,200 years old. This ancient, thriving ecosystem, with its unrivalled vitality and fragility, was granted protection as a World Heritage Site in 1981.

Virginia Chadwick,
Chair, Great Barrier Reef Marine Park Authority

---

the current with their mouths open. Many other reef animals have adopted a life of grazing, scavenging for detritus, or gorging on the mucus strands laid out by corals.

There is an almost equally rich and varied range of secondary consumers, some of which patrol by day, many at night. After the day shift retires, seeking shelter in holes and hollows, beneath overhangs, or even, like the parrotfishes, by building their own mucus cocoons, an army of nocturnal hunters swarm out from their daytime lairs. For it is during the relative safety of darkness that the zooplankton rise from the reef to feed. Immediately, they are faced by millions of stinging tentacles— those of coral polyps, sea fans, and sea anemones—and the traps, nets, and arms of countless other invertebrates.

The diverse habitats of the reefs are home to a greater number of fish species than anywhere else on Earth, of which most are unspecialized carnivores, preying on whatever comes their way. In response, most reef invertebrates have developed cryptic coloration and defensive poisons. Among the reefs' top predators, sharks are always on the alert, using their keen eyesight by day and their extraordinary sense of smell and electrical awareness by night. There are other hunters, equally effective in their techniques—octopuses with their reaching arms and deadly venom, moray eels that can make even the smallest of crevices an unsafe place to hide, barracudas and groupers that lurk in the shadows and strike with speed and precision. In all, coral reefs remain the ultimate testing ground for new survival strategies.

SEE ALSO

PATTERNS & CYCLES
Islands in Time   60–63

MARINE LIFESTYLES
The Marine Environment   172–73
Making a Living   182–87
Sexual Encounters   188–91

FRAGILE ENVIRONMENTS
Habitat Destruction   228–33

In a vast blue expanse apparently without barriers or places of refuge, athletic hunters, perfectly synchronized schools of fishes, and plankton of infinite variety engage in a never-ending game of hide-and-seek.

COMPLEX
COMMUNITIES

The Open Ocean

Synchronized swimming is an essential skill for the great schools of fish that swim the gauntlet of the open ocean. Many species, such as these herring, find relative safety in numbers. When under attack, the school responds instantly by closing ranks into a spinning ball, which shifts and changes shape in an attempt to confuse and elude predators.

The open ocean covers an area of 361 million sq km (139 million sq miles)—almost three quarters of the surface of the globe, and yet this vast expanse is sparsely inhabited. River-borne nutrients do not reach the oceanic realms and primary productivity is low, particularly in the central tropical gyres where there is little overturn of water masses. But there is plenty of food, if you know where to look, and there are creatures prepared to travel great distances to find it.

Although there might seem to be no frontiers, the open ocean in fact contains unseen physical barriers that divide it up and control the distribution of its biota. These invisible demarcation lines stem from the vertical stratification of the water column by light, salinity, and temperature, and from the horizontal variations induced by ocean currents, together with the supply of dissolved nutrients from land and local upwelling. It is these subtle differences—a few degrees of temperature here, a few parts per ten thousand of chemical concentration there—that determine the patterns of primary productivity. A cocktail of microscopic photosynthesizing organisms—cyanobacteria, coccolithophores, and dinoflagellates—is at the base of nearly all the open-ocean food webs. It is grazed by a complex zooplankton community, whose members are arranged in a hierarchy of relative size and who are in turn consumed by higher predators.

Hide-and-seek Where the ocean waters are stirred and overturned, there the phytoplankton are most abundant. Animals that feed on the plankton must track down the regions of plenty amid far greater areas of scarcity, but by gathering in localized concentrations the hunters may in their turn become hunted, in a feeding frenzy that can cost them their lives. All but the very top open-ocean predators, therefore, devote considerable effort to avoiding predation.

The defense strategy of many planktonic species is to be invisible. Single-celled diatoms and radiolarians construct tiny, see-through skeletons from pure silica glass, while the hugely abundant copepods, strange, barrel-shaped salps, and voracious arrowworms have all similarly developed transparent bodies.

The plankton community's most remarkable survival strategy, however, results in the largest mass migration on earth, involving millions and millions of tons of animals in an extraordinary daily journey. By day, the top 30m (100ft) or so of ocean are busy with phytoplankton engaged in the vital but largely invisible business of photosynthesis. But only about a quarter of the zooplankton that eat the phytoplankton risk feeding in this sunlit layer, where they can be too easily seen by diurnal, visual predators, such as various fishes, cephalopods, and birds. Instead, they come at night, journeying up from the twilight zone 1,000m (3,300ft) below, and returning at dawn to the relative safety of near-darkness. There are probably several other factors influencing vertical migration besides the avoidance of predation. Many potential predators are also migratory and closely shadow the zooplankton, while many of the migrators are bioluminescent, as visible in the dark as in the light. Conversely, many of the migrating zooplankton have transparent forms, and would be at no greater risk in daylight.

Safety in numbers The creatures that prey on the plankton must themselves in their turn seek protection, in their case from the larger ocean predators that feed on them. Small, plankton-eating fishes such as herrings have, therefore, adopted several different strategies for survival. Along with

countershading, which makes them difficult to spot, and the daily migration to the twilight zone, they also seek safety in numbers through schooling. When under attack, the school quickly closes together into a dense, swirling mass sometimes hundreds of thousands strong, twisting, turning, and darting in an attempt to confuse and evade their pursuers.

Yet schooling carries dangers of its own. Although a spinning ball of fishes may confuse predators for a while, the commotion risks attracting further unwanted attention from such higher predators as mackerel, marlin, tuna, sharks, and Killer whales, all of which may end up joining in the bloody fray. Once a school becomes the object of a feeding frenzy, the individual fishes that make it up often find themselves with no way of escape. Some may even leap clear of the water in search of a way out, only to find auks, gulls, and albatrosses attacking from above. If the school dives to take refuge in the ocean depths, dolphins, dogfish, and Killer and Humpback whales, sensing the trail of proteins and oils it leaves behind, may dive deep to find it, perhaps surprising the fishes from below or suddenly stunning part of the school with a mighty thwack of the tail.

Many of the large predators that patrol the open ocean have learned to follow the patterns of vertical migration adopted by so many of the plankton community, although they must find some way of staying warmer and therefore more alert than their quarry in the icy depths. These highly efficient hunters rely on speed, acute senses, and a deadly arsenal of weapons. Sharks have rows of razor-sharp teeth and swordfish their evil-looking toothed bills, while the great billfish such as the marlin are famed for their extraordinary long snouts.

Most of the top predators hunt alone, prepared to range over huge areas in search of food. Some, however, have learned to hunt in packs, for the combined sensory abilities of a large group dramatically increase the area searched as well as the strike rate in detecting prey. Tuna hunt in large schools, sweeping a broad area and sometimes adopting a swimming-up and gliding-down motion that covers a vertical range of 150m (500ft) or more. The Striped marlin will also work in packs, usually between 10 and 30 strong, collectively herding and trapping schools of mackerel and then taking turns at feeding.

Marine mammals, especially dolphins, are extremely social animals that often live in large herds of hundreds or even thousands of individuals, spending hours together in rest, play, and sexual activities, and communicating over long distances. They are also voracious pack hunters that combine speed and cooperation with the most sophisticated of all sonar devices. They work as a team, keeping their prey penned at the surface, and blowing a net of bubbles around them. Because of their uncanny success at locating prey in the midst of the ocean, they are often shadowed by sharks, tuna, and other fishes below the surface and also by shearwaters, albatrosses, and petrels above.

Aerial elegance The grace, beauty, and aerodynamic perfection of seabirds that spend their lives above the open ocean is legendary. The Wandering albatross has the greatest wingspan of any bird (nearly 3.5m, or 11ft) and is truly master of the ocean skies, gliding just above the turbulent waves for 90 percent of its flying time, reaching top speeds of around 90kph (56mph), and covering hundreds of kilometers per day with the minimal expenditure of energy.

The behavior of albatrosses and other oceanic birds epitomizes the nature of the open-ocean habitat. All creatures above or below the waves must keep on the move, constantly hunting for food, covering long distances to find the best feeding opportunities, and, ultimately, responding to the mating call, which may take them halfway round the world to the particular shoal, shelf, or island that is their traditional breeding ground. Most amazing of all is the Arctic tern, which makes seasonal migrations from north of the Arctic Circle to Antarctica and back again every year, spending over 10 months in continuous daylight.

Spinner dolphins on a mission. These intelligent and fast-swimming mammals are among the most accurate at locating prey in open waters, in large part aided by their highly sophisticated echolocation apparatus.

Most of our planet lies beneath the deep seas. Below the sunlit surface waters are vast realms of total and permanent darkness. Yet a great diversity of life flourishes even in these inhospitable regions.

The Deep-Sea Floor

Of all the marine communities, life in the deep sea is the most remarkable and still the least known. Even in the clearest ocean water, there is not enough light to sustain photosynthesis below about 200m (650ft); deeper still, the entire ocean becomes a zone of perpetual darkness. Heat energy is also rapidly absorbed by surface waters so that bathyal to abyssal temperatures remain near to freezing at 2–4°C (35–39°F). Added to these life-inhibiting factors is tremendous pressure. At sea level the pressure is 1 atmosphere, but since seawater is so much denser than air, the pressure increases with depth at a rate of 1 atmosphere every 10 meters (33ft). At 1,000m (3,300ft), therefore, the pressure exerted on an organism is 100 atmospheres, and at 5,000m (16,400ft) it is 500 atmospheres, or some 2.5 tons per sq cm (16 tons per sq in). A polystyrene cup lowered to this depth would return to the surface the size of a thimble.

In this inhospitable environment, there is an extreme scarcity of food. Even so, thousands of species have adapted to the harsh conditions, and as exploration continues many more will no doubt be found. Marine biologists studying the tiny animals of the seafloor, many of them hidden from sight in the pore spaces of marine sediment, have suggested they may represent a species diversity at least equivalent to that of tropical rainforests.

The twilight zone Light fades gradually through the upper ocean so that, even in the aphotic zone below the level where photosynthesis occurs, there is still some vestige of illumination down to about 600m (2,000ft) in the clearest waters. This eerie twilight soon passes into the complete darkness of the lower bathyal zone.

On the seafloor, the number of suspension feeders (sea pens, colonial octocorals, sponges) decreases rapidly downslope as the abundant organic detritus from the plankton and the constant wash of currents also diminish. They are replaced by bottom-scavenging deposit feeders such as squat lobsters and sea cucumbers. These last are one of the most common deep-sea groups—the vacuum cleaners of the ocean floor. They spend most of their lives ingesting large quantities of bulk sediment as they move slowly across the muddy substrate, extracting small, dispersed particles of organic matter and excreting the rest.

In a very few places this rather monotonous and widely dispersed benthic fauna gives way to a diversity and profusion previously thought impossible in such deep waters. In the Porcupine Bight off western Ireland, for example, the slope at about 1,000m (3,300ft) is a living, prickly carpet of densely packed hexactinellid sponges. These harbor a variety of smaller organisms and attract a range of more mobile predators. An even more unusual and less expected discovery was made in the 1990s as the oil industry began exploration on the slopes off northwest Scotland. Cold, deep-water coral communities that have been likened to the fauna inhabiting Australia's Great

There are more than 100 species of anglerfishes in the oceans. Many start out their lives near the surface, and by adulthood have moved down the water column, progressively losing the use of their eyes as they become true inhabitants of the deep sea. This Hairy anglerfish, with its squat globular shape, has a sedentary lifestyle. It simply waits for prey to arrive, forewarned by its numerous thread-like sensors.

Abyssal zone

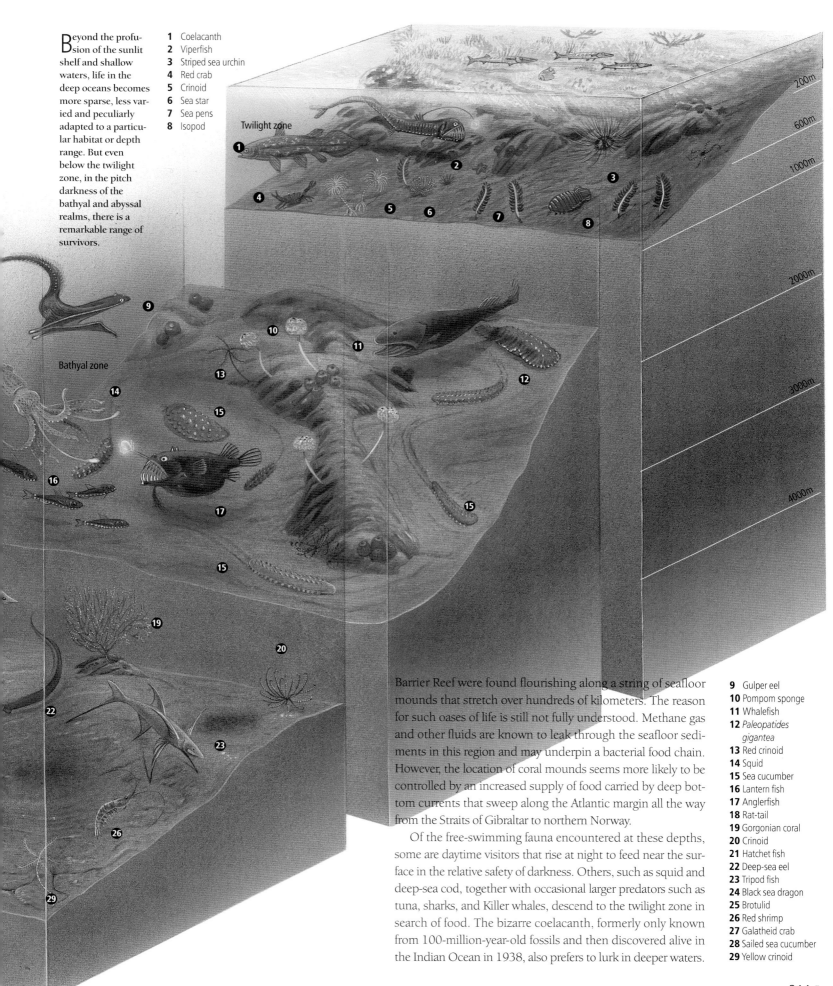

Beyond the profusion of the sunlit shelf and shallow waters, life in the deep oceans becomes more sparse, less varied and peculiarly adapted to a particular habitat or depth range. But even below the twilight zone, in the pitch darkness of the bathyal and abyssal realms, there is a remarkable range of survivors.

1 Coelacanth
2 Viperfish
3 Striped sea urchin
4 Red crab
5 Crinoid
6 Sea star
7 Sea pens
8 Isopod

Twilight zone

Bathyal zone

200m
600m
1000m
2000m
3000m
4000m

Barrier Reef were found flourishing along a string of seafloor mounds that stretch over hundreds of kilometers. The reason for such oases of life is still not fully understood. Methane gas and other fluids are known to leak through the seafloor sediments in this region and may underpin a bacterial food chain. However, the location of coral mounds seems more likely to be controlled by an increased supply of food carried by deep bottom currents that sweep along the Atlantic margin all the way from the Straits of Gibraltar to northern Norway.

Of the free-swimming fauna encountered at these depths, some are daytime visitors that rise at night to feed near the surface in the relative safety of darkness. Others, such as squid and deep-sea cod, together with occasional larger predators such as tuna, sharks, and Killer whales, descend to the twilight zone in search of food. The bizarre coelacanth, formerly only known from 100-million-year-old fossils and then discovered alive in the Indian Ocean in 1938, also prefers to lurk in deeper waters.

9 Gulper eel
10 Pompom sponge
11 Whalefish
12 *Paleopatides gigantea*
13 Red crinoid
14 Squid
15 Sea cucumber
16 Lantern fish
17 Anglerfish
18 Rat-tail
19 Gorgonian coral
20 Crinoid
21 Hatchet fish
22 Deep-sea eel
23 Tripod fish
24 Black sea dragon
25 Brotulid
26 Red shrimp
27 Galatheid crab
28 Sailed sea cucumber
29 Yellow crinoid

Hydrothermal vents: life powered by chemicals

Hydrothermal vents form around mid-ocean spreading centers where seawater penetrates the ocean crust. There it becomes superheated and also enriched with minerals leached from the rocks. These are precipitated around the vents when the hot fluids return to join the icy waters on the seafloor. The mineral clouds around the vents have led them to be called black, or white, smokers.

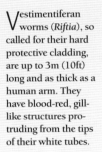

Vestimentiferan worms (*Riftia*), so called for their hard protective cladding, are up to 3m (10ft) long and as thick as a human arm. They have blood-red, gill-like structures protruding from the tips of their white tubes.

Mid-ocean ridges, with their rugged basalt outcrops and general lack of sediment cover, are the true deserts of the ocean. They lie in darkness and silence, far from any food supply. Yet in the late 1970s oceanographers discovered two thriving communities, one in the Galápagos Rift Zone and a second near the crest of the East Pacific Rise. Not only were these busy centers of life quite unexpected in such a barren seascape, but the creatures found there were among the most remarkable on the planet.

These oases are centered on hydrothermal vents that spew hot, mineral-rich waters directly onto the ocean floor. Such vents are formed at spreading centers when seawater seeps down through cracks into the ocean crust. Where it comes into contact with hot basaltic lava (near 1000°C/1,800°F), there is a violent exchange of minerals and heat. Superheated water, enriched in sulfur, iron, copper, zinc, and other metals, returns to the seafloor. On contact with near-freezing seawater a variety of minerals precipitate rapidly out of solution, building up tall, chimneylike structures.

Hydrothermal vent fluids are mainly very hot (300–450°C/560–840°F) and are full of substances normally highly toxic to life. Yet rich and complex marine communities surround the vents for a few meters in all directions. The fauna includes clams, mussels, sea anemones, barnacles, limpets, crabs, worms, shrimps, and fishes, most of them unique to vents and new to science. Both the mussels (*Bathymodiolus*) and clams (*Calyptogena*) are very large, the latter having the fastest known growth rate in the deep sea (4cm/1.6in per year).

Life positively seethes in close proximity to the hot discharge, but not all vents host the same species. In some of the Atlantic discoveries, there are no tubeworms, clams, or mussels; instead, a small species of white shrimp occurs in swarms of hundreds of thousands of individuals. Research has shown that the shrimps have a heat-sensitive spot located just behind the head, in place of light-sensitive eyes, using it to detect infrared emissions from the hot, sulfurous waters they need for their strange lifestyle, which involves harvesting gardens of

bacteria that grow on their carapace, on the lining of their gill chambers, and on their mouthparts. Other vent sites are covered in swarms of tiny white crabs or amphipods.

In any vent community the primary producers are chemosynthetic bacteria growing on and within the tissues of many different organisms. These are capable of oxidizing normally lethal hydrogen sulfide, thus providing energy to manufacture organic compounds from carbon dioxide. The entire community, therefore, is based on chemical and heat energy derived from within the Earth, rather than on the external solar energy that drives photosynthesis.

Some animals feed directly on the bacteria, while others absorb organic molecules released from the bacteria when they die. The tubeworms have symbiotic bacteria making up some 50 percent of their body weight and so have no need for a mouth, anus, or digestive system. Clams and mussels may comprise as much as 75 percent bacteria, but still retain a filter-feeding capability and a rudimentary digestive tract. Eelpout fishes nibble at the worms and clams.

Research continues, and new species are discovered with remarkable frequency. One outstanding puzzle relates to the fact that hydrothermal vents appear to be active for only 15–20 years, after which they shut down, leading to rapid collapse of their associated life. Because individual vents may be separated by hundreds of kilometers, it is not yet clear how the various animals distribute their eggs and larvae effectively, allowing their species to survive.

Professor Paul Tyler,
Southampton Oceanography Centre, UK

Deep-sea sharks have remained almost unchanged for even longer, and many of them live out their entire lives between twilight and the zone of perpetual darkness below. The largest is the sleeper or Greenland shark, growing up to 7m (23ft) long, which patrols slowly and constantly over the seafloor. The smallest is the Pygmy shark, the size and shape of a fat Cuban cigar, while one of the most fearless and aggressive is the Cookie-cutter shark that swims fast and ferociously toward any larger animal, sinking its razorlike teeth into the flesh and taking a semi-circular bite. Many of these deep-water predators are covered with bioluminescent organs, especially over their ventral surfaces, so they shimmer ghostly green through the darkness.

Into the abyss The true inhabitants of the bathyal slopes and flat abyssal floors have body temperatures close to that of the ambient water, and their metabolic rate is very slow. They move and grow more slowly, reproduce less frequently, and live longer than similar species from the surface. Most deep-water species are also smaller, but a few are true giants, perhaps due to their longevity. Sea urchins can measure up to 30cm (12in) in diameter—five to ten times the size of their shallow-water relatives; hydroids and sea pens off the coast of Japan reach amazing heights of 2.5m (8ft), while isopod crustaceans up to 20cm (8in) long crawl over the muddy seafloor. Bright red shrimps, some 20 times the size of those in our prawn cocktails, have antennae twice as long again, which they use to capture small prey.

The most spectacular giant of the deep ocean is the rare Colossal squid. Living at depths of over 1000m (3000ft) in the near-freezing waters of the Antarctic, this mollusk is the largest-known invertebrate, with the biggest eyes in the entire animal kingdom. An enormous beak, and swivelling, sharp hooks on its tentacles make it a formidable opponent for all but the deep-diving Sperm whale. Although no complete, mature specimen has been recovered, it is thought that a full-grown animal could reach a total length of 12m (40ft), larger even than the better-known Giant squid. Giant squid are mostly found in the twilight zone, but they have a strange and distant relative living in deeper waters that grows no larger than 20cm (8in). This is the Vampire squid, black-colored and with webbing between its tentacles, which appears to have descended from a group of mollusks intermediate between octopuses and squid that disappeared from the fossil record 100 million years ago. Other archaic forms that have survived in the deep sea include some sea lilies, sea urchins, and mollusks, previously known only from fossils.

For many deep-sea animals, especially the sessile ones, mealtimes are simply a matter of waiting. Every year there is a cascade of dead and dying organic matter through the water column, usually following a spring plankton bloom. Most of the debris is intercepted and recycled by organisms in the bathypelagic realm, but some does eventually reach the abyssal floor, where it is rapidly attacked. The first to arrive are the amphipods, followed by larger crustaceans, mollusks, and echinoderms. Faster-swimming predators are also attracted, such as the Rat-tail, deep-sea sharks, and the large penaeid shrimp.

Adaptations for survival Food is always scarce, so successful predators have evolved a range of useful adaptations. Deep-sea fishes such as Gulper eels and stomiatoids (dragonfishes and their relatives) possess luminous organs, elastic stomachs, and large, gaping, tooth-filled mouths. When they do locate prey, they can swallow animals much bigger than themselves. The Black sea dragon has a fleshy projection dangling from its throat, probably to act as a sensor or lure. Anglerfish have developed an array of adaptations: oversized mouths bristling with curved, needle-sharp teeth—which in some species are depressible, allowing the food to pass down and then, trap-like, springing back—expandable stomachs, bioluminescent lures that resemble fishing-rods, and several lateral lines to sense pressure disturbances caused by movement.

Avoiding predation is equally important. For many creatures of the twilight zone, near-transparency is a distinct advantage. Deep-sea amphipods and jellyfish are masters of camouflage, as are some mid-water squid and octopuses. Heteropods are small pelagic snails with transparent bodies and only the smallest vestige of a shell. They are also particularly adept predators, having rotary sets of razor-sharp teeth mounted on the end of a mobile proboscis—rather like having a mouth at the end of a trunk.

In the deep ocean light is a vital asset, and many animals have evolved ways of producing their own in the form of bioluminescence. Some have luminescent organs; others play host to bioluminescent bacteria. Lantern fishes use lights to identify their own species and gender, helping them to attract a mate, and also to spot prey. In other cases light is used defensively; some deep-sea squid and the Red opossum shrimp both release fluids that burst into clouds of bioluminescent lights, confusing predators. Most amazing of all are the siphonophores—colonies of jellyfishes that may extend over 40m (130ft) across, with dragnets comprising hundreds of stinging tentacles. For reasons that are not yet clear, they flash furiously when disturbed, producing the most spectacular bioluminescent displays ever witnessed.

Anglerfish using a bioluminescent lure to attract its prey of red shrimps. As no red light penetrates to these depths, the red color normally appears black and so the shrimps are more or less invisible. However, another predator—the small black *Malosteus* fish —also captures and eats these delicacies of the deep by locating them with a beam of red bioluminescent light.

OCEAN
LIFE

Each year, as the Antarctic ice begins to melt, Chinstrap penguins are among the first to arrive on the floes. They are waiting for the spring plankton bloom and attendant bounty of food after a long, tough winter. This blue ice has been sculpted into pinnacles by wave action. But each year, as the mean global temperature continues to rise, a little more of the ice shelf melts. We cannot yet tell what long-term effects human-induced climate change will have on the delicate balance of life here in the high Antarctic.

FRAGILE ENVIRONMENTS

CHAPTER 10

Sustainability and preservation in the oceans

From time immemorial we have looked to the ocean for food and sent our children to play along its shores. We have plied the seas for trade and war, extracted energy and mineral wealth in abundance, and returned our waste as though to a bottomless pit. Only over the last century have we become increasingly aware that the ocean world is limited both in size and resource, and that it is every bit as fragile as it is beautiful, as vulnerable as it is awesome.

The seas provide the main source of protein for more than 1 billion people, and are vital to the livelihood of island and coastal populations worldwide. Traditional fishing methods are fast giving way to industrial-scale fisheries, where a single expedition can involve a fleet of several hundred vessels that may circumnavigate the globe and remain at sea for 10 months. Under this kind of onslaught, fish stocks are dwindling, many species are now endangered, and prime fishing grounds are in crisis.

The ever-growing world population also generates copious volumes of waste. Sewage, agricultural runoff, and industrial waste account for 75 percent of marine pollution and are the main cause of damage to marine habitats. Offshore oil production, pollution from ships, and waste disposal also contribute. With more than 3 billion people living in close proximity to the ocean, the pressures on our coastlines are without precedent. Many of the world's mangrove forests and coral reefs, temperate estuaries and salt marshes, have either been destroyed or severely damaged. It remains to be seen whether we can achieve sustainable management of the ocean environment before it is too late.

Marine life has become an extremely valuable commodity.
Both hunted excessively and nurtured commercially, it is still only
partially understood and poorly managed.

The Living Resource

KEY TERMS

BYCATCH

DRIFT NET

PURSE-SEINE NET

TRAWL NET

UPWELLING

EL-NIÑO

EXTINCTION

MARICULTURE

The total quantity of fish (freshwater and marine) caught worldwide reached its highest level ever in 2000 at 94.8 million metric tons. China and Peru were the top producers, with China accounting for (a possibly overestimated) 18 percent. Marine fish, led by the anchoveta—which is profoundly affected by environmental factors such as El Niño—represent over 90 percent of the total haul.

The ocean has provided for humans ever since their appearance on Earth. Our early ancestors gathered seaweed and shellfish along its shores, caught fish, and hunted seals. Before long they were fishing from boats. For countless coastal communities, a maritime way of life grew up over the millennia around small-scale subsistence fishing. An ingenious array of techniques evolved and was passed from generation to generation.

In one short century, however, these traditions have been dramatically eroded. The steep rise in commercial fisheries through the 20th century, and in particular the exceptionally rapid growth in fish catches between 1950 and 1990, have completely changed the face of the industry. Today, fish and shellfish provide a global average of 10–20 percent of all animal protein in human diets. At one time a heady optimism even arose that the oceans might provide unlimited food supplies for a burgeoning world population. That optimism was short-lived. There is little doubt that fish stocks are dwindling, many species of commercially caught fish, crustaceans, and mollusks are now endangered, and once-prime fishing grounds are in crisis.

Traditional fisheries For more than a billion of the world's people, seafood provides the principal source of protein, and fish are also rich in vitamins B and D, and in important natural oils. In some developing countries like Bangladesh, fish provide 80 percent of total nutritional needs, while the inhabitants of some small island states survive almost wholly on protein from the sea. In such countries fishing is still mainly a small-scale affair, for subsistence and for the local market, and it rarely poses a threat to fish stocks. Despite the large numbers of people fed, these operations collectively account for only about 10 percent of the global catch.

In such communities, traditional methods—spears, bows and arrows, rod-and-line, hand-thrown nets or hand-made traps—are still used in the timeless manner of centuries past. Shells, feathers, wood, and spiders webs are all used as lures or bait, as are worms, small fishes, shellfish, and (for night fishing) bright lights. In some communities, strong-smelling substances such as aniseed, musk, castor oil, or female salmon ready to spawn are used to attract fish, playing on their strong olfactory senses. The skills and secrets that are handed from one generation to the next include an intimate knowledge of the lifestyle and habits of prey species—their migration pathways and reproductive patterns, their feeding and spawning grounds.

Fishers in Oceania beat the sea with coconut bats to attract sharks, and even harness remora as living fish-hooks to capture turtles and larger sharks. Cormorants have been trained by the Japanese since at least 813 AD to dive, catch fish, and return to their masters, with their beaks or throats partially tied so as to prevent the birds from swallowing the larger prey. Chinese fishermen have traditionally used otters to round up schools of fish rather as shepherds use sheepdogs.

The maritime nations of Europe and the seaboard states of North America have also long plied their coastal waters. From as early as the 16th century, Basque fishermen made the transatlantic crossing to exploit hugely profitable cod fishing grounds off Newfoundland, and they were soon joined there by the Portuguese, Spanish, English, and French.

Commercial enterprise For the world's richer nations, however, such small-scale enterprise has very largely vanished. The advent of steam, diesel, and progressively larger sea-going fishing vessels, together with an expanding

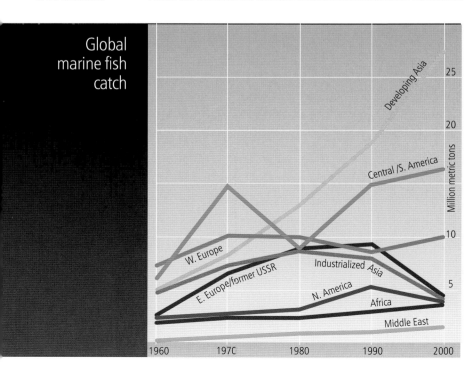

Global marine fish catch

Traditional fishing methods are still the lifeblood of many island nations and coastal communities and, for the most part, are fully sustainable. However, competition from large-scale commercial fishing enterprise is a constant and growing threat. Here at Fujaeira in the United Arab Emirates, the daily haul of sardines in these gill nets must also survive increasing oil pollution in the Arabian Gulf.

The principal fishing grounds are actually quite few in number. Half the world's annual catch is taken from the North Pacific, the North Atlantic, and the west coast of South America. In these regions, upwelling currents bring cool, nutrient-rich water to the surface, encouraging the rapid growth of phytoplankton during the spring and summer months. These microscopic organisms form the base of a rich food chain on which larger zooplankton, crustaceans, and small fishes feed and breed in abundance. It is the species that feed on these animals that are most often the target for commercial fishing. Fishing fleets are often guided by the presence of the fishes' natural predators—dolphins, seals, diving birds—but also have sophisticated technology available in order to pinpoint their quarry with accuracy. Sensitive sonar equipment is used to reveal large schools of fish just below the surface or in the twilight zone. Satellite remote-sensing techniques can provide immediate data on such parameters as phytoplankton abundance—and hence specific information on the most likely feeding grounds that day.

A catch of Orange roughy aboard a New Zealand trawler. This fish began to be commercially exploited in 1984 and soon became popular, with 63,000 metric tons being landed in 1988. However, before long catches fell, despite the opening up of new fishing grounds. Unlike most commercial species, the Orange roughy is very long-lived, taking 20 to 30 years to reach sexual maturity and reproducing only slowly, so demand rapidly outpaced supply.

global market, catapulted the fishing industry into a new world. Commercial fishing now accounts for 90 percent of the global catch. This grew from around 20 million metric tons per year in the 1950s to around 80 million metric tons by the late 1980s, but now appears to have reached a plateau. Just over two-thirds of the fin fish and shellfish taken are sold directly for food, while the rest is ground up to make fish meal and oils that are used in animal feed, as fertilizer, and in the manufacture of paint, glue, and soap. A substantial but unrecorded tonnage of species that are either unwanted or that are too small to be commercially viable—possibly between 20 and 30 million tons in all—is caught as bycatch and subsequently discarded at sea.

The top producing nations in the world today are China and Peru, followed by Japan, the United States, Chile, Indonesia, Russia, and India. Their operations are far-flung and astounding in scale. Fishing fleets that may number several hundred vessels, each accompanied by one or more gigantic factory ships, operate throughout the year in every ocean. They fish with gaping, open-mouthed trawl nets dragged above and across the seafloor, with giant purse seines, and, until fairly recently, with curtain-like drift nets that stretched over tens of kilometers. Large fishes such as tuna are sometimes caught using hundreds of baited hooks strung out on very long lines. Unfortunately, the size of catches is often matched by the scale of waste they create. Supertrawlers can scoop up 90 tons of bottom-dwelling fish in a single trawl, but then reject a quarter of the total. Where drift nets are left out for days at a time, creatures such as dolphins, whales, turtles, and seals will often become entangled in them and drown. The extensive bycatch of marine mammals and birds by drift-net fishing finally led to a United Nations ban in 1992; despite this, the practice continues illegally.

Because the fleet stays at sea for weeks or months, the catch must be processed and packaged on board. Just one factory ship is capable of salting 200 metric tons of herring, processing 150 tons of mixed fishes to fish meal, filleting and freezing 100 tons of flat fish, and manufacturing 5 tons of fish oil, all in one day.

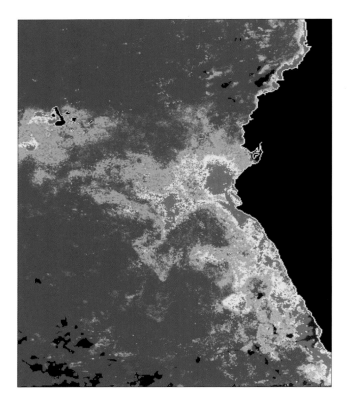

Atlantic cod has long been a favorite capture species because of the large size of individual fish and the potential catch tonnage, but both overfishing and regional warming in the northwest Atlantic have depleted stocks. From a peak in 1968, catches declined rapidly. There was a subsequent recovery from 1982 to1989, followed by collapse to the point of closure of Canadian fisheries.

Satellite image of the coastal waters off Peru and Ecuador (*right*), with artificial coloration showing the distribution of phytoplankton in the surface waters, from red (most dense) to blue (least dense). The image was produced from data acquired by the Coastal Zone Color Scanner on NASA's Nimbus 7 satellite. Such data can be relayed in real time to commercial fishing operations in the area, so that they can better determine likely fishing grounds.

Changing fortunes Several hundred different species of fish and shellfish are harvested from the sea, although relatively few types dominate the global catch. The species most frequently taken depend on the fishing grounds that are open or active, on the availability of particular stocks, and on market demand.

Through the 1950s and 1960s, cod, mackerel, anchoveta, haddock, and herring were dominant. During the 1970s and 1980s, a continued decline in the population of larger fish such as cod forced attention toward other fish stocks, including anchovies, sardines, pilchards, and capelin. Although smaller in size, these species were available in extremely large numbers. Within the last decade, the markets have seen a markedly increasing share for pollock and tuna, as well as revived trade in herring, mackerel, and anchoveta. Other catches that measure in the millions of tons include jackfish, salmon, redfish, halibut, and whiting, as well as several types of shellfish. Some of the highest earners are crustaceans, mollusks, and salmon.

One very important influence on the fishing industry is climate, as dramatically illustrated by the changing fortunes of the Peruvian fisheries over the past 50 years. Catches declined from over 14 million metric tons (10 percent of the world total) in the late 1960s to less than 2 million tons by 1984. This decline mirrored a global trend toward warmer sea-surface temperatures that was suddenly accelerated in the equatorial East Pacific by an especially severe El Niño event in 1982–83. The Peruvian anchovy fishing industry simply collapsed in the following year, with harsh consequences for hundreds of coastal communities. Interestingly, the Chilean fish catch expanded as a direct consequence of the southward shift in upwelling caused by the El Niño phenomenon.

Anchovy and sardine populations seem to alternate in dominance in such upwelling regions in response to climate and water temperature. A detailed study carried out off California looked at the type and number of fish scales preserved in laminated sediments in the Santa Barbara basin, stretching back as far as the year 300 CE. It was demonstrated that sardines prefer the warmer episodes and anchovies the colder ones.

A similar climate control is evident from historical records of the North Atlantic herring catch. During cool periods, coastal ice remains for longer off Iceland and Scandinavia, and herring stocks are forced to move south into the North and Baltic Seas. During warmer periods, the herring spread out across the North Atlantic between Norway, Iceland, and Svalbard.

Reaching the limits Climate change can drive particular fish stocks to different parts of the ocean, and even result in significant short-term decline where whole upwelling systems are temporarily shut down. Of greater concern, however, is the overfishing of particular species as well as of whole fishing grounds brought about by improved technologies and larger fleet sizes. If fish are being caught more quickly than they can effectively breed and so be replaced by younger stock, then their numbers gradually decline. Overfishing may also result in changes to the ecosystems in which the target fish reside, through alteration in prey–predator relationships.

It is not always easy to distinguish natural decline, or change due to climate and other factors, from overfishing. The sardine and anchovy stocks in the Pacific have probably suffered from both. Replacement of the drift net with the purse-seine net by European fishers in the 1960s led directly to overfishing of herring stocks, followed by various fishing bans in the North Sea. The embargoes fostered a massive rise in commercial fishing for sand eels, which duly suffered from overfishing in their turn. Years of overfishing in the northwest Atlantic caused the collapse of cod and haddock fisheries in the early 1990s, and the forced closure of North American fishing grounds. Although limited cod fishing started again in 1999, the population has not recovered as expected—a result, perhaps, of competition from dogfish or of steadily warming seas.

Whatever the reasons for their decline may be, the World Conservation Union listed more than 100 species of commercially caught fish and shellfish as threatened or endangered in 1996. The United Nations Food and Agriculture Organisation further declared, in 2002, that 13 of the world's 17 main fishing grounds were currently either overfished or being fished to the limit of sustainability. Managing fish resources first requires a very precise understanding of these limits, as well as of the long-term pattern of changes and their causes. These are admittedly complex but realizable goals of fisheries science. The second requirement is to achieve agreement among the principal fishing nations about what stocks can be taken, from which areas, and in what quantities. Setting and then policing these regulations is much more fraught with difficulty and dispute.

Hunting to extinction

The ignominious demise of the whaling industry offers a prime example of the results of overexploitation. The hunting of whales dates back many centuries, and the capture of these large mammals was particularly prized in the past. Their enormous bulk yielded many tons of high-protein meat, their thick insulating blubber was boiled down for oil to burn in lamps and to lubricate industrial machinery, baleen whalebone was used in dresses and corsets, and the nasal oils from Sperm whales served to make candles and cosmetics. But the long breeding cycles of whales, coupled with vastly improved hunting technology, has led to a very serious decline in whale populations globally.

Both the Right and Bowhead whales proved easy targets for early whalers in the North Atlantic and Arctic, so that by the late 18th century stocks were vastly depleted. Whalers then turned their attention to other species and started to use larger ships mounted with heavy guns that fired explosive harpoons, travelling as far afield as the whales' feeding grounds in the Antarctic Ocean. Whaling stations were set up on remote islands such as South Georgia, 1,300km (800mi) east of the Falkland Islands. Then came factory ships, which could process the whales while still at sea. Soon, the favored large species—the Right, Blue, Fin, Humpback, and California Gray whales—were hunted to commercial extinction. From an estimated quarter of a million Blue whales that populated the Antarctic Ocean 100 years ago, there remain only a few hundred individuals. It is uncertain whether these trends can ever be reversed. Sperm whales proved more difficult to catch so their numbers have been reduced less, while the smaller Sei and Minke whales fared better.

The International Whaling Commission was set up in 1946 to control the hunting of whales, and quickly set quotas on the numbers of whales each country could kill in a year. Policing was always difficult and some nations, such as Japan, Norway, Iceland, and Russia, consistently ignored the quotas. When a moratorium on commercial whaling was introduced in 1986, the effect was more successful, although several nations still refused to comply. The creation of whale sanctuaries around Antarctica and in the Indian Ocean should further help to conserve and renew whale populations.

A dead baleen whale has been hauled out of the water at the Hvalfjorden whaling station in Iceland; a valuable prize, it will be cut up for blubber, meat, and other products. Iceland left the International Whaling Commission (IWC) in 1986 in protest at its worldwide ban on commercial whaling. It was readmitted in 2002, having agreed to suspend commercial whaling temporarily. However, in 2003, to international criticism, it resumed "scientific whaling." Policing the oceans and ensuring that set quotas are adhered to seems to be an almost impossible task. But if firm action is not taken, whale stocks will continue to decline.

The blue revolution? By analogy with the 1960s green revolution in land-based farming, which promised and delivered so much for increased agricultural production in developing countries, the recent upsurge in mariculture (as marine fish farming is known) has been heralded as the "blue revolution." In fact, there is nothing new about the concept; the Chinese have been farming fish, shellfish, and seaweeds for some 4,000 years, and a Chinese author even wrote the first textbook on the subject as far back as 475 BCE, describing the farming of freshwater carp, brackish-water milkfish, and marine mullet.

The practice has continued through the centuries, especially in south and east Asia, where shallow ponds are dug in coastal areas in such a way that they are readily replenished by the incoming tides. These ponds are then seeded with the eggs of fish and crustaceans; in due course the larvae or juveniles are transferred to larger ponds where they are fed on larvae and plankton. Oysters and mussels also have a long history of cultivation in many parts of the world.

World mariculture production stood at around 15 million metric tons in 2001, and has now become commercially more important than freshwater aquaculture. Oysters, scallops, carpet shells, and mollusks are produced in vast quantities, as are prawns, shrimps, and lobsters. Salmon, amberjacks, and freshwater carps are the principal farmed fish species, with Japanese

Artificial reefs: life from the scrap heap

The semi-enclosed spaces within old tires provide ideal shelter for these squirrel fish on an artificial reef.

In natural reefs, corals, coralline algae, and other lime-secreting organisms build their own elaborate constructions, which then provide shelter and feeding opportunities for a variety of marine life. Human-made structures, such as harbor breakwaters and bridge supports as well as incidental debris discarded at sea, provide alternative hard substrates that are rapidly colonized. They offer new habitats for life across an otherwise barren seascape, acting as magnets for many types of fishes, shellfish, and other organisms.

It was this observation that first provided the impetus for the deliberate construction of artificial reefs in shallow seas—a practice that dates back many centuries. Today, tens of thousands of such structures exist, from small-scale reefs built by artisanal fishermen using locally available materials (rocks, trees, bamboo) in tropical countries to the creation of entire fishing grounds by the Japanese where there had been none before. Both the Japanese and the Americans are world leaders in the field, with well-advanced national development plans and substantial government investment in both research and construction.

In Japan, offshore Taiwan, and off the eastern Florida coast, artificial reefs are made from large open-work blocks of fly-ash waste, taken from oil- or coal-fired power stations, mixed with concrete. This practise, which is now becoming more common throughout the world, has been extensively tested by ocean scientists in Poole Bay, southern England. Another waste product with good potential for artificial reefs is the rubber tire. Almost a quarter of a million scrap tires are disposed of annually in the USA alone. Some have been used for artificial reefs, while others are incinerated to generate electricity. Malaysia, the Philippines, and Australia all use scrap tires for reef construction.

As the oil industry has matured in offshore areas, progressively more obsolete drilling platforms are in need of disposal. The states of Louisiana and Texas are actively involved in utilizing such platforms as artificial reefs in the Gulf of Mexico. Over 4,000 have been either toppled *in situ* or else relocated to shallow water, where they are being successfully managed for recreational fishing.

Artificial reefs have been put to many and varied uses around the world. For example, mussels and oysters have long been farmed on artificial structures. Israel, Canada, and the UK have focused on special sites for enhanced lobster fishing. Japan has designed reefs for the culture of kelp, sea urchins, and abalone. For the enhancement of wild fisheries, as well as for more intensive mariculture; in the restoration and protection of marine habitats, as well as for biodiversity management; and for the creative disposal of a wide range of inert waste materials, artificial reefs have an important role to play in the future management of our oceans.

Dr Anthony Jensen, Southampton Oceanography Centre, UK

Dr Ehud Spanier, Center for Maritime Studies, University of Haifa, Israel

Seaweed farming throughout the world now yields around 6 million metric tons each year, the greatest tonnage of all mariculture products. It is most widespread through East and Southeast Asia, but of growing importance also in India, Africa, and Latin America. It was introduced to the island of Zanzibar (*far left*) in 1989, where it is now a signficant part of the local economy. Most of the farmers are women, many earning more than their fishermen husbands.

Salmon is one of the principal farmed fish species, fetching a high price in the marketplace and destined for supermarkets and restaurants the world over. The fish are cultured from the fjords of Norway and Scotland to large coastal embayments, as here in Tasmania (*near left*). Salmon ranching, in which juvenile fish are released into the sea to be caught at a later stage, has become commercially viable in Alaska and Japan.

yellowtails, Florida pompano, and certain flatfish also playing an important role. The most valuable items in monetary terms are certain exotic species such as Giant tiger prawns, Pacific cupped oysters, and sea-urchin roes, although in terms of total tonnage seaweeds actually yield the greatest harvest. Not only are seaweeds consumed as food throughout Asia and elsewhere in the world, but they are also widely used to create the smooth texture of ice-cream, toothpaste, and paint, and to keep the foaming head on a glass of beer.

Recent years have seen a boom in mariculture production as a proportion of the global fisheries catch, and this trend will almost certainly continue in the decades to come, with China and then Japan leading the way. There is still much potential to be realized; in recent years, for instance, genetic engineering initiatives have been undertaken to enhance weight gain in certain species, to improve flavor, and to produce strains that are more resistant to disease.

But the blue revolution is not without its critics. The rights and wrongs of genetic engineering itself always spark debate. Intensive farming also increases the potential for the rapid spread of diseases, and of parasites such as the fish lice commonly found on salmon farms, and these may even spread outward to infect wild fish stocks. The use of antibiotic supplements in fish foods may counter the threat of disease but also creates its own type of pollution. Furthermore, the production of fish food itself consumes more biomass (in the form of small crustaceans and fishes turned into fish meal and oil) than is produced in the form of shrimps, salmon, and other table fishes.

With the global population exploding and inadequate management strategies in place, the vast waste receptacle that we call the ocean appears ever smaller and more vulnerable.

Marine Pollution

The sheer scale of waste generated across the globe is staggering. World population now exceeds 6 billion people, and seems set to top the 10 billion mark well before the middle of the century. Collectively we produce very nearly 6 billion metric tons of domestic waste each year, unevenly distributed between the profligate developed and more frugal developing worlds, and approximately 10 billion tons of (wet) sewage sludge. Industrial and agricultural operations, engineering works, and energy production yield waste products in equally enormous amounts.

Although a large proportion of the waste from these various sources is disposed of or treated on land, marine dumping takes its share, particularly along the coastal zone in highly populated regions of the world. Even the high seas are tainted by the human waste machine. It is estimated that close to 4 million metric tons of oil are spilt into the sea every year, stretching out along ocean shipping lanes. A remarkable 150,000 tons of plastic nets and other fishing gear are abandoned or lost at sea annually, while plastic refuse can be found everywhere from the most isolated Pacific islands to the shores of the Arctic Ocean.

The global environmental focus today looks more toward the problems of the skies than of the sea. We are understandably mindful of the 6 billion metric tons of carbon dioxide emitted into the atmosphere each year from burning fossil fuels. While burgeoning car numbers are a recognized area of concern, few people realize that sulfur dioxide emissions from the world's shipping exceeds that from all of Europe's industry. However, the oceans and the atmosphere are inextricably linked through the sea-surface microlayer. This invisible layer, rich in organic chemicals and micro-organisms, forms an oily "skin" that moderates air–sea gas exchange, but it is also highly vulnerable to airbone contaminants and other pollution .

Pollutant persistence It is not only the absolute volume of waste product that is important in determining its effect on pollution, but also the length of time that it remains in the marine environment, and the natural coping mechanisms of that environment. Organic substances, such as sewage sludge, can be dealt with by natural processes of bio-degradation and recycling in a few days or months, provided that the environment is not overloaded. Thermal pollution caused by industrial emissions and cooling plants located along the coast is also rapidly dissipated, especially where the tidal range is high and currents strong. Marine organisms have also developed effective means of dispersing and removing oil when it is present in small amounts, although large spills may persist for months to years and can cause considerable damage to marine life.

The movement of sediment—mud, silt, sand, and gravel—is a natural process occurring throughout the oceans on a daily basis. Sediment only becomes a potential pollutant when human interference markedly affects its movement or composition. Large-scale deforestation and farming may lead to sudden, excessive sediment discharge into the sea via river runoff.

Large amounts of garbage are dumped into the sea each year, from shore and ship. This Hawaiian monk seal has been caught in fishing tackle off Kure atoll, in the Pacific. It has been estimated that at least 100,000 marine mammals die each year by drowning or exhaustion after being trapped in discarded plastic waste such as this.

Principal pollution types and sources

Type	Examples	Human sources	Effects
NUTRIENTS	Nitrogen, phosphorus	Sewage outflow, agricultural runoff, airborne oxides from burning fossil fuels	Proliferation of algae leading to oxygen depletion and toxicity, harming marine life
PATHOGENS (bacteria, viruses, parasites)	Hepatitis, cholera, typhoid, diphtheria	Sewage, livestock	Contamination of beaches and shellfish, leading to spread of disease
SEDIMENTS		Erosion from mining, forestry, farming, etc. Also, dredging, water engineering	Clouding of water, leading to reduced photosynthesis and smothering of ecosystems
TOXINS (heavy metals, organohalogens)	Mercury, lead, cadmium, arsenic, DDT, PCBs (polychorinated biphenyls)	Industrial effluents, pesticides, urban wastewater discharge	Poisons coastal marine life; accumulates in food chain to cause disease and disrupt reproduction in top predators
HYDROCARBONS	Petroleum (crude oil)	Urban runoff, oil transport and shipping, refineries, atmospheric fallout, tanker accidents, offshore production	Oil spills in coastal areas devastate local biota; prolonged low-level contamination can cause disease and reduce productivity
RADIONUCLIDES	Uranium, plutonium, caesium	Discharges from power stations, atmospheric fallout, discarded military waste	Contamination of shellfish leading to diseases such as leukemia in humans
LITTER	Plastic bottles, fishing gear, disposable diapers, aluminum cans	Fishing, shipping, tourism, industrial and landfill wastes	Animals ingest plastic items or become entangled; debris can remain on beaches for 400 years

The cloudy waters that result may only remain opaque for a few days, but they can impede photosynthesis, clog the gills of fish, and smother coastal ecosystems, particularly coral reefs. In addition, dredged spoils from harbors and estuaries account for the transfer of huge amounts of sediment offshore and its continual redistribution within coastal waters. Although the dumped sediment will settle into its new site within days, dredging can also release buried organic matter and heavy metals into the water column. Other bulky, more-or-less inert wastes include fly ash from power stations and mining tailings.

Many artificial products, the detritus of our 21st-century lifestyles and of industrial proliferation, are longer-lived than natural substances and are more dangerous to the marine environment. Being synthetic and only recently introduced into the natural world, they are more resistant to degradation by bacteria and other micro-organisms, which have not had time to evolve the biochemical systems needed to cope with them. Certain pesticides, detergents, and other organo-chemicals are very persistent over periods ranging from years to many decades, while the effects of radioactive isotopes and toxic metals can last for centuries or even longer.

The sewage debate Sewage is a rich cocktail of dissolved and solid material—organic and nitrogen compounds, ammonia, and concentrations of heavy metals that are often many times higher than those found in the surrounding medium. It also contains and attracts an army of disease-causing organisms such as viruses, bacteria, and parasites, while the natural bacterial decay of the sewage can lead to oxygen depletion and water toxicity.

Historically and regionally, sewage has been subject to a variety of regimes—treated or untreated, disposed of on land or at sea. Until recent times, major urban populations—for example those of the Los Angeles Bay region or the New York East Coast conurbation—have dumped countless millions of tons of sewage, much of it quite untreated, into their coastal waters. The same has been true for the European cities that surround the North Sea, and for the growing population fringing the Mediterranean. Sewage input into this inland sea is especially heightened during the tourist season.

The recent setting of pollution controls through Europe and much of North America requires better sewage treatment and the progressive cessation of most marine dumping. Yet the beautiful harbor city of Halifax in Nova Scotia, Canada, still pumps its untreated waste directly into the Northwest Arm inlet adjacent to the main downtown area, maintaining that natural processes adequately cope with the effluent. The majority of developing-country coastal cities do exactly the same, although the reasoning is generally economic rather than environmental. Inland towns and cities often pump sewage directly into the river systems, and some of the partially degraded solids and dissolved chemicals find their way ultimately to the sea.

Sewage is a natural organic waste product, which is rapidly decomposed by microbial action. Controlled marine dumping of pre-treated sewage sludge in offshore areas, therefore, poses little threat to the environment. But for the children (*above*) playing in raw sewage on Gaza beach, Palestine, there are very serious health hazards from the many disease pathogens that flourish in such conditions. The same is true for large coastal populations in many developing countries.

Although the worst zones of marine pollution border urban centers and industrialized areas, human contamination is gradually spreading into the open ocean, particularly along the major shipping lanes. Shipping accounts for about a third of oil pollution in the ocean, although it is the more spectacular tanker accidents that grab the headlines.

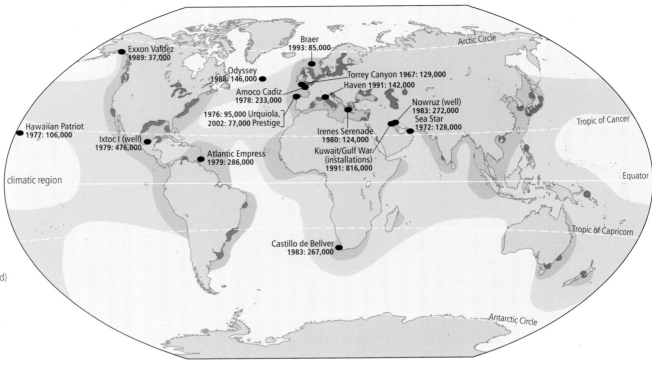

Exxon Valdez
1989: 37,000

Braer
1993: 85,000

Arctic Circle

Odyssey
1988: 146,000

Torrey Canyon 1967: 129,000
Haven 1991: 142,000

Amoco Cadiz
1978: 233,000

1976: 95,000 Urquiola,
2002: 77,000 Prestige

Nowruz (well)
1983: 272,000
Sea Star
1972: 128,000

Hawaiian Patriot
1977: 106,000

Irenes Serenade
1980: 124,000

Tropic of Cancer

Ixtoc I (well)
1979: 476,000

Atlantic Empress
1979: 286,000

Kuwait/Gulf War
(installations)
1991: 816,000

Equator

climatic region

Castillo de Bellver
1983: 267,000

Tropic of Capricorn

Antarctic Circle

major oil spill, with
date and tons lost
(tanker unless indicated)

marine pollution levels
- constant and high
- frequent and moderate
- intermittent and low
- minimal

Workers in the Gulf of Mexico attempt to contain a fire and oil-spill on the Norwegian oil tanker *Mega Borg*, in 1990. Such spills are dramatic and can be devastating in their local effect. But more significant long-term and worldwide are the chronic oil pollution problems due to routine ship operations and run-off from land.

The consequences of overloading the marine environment with sewage can be quite dramatic. The influx of abundant nutrient elements creates a flourishing micro-cosmos of competing organisms. and a consequent rapid reduction in oxygen levels. Water eutrophication occurs, together with reduced biotic diversity, shellfish poisoning, and an increased likelihood of toxic phytoplankton blooms. These "red tides" release toxicants into the water that may spell catastrophic death for marine mammals, fishes, and invertebrates, as well as encouraging the emergence of certain common disease pathogens. Hepatitis, typhoid, dysentery, and enteritis are all associated with sewage contamination, as is a worrying new form of cholera that has appeared in several Asian nations.

Oil pollution Until the coming of the age of hydrocarbons in the latter part of the 20th century, oil seepage into the marine environment came almost entirely from natural sources. It was readily dispersed and biodegradable. But the world now is a very different place. At the outset of the 21st century, oil and gas are being consumed at the alarming rate of 6 billion metric tons per year—roughly the amount it has taken nature 1 million years to produce! Against this figure the spillage of a few million tons a year into the vast ocean reservoir may seem small, but the released oil's concentration along shipping routes and coastlines, in harbors and areas of major offshore oil production, is certainly cause for concern. Such long-term, low-level contamination, particularly by refined oil, can kill larvae in coastal nurseries, cause a variety of diseases among marine organisms, and generally lead to negative changes in population structure and dynamics.

The majority of oil spillages result from routine operations such as loading, discharging, or bunkering, coupled with runoff from cars, heavy machinery, industry, and other land-based sources, yet this chronic pollution never makes headlines. At worst, tar balls wash up on nearby beaches and tourism suffers.

Far more newsworthy and immediate in their effect are the intermittent accidents associated with production and transportation—well-site blow-outs, or collisions, groundings, hull failures, fires, or explosions involving the supertankers that ply the energy arteries of the world. The enormous size of these vessels results in spills of staggering proportions. The largest ever was caused by the collision between the *Atlantic Empress* and the *Aegean Captain* off Tobago in 1979, when some 286,000 metric tons of crude oil emptied into the Caribbean Sea.

Although several other disasters have approached this in scale, it was the *Exxon Valdez* incident in 1989 that alerted public opinion to the potential risks. One small error by the captain

Living organisms are largely composed of just four elements—hydrogen, carbon, nitrogen, and oxygen—together with a number of minor trace elements present in only very small amounts. Many of these are metals, including calcium, phosphorus, magnesium, iron, copper, cobalt, and zinc, and all are essential for the normal metabolic functioning of the organism. They are active components of metallo-enzymes, enzyme activators, vitamins, proteins, and respiratory pigments. Deficiency or excess may lead to depressed cell function and death. Other trace elements known as heavy metals—mercury, lead, arsenic, and cadmium—are highly toxic to most life even in very small doses (often less than 1 part per million).

All these elements are normal components of seawater, introduced from the natural weathering of rocks, volcanic emissions, and hot fluid discharges at mid-ocean ridge vents. Problems can occur for living organisms where concentrations of toxic heavy metals have been significantly elevated, generally as a result of the introduction of contaminated waste materials into the oceans. Disposal occurs both from untreated sewage and from treated sewage sludge, in widespread dredging operations that inadvertently recycle metals safely buried in sediment, and in runoff from mining or industrial operations. Heavy metals, such as methyl mercury, have a further tendency to build up in the food chain, forming stable organic compounds that collect in the tissues of organisms (bioaccumulation). They are then increasingly concentrated at each successive trophic level (biomagnification), passing from microbes and other single-celled organisms, through filter and detritus feeders, to progressively higher predators.

That mercury is toxic to humans is well known. "Mad" hatters in Victorian England got their name from neurological symptoms due to chronic poisoning by the mercury employed in the treatment of felt used for the making of hats. But the origin of diseases related to mercury and other heavy metals, as well as to a variety of trace elements, has only been systematically studied in the last half century. Minamata disease was first reported from the small coastal town of Minamata in Japan in 1953, apparently confined to fishermen and their families. The symptoms were numbness, speech and hearing impairment, failing eyesight, and loss of coordination, all due to progressive disintegration of brain cells, and culminating in coma and death. It was not until several years later that its cause was eventually traced to the effects of methyl mercury introduced into the bay from a plastics factory. At least 400 people died and more than 2,000 were left with permanent brain disorders.

Itai-itai disease was also first reported from Japan in the 1950s. It is an intensely painful kind of osteomalacia (softening of the bones) caused by an excess of cadmium in the diet. Mollusks and squid are known to accumulate large amounts of cadmium. Lead is notorious as a harmful toxin on land, leading to reduced neurophysiological functioning, and it has been found to bioaccumulate selectively in certain marine organisms. There are instances of whole communities facing slow poisoning from arsenic contamination of drinking water, for example in West Bengal, India. The toxins induce a variety of disorders including cirrhosis, anemia, fibrosis, and painful skin ulcers. However, lasting or demonstrable effects of lead and arsenic contamination via seawater are very rare.

There is still a great deal to be learned about the distribution and effects of heavy metals and other trace elements in the oceans. Different chemical forms of the same element have very different effects: methyl mercury is, for example, far more toxic than metallic mercury, but its toxicity is lessened by complexing with selenium; chromium (III) is an essential element in humans, whereas chromium (VI) is toxic. Numerous diseases are now known to be caused by a deficiency of important trace metals, for example selenium, copper, zinc, chromium, magnesium, and others. These are exciting times for biochemists in the medical profession working alongside ocean scientists.

Professor Bill Fyfe,
University of Toronto, Canada

Heavy metals: small but deadly

Effluent from a pulp mill washes up on the coast of northwest Tasmania. Methyl mercury, a deadly poisonous heavy metal, is one of the by-products of the paper-pulp and chlorine-alkali industries, as well as being an ingredient in antifouling paints used on ships, in pesticides, and in some pharmaceutical products. It resists degradation by microorganisms and remains in the environment for a very long time.

of this 274m (900ft) supertanker shortly after leaving the Valdez oil terminus in southern Alaska led to the loss of 37,000 tons of oil at sea. Over 2,400km (1,500mi) of previously unspoiled high-latitude coastline became black and desolate almost overnight. An estimated 4,000 Sea otters, 500 seals, and 20 Killer whales perished, together with as many as 1 million seabirds. Exxon-Mobil was fined $5 billion (later ruled as excessive) and paid over $2 billion for clean-up operations. The long-term biological cost is still being calculated.

Accidents to oil wells at sea are rarer but potentially even more damaging. Blow-outs occur when a highly pressured reservoir of oil several thousands of meters below the seafloor is penetrated with insufficient weight of drilling mud to keep the oil in place. The worst disaster of this kind occurred beneath the drill rig Ixtoc I in the Gulf of Mexico on 3rd June 1979. The resultant outflow was not brought under control until 24th March 1980, almost 11 months later. During this period 476,000 metric tons of oil escaped; oil slicks covered large tracts of tropical sea and washed up on beaches in southern Texas.

Probably the worst single oil disaster ever was, in fact, a deliberate act of ecological terrorism by Iraq's former President Saddam Hussein as his troops retreated from Kuwait in 1991. More than 600 well heads were set ablaze by the Iraqis. Plumes of black smoke filled the air for many months, and thick layers of soot and sludge were found more than 3,200km (2,000mi) away in the Himalayas. Oil refineries along the Persian Gulf were sabotaged, creating one of the biggest oil slicks ever known and causing catastrophic environmental damage in a region of the world already under severe stress from chronic oil pollution.

While there is little doubting the obvious and immediate effects of major disasters such as these in terms of the marine life that perishes, the beauty spots that are tarred and beaches closed for tourism, and the fishing stocks that are poisoned and depleted, the long-term effects on marine ecosystems are still poorly known. In regions of persistent oil pollution, however, including many inland and marginal seas as well as the coastlines adjoining heavily populated or industrialized areas, there is little doubt that marine life is permanently affected.

DDT concentration (parts per million)

Zooplankton 0.04
Minnows 0.5
Needlefish 2.0
Ospreys 25.0

DDT contamination decimated the osprey population during the 1950s and 1960s. Marine predators at the top of their food chain, ospreys in Long Island Sound were found to contain over 600 times the amount of DDT found in zooplankton, near the base of the food chain. Biomagnification of toxic substances is, unfortunately, common in marine ecosystems.

Organohalogens Two miracle chemicals were introduced in the 1950s and 1960s and were lauded at the time for their beneficial properties. These were DDT (dichloro-diphenyl-trichloro-ethane), the chief component of pesticides widely used in agriculture, and PCBs (polychlorinated biphenyls), used in paints, plastics, adhesives, hydraulic fluids, electrical appliances, and aerosols. Each has been widely distributed via the atmosphere into rivers and seas throughout the world.

Unlike petroleum hydrocarbons, both of these chlorinated hydrocarbons are not easily degraded by bacterial or chemical action and are very long-lived in the marine environment. They become attached to particles of silt, which are then ingested by filter feeders. Relatively insoluble in water, they dissolve and remain in the fatty tissues of living organisms and become increasingly concentrated along the food chain as one animal eats another—a process known as biomagnification. They inhibit normal growth and cause infertility in many vertebrates. Alarming concentrations of the chemicals have been found in many different organisms around the globe, from the penguins and whales of the Antarctic Ocean to Arctic seals and Polar bears, and from seabirds and dolphins at the ocean's surface to the rat-tail fish of the abyssal depths.

The toxic effects of DDT were soon recognized and graphically portrayed to an unsuspecting public in Rachel Carson's poignant book *Silent Spring* (1962). Its use was banned first in Hungary in 1968, and then by the United States in 1972; other developed countries soon followed suit. PCB usage has also declined considerably, but both chemicals are still actively sold in developing nations, so their transference to the global ocean continues. Their toxicity is not in question, although scientific debate continues as to the amount of each the oceans can absorb before their adverse effects become irreversible.

Unfortunately, dozens of similar organic chemicals (generally linked with one of the halogen elements, such as chlorine) are newly introduced every year before their long-term environmental effects are fully known. Tributyn (TBT) was one such; extensively used in wood preservatives and as an anti-fouling agent in marine paint, it too has been shown to cause serious damage to marine life such as shellfish, and its use has now been banned or curtailed in some countries. A variety of synthetic fertilizers are considered essential in many parts of the world for maximizing the yield of food and other crops. Their application, together with that of insecticides and herbicides, all contribute to organic runoff that eventually finds its way into the neighboring seas.

Industrial plants contribute all manner of other chemicals, including large amounts of nitrogen and phosphorus compounds, fluorides, cyanides, acid and alkaline cleaners, solvents, pigments, plating salts, and phenols. In the USA alone, approximately 1,000 new chemicals are marketed each year, and around 60,000 chemicals are currently on the market. Over half of these are officially classified as being definitely or potentially hazardous to human life.

Interestingly, a number of lawsuits have recently been brought against large companies for polluting the oceans. A notable and successful one was initiated by the Los Angeles Surfrider Foundation against the Louisiana–Pacific Corporation and the Simpson Paper Company, whose industrial effluents included dioxins and other toxic compounds from their paper mills near Eureka. In addition to paying a large fine, the companies have agreed to invest over $50 million in reducing the effects of such dangerous pollutants.

Dumping at sea The oceans have always been the ultimate repository for human waste, as they are for sediment, the natural detritus from the continents. It is only as civilization has "advanced" and populations have expanded that we have become more acutely aware of how directly we influence our environment: land, ocean, and atmosphere. For several centuries now, the seawater canals of Venice, the narrow waters of the Bosporus that have long divided Asia and Europe, the headwaters of the Persian Gulf, and several of the major estuaries and bays of China have been recognized as posing particular health hazards for the local populations. The number and intensity of such pollution hotspots is rapidly increasing.

Dumping at sea is one of the principal reasons for this upward trend. Radioactive waste and toxic chemicals are particularly dangerous and long-lived, and marine dumping of the former was discontinued in 1983. A new menace currently threatening the high seas and coastal regions alike is the global obsession with plastics and packaging. Countless plastic containers and garbage items are tossed overboard, or find their way into the marine environment from beach litter and landfill sites. Plastic does not degrade easily; sea creatures either eat it or become hopelessly entangled. As many as 1 million sea birds, and 100,000 seals, dolphins, and whales, are thought to die each year as a result.

Managing waste It should not be difficult to manage our waste products effectively, if only we have the intellectual and economic will to do so. Attitudes to waste management have evolved in the developed countries since the beginning of the Industrial Revolution, varying markedly over time. For the first hundred years or so of the industrial era, the tendency was to dilute and disperse the refuse. As volumes and toxicity increased, however, the policy became one of concentration and containment, although leakages were often difficult or costly to prevent. Now we are slowly being forced to consider fresh strategies again: the most ecologically sound option is to reduce the volume of waste produced, to recycle an increasing proportion of materials, and to re-use as much as possible.

A key element dictating policy with regard to waste is economic. Waste management, especially recycling and re-using, is extremely costly, and in purely economic terms is generally non-viable. In order to ensure that we preserve the environment for future generations, there is a clear need for strong national directives and international treaties. Many such agreements have indeed come into being since the early 1970s, banning, phasing out, or controlling the dumping of many waste materials at sea. Groups of countries have adopted regional plans to start cleaning up their common seas—the Baltic, North, Caribbean, Mediterranean, and South China seas, for example. Their success is just beginning to be seen in the marine world. There is reason for cautious optimism, but we cannot afford to take our eye off the ball.

Litter lies on the formerly pristine shores of Svalbard—the inevitable result of our global dependence on plastics and packaging. Far within the Arctic Circle, even these remote islands receive their share of garbage. Not only is this kind of waste unsightly, it is also potentially harmful to the marine creatures and seabirds that may consume it.

For marine scientists, there are still many questions that need urgent answers, especially with regard to the potential use of deep-ocean sites for waste disposal. For example, how would oxygen concentrations be affected by the deep disposal of organic-rich waste? How much damage might be caused to deep-sea life by the introduction of toxic chemicals or radioactive waste onto the deep ocean floor? Should we consider the introduction of excess carbon dioxide into deep water as a means of slowing down global warming, and what effects might this have? These are just some of the very complex issues requiring further research in order to better advise on the future role of the oceans in waste management strategy.

Effluent from a chemical plant being discharged into the North Sea at Hartlepool, on England's northeast coast. An increasing amount of industrial waste, which pours daily into the marine environment, is classified as conservative pollutant. This is non-biodegradable and not readily dissipated, and so tends to accumulate in marine organisms.

The oceans are vast, but their richly varied coastal fringe is fast showing signs of environmental stress under the growing weight of urban sprawl, industrial development, and marine tourism.

Habitat Destruction

The Marine Species Population Index charts the changes in population of 217 species of marine animals. Between 1970 and 1999 there has been an average decline of 35 percent. That this appears more pronounced in the southern oceans is due to the fact that the major losses and damage in the industrialized north had already taken place.

History is littered with the excesses of human activity, just as it is punctuated by magnificent achievement. Our apparent determination to dominate the natural world is tempered only slightly by an increased understanding of the environmental issues we trample across with such abandon in pursuit of our goals. The voice of reason and restraint, born out of scientific insight and ecological awareness, is all too often drowned out by society's need and greed. The world order is currently dictated by political and economic imperatives rather than by environmental concerns.

Set against the remarkable progress made by humans is the often appalling mismanagement that has blighted great tracts of the world's land area. Continental habitats that are now facing the unknown effects of global warming have already suffered irreparably from a variety of different blights: marching desertification and deforestation; urban sprawl and the industrial wastelands it has created; soil erosion and biodiversity loss; and the ill effects of a century marred by warfare. Now the oceans also are under threat as never before. There is a growing concentration of population and industrial development along the coastline, coupled with a rapid expansion of marine tourism and with rising pollution of already stressed marginal seas. The net result is increased environmental degradation and habitat loss.

The oceans are vast. Unbelievably large expanses of the middle-ocean and deep-sea worlds are still little known and are quite untouched by human hand or deed. It must be the firm duty of ocean scientists not only to continue to explore and understand these wild frontiers, but also to argue strongly for their cautious use and careful stewardship.

Coastal concerns The coastal zone embraces a wide variety of habitats including rocky shores and sandy beaches, barrier islands and sheltered lagoons, muddy estuaries and fertile delta plains, salt marshes and seagrass beds, coral reefs and mangrove swamps. Together, these regions boast a biodiversity second to none, play host to marine productivity of the first order, and are the most gravely threatened ecosystems. About 90 percent of the world's total seawater fish catch reproduces in coastal areas. At least 60 percent of the world's population—over 3 billion people—live close to the sea, and two-thirds of all cities with populations in excess of 2.5 million have been constructed near estuaries or deltas. The population of the coastal zone is expected to double by the year 2020.

There is no doubting the changes already wrought by human development in coastal regions, nor can there be reason to question the increased ecological stress that a growing population density will cause. Seasonal excesses due to the booming tourist industry have an especially deleterious effect in many parts of the world—mostly where sunshine and golden sand is on offer, but now also in some more remote corners of the globe, including lesser-known tropical shores and ice-clad polar regions. Tourism is one of the world's boom industries, with an annual turnover of $4 trillion. It is also, collectively, the world's largest employer, estimated to generate, directly or indirectly, nearly 200 million jobs, or around 10 percent of global employment. A sizeable proportion of those jobs, and of the estimated 1 billion tourists, is focused on or close to the sea.

Inevitably the urban, recreational, and industrial developments involved have brought about significant habitat change and destruction. Hotels, villas, airports, roads, and causeways all

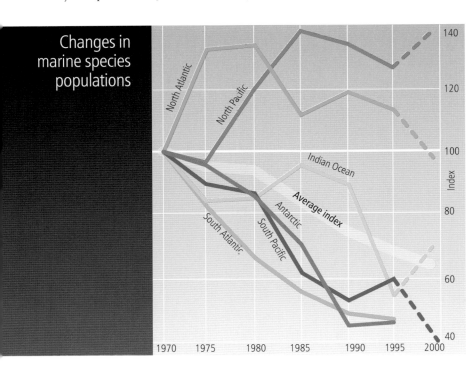

Changes in marine species populations

North Atlantic

North Pacific

Indian Ocean

Average index

Antarctic

South Atlantic

South Pacific

Index

140

120

100

80

60

40

1970 1975 1980 1985 1990 1995 2000

Crowded beaches are a common sight throughout the world, as hundreds of millions of tourists flock to the coast each year. This booming industry poses the single largest threat to coastal habitats globally. For the big city beaches—Los Angeles, Rio de Janeiro, or Sydney, for example—the sheer quantity of raw sewage and industrial and urban runoff presents the greatest problems, but resorts everywhere face a growing concern.

d'Ivoire, with Sierra Leone not far behind. In Southeast Asia, they have been decimated to make way for the urban development of Singapore, and have been progressively reduced by around 80 percent over the past century in the Philippines and Myanmar. In Thailand, Malaysia, and Indonesia, the decline has been more recent, caused mainly by the development of mariculture, as well as by felling for the woodchip industry. The Sundarbans, fringing the shoreline of the Ganges and Brahmaputra deltas, are the world's largest remaining mangrove ecosystem. Much of the area has been protected for over 125 years as a key habitat for many different bird, reptile, amphibian, and mammal species, including the Bengal tiger. The mangroves are also vital for the regional fishing industry and the livelihoods of millions of people.

There are now serious conservation efforts underway in some countries, together with major replanting programs. Some of the islands offshore from the Ganges Delta are being replanted in this way, as are many kilometers of the Vietnam coastline. The largest destruction of mangroves in that country resulted from spraying with the herbicide Agent Orange during the Vietnam war; there has been virtually no natural regeneration since. The Everglades mangroves and other wetland regions of Florida suffered badly in the past, although they are now protected in part by World Heritage Site status and in part by careful management and regeneration programs.

These mangroves (*below*) have been killed as a result of road construction along the northern shores of Colombia, which resulted in a significant change in drainage patterns and water salinity. The loss of mangrove swamps worldwide is a serious problem, not only because of habitat loss and biodiversity reduction, but also because they are a very valuable national asset when managed properly. Where preserved, they provide revenue and local jobs in forestry, fisheries, and tourism.

take space as well as using up local construction materials, as does housing for the influx of workers. Groynes, breakwaters, and jetties are specially built to modify the coastline, preserving sandy beaches in one area but leading to sediment removal elsewhere. Over 40 percent of Japan's very extensive coastline has already been modified by human activity. In Bali and Sri Lanka, there has been much use of coral-reef limestone as a raw material for the cement needed for the tourism-related construction industry. There are countless other examples.

Human activities farther inland can also seriously affect coastal habitats. Widespread deforestation and land cultivation lead to increased sediment discharge via river systems, as is the case for many rivers in China and Southeast Asia, where excess sediment has been known to damage and even destroy estuarine habitats such as saltmarshes and seagrass meadows. Off the Malaysian coast, tin waste from inshore mining operations is carried with the sediment, blanketing over local coral reefs. The construction of large dams across major rivers has the opposite effect, reducing the outflow of sediment, which can lead to serious problems of coastal erosion, as has occurred in the Nile Delta and with the Colorado River in the USA.

Mangroves and wetlands Mangrove forests are the intertidal wetland ecosystem most under threat in tropical and subtropical regions. These unique environments, with their salt-adapted trees and tangled mats of spreading roots, are among the most fecund and productive in the world. Not only are mangroves vital spawning grounds and nursery areas for fishes and shellfish, but also the feeding grounds of many seabirds and the home of many other organisms.

They have suffered catastrophically, however, as a result of clearance for agriculture or urban and industrial uses, as well as being cut down for firewood and charcoal. In West Africa, they have been almost completely cleared from Ghana and Côte

The coral reefs of Hawaii (*right*) are a diver's paradise and so-called "eco-tourism" has become a mainstay of the islands' economy. If the beauty and wonder of these and other marine habitats are to be preserved for future generations, the term must become more than just a fashionable label. We have to achieve sustainable tourism guided by ecological awareness.

Measures to conserve the wetlands are not purely altruistic in their aims. In fact, cost–benefit analysis of the Thai and Malaysian mangroves has demonstrated clearly that undamaged forests are between two and ten times more valuable, when carefully managed for traditional fisheries, sustainable forestry, and ecotourism, than they are when destroyed and converted for use as rice paddies or for aquaculture or oil-palm plantations. Typically, such developments also require costly engineering works to prevent coastal erosion and flooding.

Reefs at risk The most beautiful and complex of marine ecosystems are coral reefs, covering only 0.2 percent of the ocean floor but supporting around 25 percent of all known marine species. Unparalleled in their biodiversity, reefs in one form or another have been an essential part of the ocean world for more than 500 million years.

From the human economic perspective, the reefs act as an energy baffle for waves and storms, protecting the coastline from erosion and allowing sandy beaches to accumulate while also providing shelter for harbors or lagoons. Like mangrove swamps and wetlands, they play an important role in the fishing industry, and are vital to many island subsistence economies. Progressively they have come to epitomize the romance and excitement of marine tourism.

It is tourism, in particular, that has heightened awareness of the plight of these irreplaceable habitats. Coral reefs are being degraded at unprecedented rates, largely as a result of human activities. Of the 109 countries with significant coral reefs, at least 90 have reported damage. One of the principal problems is land reclamation and the building of harbors, airports, and hotels near to reef complexes. Parts of some reefs are even mined extensively to provide construction materials—surely one of the more perverse and short-sighted examples of environmental mismanagement! Another serious concern is the increasing volume of silt supplied to coastal regions, smothering large areas of the living reefs and so destroying most of the life they support and, ultimately, the reefs themselves. The silting results directly from land clearance and deforestation in the hinterland.

Coral reefs in many parts of the world are at serious risk from over-collecting, over-fishing, recreational use and other coastal development, mining, siltation from land clearance and soil erosion, and increasing pollution. Coral bleaching (*below*) is a sign of environmental stress, believed to be related in some cases to a rapid rise in sea-surface temperature.

In a typical situation, more and more of a reef may be mined for the souvenir trade as tourism escalates; giant clams, pearl oysters, and other colorful mollusks are similarly taken as attractive trophies. At the same time the reef may be broken and damaged by careless boating, while fish stocks can be seriously depleted by scuba divers and spear fishing. Seasonal pollution becomes ever more serious, to the point where the ecosystem never fully recovers between peak holiday seasons.

Global warming is also beginning to affect coral reefs. Reef-building corals are extremely sensitive to changes in water temperature. A rise in the incidence of coral bleaching—where the coral animals expel their symbiotic zooxanthellae—has been noted. It is thought to be a response to stress induced by raised water temperatures and can be fatal when severe. The rise in sea level that would accompany such warming might also spell disaster for the reefs.

Dying seas Enclosed and semi-enclosed seas marginal to the major ocean basins support their own range of marine habitats, which mirror those of the larger oceans. There are estuaries, deltas, and all types of coastal environment, as well as narrow shelf seas and broad areas of open water over shallow platforms, steeper slopes, and small basin plains. Some of the marginal seas in tectonically active areas—the Pacific Ring of Fire or the Caribbean, for example—even boast their own spreading centers, hydrothermal vents, and nearby trenches. For the ocean scientist, such areas are natural marine laboratories that enable us to understand more easily the biology and dynamics of whole oceans.

When such seas are surrounded by high populations or have become centers of tourism, however, they become subject to problems of chronic pollution. The whole suite of marine habitats they comprise is gradually put under greater and greater strain; the seas are slowly poisoned around the edges, and marine life is permanently affected. The stress results in part from very high levels of pollutants, many of them carried long distances downstream from inland centers of population or industry, and in part from the generally restricted circulation of some marginal seas. There is less ready exchange of water

masses with the global ocean, and hence less chance for renewal, re-oxygenation, and cleansing.

Completely isolated inland seas, such as the Great Lakes of North America or the Caspian in central Asia, may quickly become completely polluted and more or less dead. The Gulf of St. Lawrence, the North and Baltic seas, the Gulf of Arabia, and the Yellow and East China seas all contain some or many habitats that have been destroyed through long years of industrial, urban, or agricultural waste. The Mediterranean and Black seas have been affected both by long-term industrial and demographic growth and, more recently, by the huge seasonal influx of tourists. Parts of the Caribbean are similarly stressed.

Vast blooms of phytoplankton have become a commonplace and serious effect of excess nitrogen and phosphorus pollutants derived from sewage discharges, fertilizers, and some industrial wastes. Large amounts of these nutrient elements cause sudden growth explosions of phytoplankton, quickly followed by similar booms in the herbivorous zooplankton communities. Some of these plankton blooms are quite harmless, but others, known as red tides, produce neurotoxins that can wreak havoc with marine life higher up the food chain. In 1988 a layer of toxic plankton 10m (33ft) thick spread along the narrow straits between the North and Baltic seas, coating beaches with slime and leading to the deaths of millions of fish.

Red tides are no longer rare occurrences. The northern Adriatic is particularly prone to summer blooms, following years of chronic pollution from the River Po. In bad years, beaches of the Italian Riviera and the Yugoslav coast have to be closed and tourism suffers. Toxic blooms occur several times a year in the waters around Japan, especially in the Seto Inland Sea. Several types of edible shellfish are known to accumulate the toxins without being adversely affected themselves. Human ingestion can lead to paralytic shellfish poisoning, as well as to other neurological, cardiovascular, and gastro-intestinal symptoms.

Another effect of repeated plankton blooms in marginal seas is to slowly use up all the dissolved oxygen at depth, as the dead plankton sink through the water column and bacterial decomposition takes over. The Black and Baltic seas are both at or near oxygen minimum levels for most of the year, rendering the deeper levels more or less void of macro-organisms.

Pressure at the poles Wild and inhospitable yet strangely beautiful, the polar seas are also desperately fragile. While the Arctic's land creatures suffer temperatures of -40°C (-40°F) and less, the ocean waters rarely fall much below 0°C (32°F). As winter sea ice breaks up across the broad Arctic shelves, almost a million marine mammals swim through the Bering Strait to feed and summer in these richest of waters. At least 50 million birds migrate to their favored nesting grounds on deltas that fringe the land of the midnight sun.

The Arctic seas have long attracted human interest of an adventuring or commercial nature, drawing hunters in pursuit of prized narwhal tusks, whales, walruses, seals, and Polar bears. Steller's sea cow was discovered and exterminated within less than three decades in the 18th century. Some seabirds have also been hunted to extinction. Today, most species of mammals and birds that inhabit the far north are protected, and their numbers show a very slow recovery. The region remains one of the world's more prolific fishing grounds, and the coastal waters off Alaska and Canada are also spectacularly rich in hidden oil. Some of the most northerly land areas—Svalbard, Greenland,

Finland, and Siberia—are equally rich in mineral deposits. The exploitation of all these resources, living and non-living, is placing ever-greater environmental pressures on polar ecosystems.

It is fortunate, therefore, that the world still respects the Antarctic Treaty, first signed in 1959, which sets the continent aside for scientific investigation, suspends territorial rights, prohibits economic exploitation, and seeks to protect the environment. It is vitally important that we should continue to do so, although the waste left by research bases is already fouling the nest on a small and local scale. The Antarctic Ocean, however, has suffered rather more. First seals and then whales (Blue whales, Humpbacks, and Fin whales) were hunted nearly to extinction before protection measures were introduced.

The Antarctic Ocean is immensely fertile. Spring and summer blooms of phytoplankton feed countless billions of shrimp-like krill, whose gigantic swarms color the waters pink by day and shimmer with phosphorescence by night. The world's major fishing nations ply their trade in these rich waters, turning now to krill harvesting in addition to fish, as certain stocks are being pushed beyond the limits of sustainability.

Agents of change The destruction of natural habitats is not, of course, a uniquely human achievement. Cyclones in the Indian Ocean can wreak terrible damage on the mangrove forests of the Sundarbans in just a few hours. As hurricanes unleash their power across the Caribbean, large tracts of coral-reef communities suffer in their wake. Submarine earthquakes may cause massive undersea landslides, tearing asunder whole segments of deltaic ecosystem and

Cruise ships are a growing part of marine tourism. They can help create a respect for and awareness of the ocean world through their choice of destinations and through onboard educational programs. This cruise liner, anchored in Neko Harbor on the Antarctic Peninsula, is one of many that now sail to far-flung parts of the globe, bringing a sense of awareness as well as wonderment to thousands of passengers at a time. But such ships are like mini-cities, with the potential to leave damage and waste in their wake.

Ecotourism is the single fastest-growing sector in one of the world's booming multinational industries, and yet its credentials are fast slipping. So-called ecotours are now available to the most remote parts of the Earth's wilderness areas—the snowfields of Antarctica and the Himalayan roof of the world, for example—as well as to some of its most popular destinations, such as the pyramids of Egypt or the wildlife reserves of East Africa. Even lunar tourism is now being proposed. Whereas mass tourism usually spells environmental stress and ultimate disaster, ecotourism is supposedly more sensitive and environmentally friendly—but does it work and can it be sustained?

The marine environment is particularly popular with aspiring ecotourists. Remote island locations, Galápagos wildlife oddities, Australia's Great Barrier Reef, whale-watching cruises, and Red Sea diving are among the principal attractions on offer. But there are still very few operators who offer genuine marine tourism with a conscience.

Ecotour operators should be mindful of the ecosystems in which they are operating, and of the health and protection of the global ocean environment. They need to follow sustainable practices ensuring minimum waste production and the economic well-being of the local community. They must raise the environmental awareness of their clients, at the same time as introducing them to some remarkable aspect of the marine world. It is a delicate balancing act to achieve all this, but even large-scale operators can go part way to achieving such aims.

Of the many hundreds of tourism companies operating in the Caribbean, only one or two dive operators can demonstrate a truly sustainable, ecological approach. Yacht charter companies may stress the environment-friendly nature of sailing in their brochures, but money and deadlines tend to take precedence. The much heavier tourist traffic to the highly stressed Mediterranean region is almost wholly without marine eco-sensitivity—the norm is for high-density beach resorts, offering convenience, luxury, and marine activities such as jet-skiing and power boating. There is every temptation for many of the Pacific and Indian Ocean island

nations to turn to tourism to satisfy their development aspirations. The challenge is for them to introduce true, small-scale ecotourism rather than the largely unplanned and unsustainable scramble for development already witnessed in, for example, parts of the Maldives.

New Zealand's people and government take marine ecotourism very seriously. Kaikoura on the South Island is a fine example. Upwelling of nutrient-rich water off the east coast results in an abundance of marine life, from tiny plankton to various cetacean species at the top of the chain. Sperm whales are resident year-round, growing up to 18m (60ft) long and weighing as much as 45 tons. Regular visitors include the mighty Blue whale as well as pilot, beaked,

minke, Bryde's, Humpback, Fin, Sei, Southern Right, and Killer whales. There are at least five different types of dolphins, large numbers of New Zealand fur seals, and occasional visits by Leopard and Southern elephant seals. Vast numbers of seabirds invade the skies—albatrosses, penguins, petrels, shearwaters, shags, cormorants, terns, and more. Kaikoura's status as the country's marine ecotourism hub is undisputed, but there has been no gold-rush style competition for trade. Instead there are effective management plans, fully involving local residents, for achieving sustainable tourism in tune with the ocean environment.

Dougall Crockett, Kaikoura, New Zealand

Marine ecotourism: a delicate balancing act

To see marine mammals, such as this Gray whale, in their natural habitat is a memorable experience. Whale watching is now a major part of the tourist industry, attracting over 4 million people per year. Surprisingly, Japan now has one of the fastest growing whale-watching industries in the world, so that before long live whales in Japanese waters might be more valuable that the dead ones taken by the whaling industry.

---

sending them cascading downslope to bury forever the deep-basin habitat below. Dormant volcanoes may suddenly erupt, blowing apart a whole island and sending giant tsunami across an entire ocean basin. Such is the ever-present power of nature, yet the devastation that it occasionally wreaks is rarely permanent; the damaged habitats can usually recover and rebuild.

Stepping back in geological time, we can see that the Earth's climate has swung dramatically between icehouse and greenhouse conditions over the past million years and more. Climate change is one of the most effective and permanent means of altering ecosystems completely. As the polar ice caps alternately contract and expand, sea levels rise and fall, causing coastlines around the world to change their position and nature. As one coastal habitat vanishes beneath the waves, so another takes its

place. These are natural cycles of profound change, played out over the millennia.

Ever since the advent of humans on Earth, but mainly in the last few centuries of our history, the style and pace of change has been influenced by our own actions. Clearly we are now contributing to very rapid global warming, which will undoubtedly lead to the melting of polar ice and a steady rise in sea level throughout the oceans. All sorts of physical and biological effects will derive from these planetary modifications. With global warming, as with all the other human activities that are altering or destroying natural habitats, we need to decide quickly and definitively the goals that we are proposing to accomplish. International agreements have to be made and adhered to if we are to stand any chance of achieving them.

CONCLUSION

FUTURE CHALLENGE

In recent times the explosive growth in human population and technological progress, coupled with a deep lack of concern for the environment, have seen our interaction with the natural world enter a new dimension. We now know that we can change climate and bring about mass extinctions, but we are not yet certain whether we can reverse such trends once set in motion. Looking ahead, we can see a time of great danger, but also one of opportunity. If scientific understanding can be coupled with wise management, we may be on the threshold of a new era—the Age of the Ocean.

The previous chapters in this book have provided a vista on our ocean planet as we know it today—from the tumultuous outpourings of molten rock that create the seafloor to the awesome power of currents and waves that swirl above it, from the colorful kaleidoscope of life teeming along its sculpted shoreline to a dark, deep-sea world that few have ever seen. Yet even though we have learned much in the relatively short period for which systematic ocean exploration has been a scientific discipline, we have also come to appreciate that all our knowledge opens only a tiny view of what the French pioneer Jacques Cousteau once called "the third infinity."

Science has barely dipped its toe in the oceans. There are submarine mountain chains that no one has climbed and shallow sea plateaus and abyssal deserts that lie unexplored. Indeed, almost the whole undersea world remains unseen by the human eye. Almost certainly, too, there are oases of life not yet found, whole encyclopedias of new species undiscovered, vital clues to the past and to the future as yet unknown, and vast untapped resources awaiting the attention of a greedy or hungry global population. The slow but steady progress made by ocean science both in discovering and in explaining the astonishing world beneath the waves is the fundamental challenge that permeates and thrills the oceanographic community.

We know that the oceans play a critical role in maintaining life on the planet and in moderating its climate. They also provide food, energy, transportation, and leisure activities for humans, as well as acting as gigantic septic tanks for our domestic, industrial, and other wastes. We are also aware that the oceans present an inescapable ecological unity. Huge, interconnected cycles of energy, climate, chemicals, and life processes are forever in motion, so that a coastal input in one part of the world can translate clear across an ocean basin to become an output elsewhere. These cycles are finely balanced, and the balance is under threat from human activities. But the timescale and degree of change

are poorly studied, and the ability of the ocean system to buffer our worst excesses is completely unknown.

Well-founded concern

Knowing that the oceans are crucial to securing our common future on the planet, we have every right at present to feel a well-founded concern. Even though systematic exploration has only recently begun and our understanding of the ocean system is in its infancy, people of all nations are already dreaming up a thousand ways in which the oceans could support us through the third millennium. The "blue revolution" in aquaculture has been vaunted as a panacea for world hunger; deep-water oil and gas have been heralded as providing short- or medium-term solutions to the fossil-fuel crisis; offshore sand and gravel resources are believed to be more than enough to construct the megacities and tourist resorts of the coming century. Some scientists have earmarked thick piles of sediment on the seafloor as future garbage containers for toxic waste, or even as repositories for surplus carbon dioxide in our battle against global warming.

The truth, however, is that we do not yet know enough to turn such dreams into reality. We are playing games with the unknown, anticipating ocean properties or ocean responses well before we can be certain that the game will play out in the expected way. Human need and human greed are outstripping the pace of understanding and backstaging the thirst for scientific knowledge. The concept of ecological unity in the earth–ocean system is fast being replaced by one of economic unity.

The use and abuse of the oceans will ultimately affect every individual on the planet, making it far too important an issue to leave in the hands of politicians. Already ocean-related topics—El Niño and the ocean–climate links, the melting of the polar ice shelves, increased storminess at sea, oil tanker disasters, coastal pollution—are becoming ever more prominent in the world's media. It is therefore a vital challenge for ocean scientists to ensure that the facts and issues are correctly reported, and that ordinary people as well as politicians have the knowledge needed for sensible decision-making. Developing a keen global sensibility is the responsibility of us all as world citizens.

A look into the future

The potential benefits to humankind from an improved scientific understanding of the oceans are immense, always assuming that they are coupled with sound management practices. So it is interesting to speculate, albeit fancifully, on some of the advances that we might possibly expect for our children's children by the end of the 21st century.

In full awareness of the many different scenarios that are equally possible, one might venture a Utopian vision and predict that the 21st century will, in fact, turn out to be the Age of the Ocean. With a global population controlled at, say, 16 billion, just over 50 percent of all the world's people will live within 1 km (0.6mi) of the coast. A simple process for harnessing wave, tide, and ocean-thermal energy will provide ample fuel for all needs. Controlled mining of gas hydrates offshore will yield abundant

clean drinking water from their ice-like lattice structure, as well as short-chain hydrocarbons for the plastics industry. All transport will be solar-powered. The tourist industry will have opted for underwater hotel chains and leisure parks rather than for space stations. Long-distance ski-noeing—a subaqueous combination of skiing and canoeing—may even become popular down the continental slopes.

Human and animal organic waste materials will be used to fertilize vast mariculture farms spread out across the continental shelves. The artificial reefs that form the framework of such farms will be constructed from solid, non-toxic waste. This blue revolution, and a planetary diet based increasingly on items low on the food chain, will have helped solve the problem of global nutrition. Toxic chemical waste-products will be fed to the chemosynthesizing organisms that surround deep-sea hydrothermal vents and will be broken down by microbial action into harmless products. Medical advances in the control of common diseases will have come from the widespread use of bacteriophages, cultivated as a byproduct at marine farms and vent sites.

The climate will be warmer and more equable. Following much squabbling between nation states about the ideal temperature range for the planet, the secret of the ocean–climate link will finally have been unraveled and the technology of climate control mastered. Restricted parts of the world—the polar oceans and Antarctic continent—will be kept cold and ice-bound, while the rest will enjoy relative warmth and adequate rainfall. The carbon dioxide tap, which controls flux between ocean and atmosphere, will be carefully monitored by a global institution named, perhaps, the World Ocean Forum.

The remarkable variety and scale of life found within and below the oceans will lead to the recognition that a pattern of simple lifeforms exists everywhere in space and that every now and then a suitable substrate such as planet Earth allows some to proliferate and evolve in complexity. Many illnesses, from common colds to Asian flu and the AIDS virus, will have been found to have extraterrestrial origins; some such may have led to the mass extinctions of the past.

Perhaps most important of all, the Age of the Ocean will see the demise of the nation state. It will finally be recognized that issues of world health, nutrition, energy, and climate, as well as the eradication of world poverty, warfare, and religious divisions, can only be managed by cooperation rather than competition. Earth itself may ultimately be renamed "Ocean."

Although some of these speculations may seem very wild, many are already possible. What is abundantly clear is that we will have to transcend the self-interest of nation states and multinational corporations in the century ahead if we are to embrace a broader common interest—that of the whole ocean planet.

CLIMATE CHANGE

There is now a general consensus that the Earth is warming rapidly under a cloak of greenhouse gases. Certainly, the world seems to be hotter at the turn of the 21st century than at any time in the last seven or eight hundred years; the 1990s were the warmest decade since reliable records began, and 1998 probably the warmest year. There is also widespread agreement that humankind is now the principal contributor to this atmospheric build-up. What is more controversial is the apportionment of blame for the process, and the steps that need to be taken if its worst effects are not to be realized over the coming decades.

Over the past 100 years, the global average surface temperature has increased by about 0.6°C (1°F), and it is predicted to rise by a further 1.4–5.8°C (2.5–10.5°F) over the next century. Even such small changes can have dramatic effects on already fragile ecosystems. Average sea level has risen by 10–20cm (4–8in) through the 20th century as polar ice sheets have begun to thin, and it is expected to rise a further 25–100cm (10–40in) by 2100 as melting continues, not least because seawater expands slightly as it warms. In a worst-case scenario, a great many island nations would vanish altogether, along with a number of major coastal cities—Shanghai, Venice, and New Orleans, for example. Some 15 percent of Bangladesh would be submerged, leaving 15 million people homeless. Coastal erosion and flooding would threaten homes, farmland, and natural habitats in many low-lying areas.

Extremes of weather are already becoming more common, and the trend can be expected to continue. In middle to high latitudes of the Northern Hemisphere, cloud cover has increased and heavy rains are more frequent. Ferocious tropical cyclones and temperate-area storms seem likely to intensify further, bringing torrential rains to some areas and drought to others. Some of these changes are positive, causing harsh climates to become more equable, greening desert areas, and providing longer growing seasons in high latitudes, but the regional patterns of change, their influence on global productivity, and the longterm ocean–climate response are still very poorly understood.

Cycles of change

The greenhouse effect is entirely natural and has been operating for billions of years. There is a constant exchange of greenhouse gases and heat energy between ocean and atmosphere, while the circulation of both serves to distribute heat from the tropics to the poles and so to moderate extremes of climate. The warmer and colder periods of Earth history that geological

evidence has unraveled are due to a complex interaction of causes, including variations in incoming solar radiation, the biological and volcanic pumping of greenhouse gases into the atmosphere and their retention in the ocean storehouse, and changes in patterns of circulation and hence in the distribution of heat energy. However, scientists are not yet able to explain the ultimate triggers for the five principal cold-planet periods of the past billion years. We can be more confident in ascribing the alternation of ice-age and greenhouse conditions within each of these epochs largely to variations in planetary orbit, affecting incoming solar radiation over cycles respectively of about 20,000, 40,000, and 100,000 years. By this reckoning, the Earth appears set to return to ice-age conditions within the next 10,000 years.

On a shorter timescale, the transfer of heat, moisture, and other gases between the ocean and atmosphere drives oscillations in climate. The El Niño Southern Oscillation (ENSO) is the best known cycle, causing climate variations over periods of three to seven years that are driven by the interaction of sea-surface temperature and trade winds along the equatorial Pacific. The effects are worldwide in their impact, causing floods in Europe and North America and monsoon failure and severe drought in India, Southeast Asia, and Australia, as well as decimating the anchovy fishing industry off Peru. The more recently discovered Interdecadal Pacific Oscillation (IPO), centered on the North Pacific, creates climate cycles with global influence over a period of one to three decades. The causes and effects of IPO are still being studied.

Both physical and biological pumps then act to remove carbon dioxide from the ocean surface, so that it remains out of contact with the atmosphere for long periods. The physical pump works because cold water, charged with dissolved carbon dioxide, is heavier than warm water and so sinks to the seafloor. The biological pump operates through the uptake of carbon dioxide by planktonic organisms to build their soft tissues and construct hard skeletal material. After death, this carbon-rich matter sinks through the water column as a marine snow that is progressively broken down by bacteria and chemical corrosion as it falls.

The oceans are in a state of perpetual motion. Heat energy, carbon dioxide, and particulate matter transferred from the surface to the abyss are all caught up in the silent stirring of the deep sea. Deep cold waters are moved relentlessly from the poles toward the tropics. Together with surface currents, these lower flows act as a gigantic global conveyor belt, shifting the planetary store of heat and so moderating climate. Even small changes in ocean temperature are known to have marked effects on ocean circulation. The climatic results are still hotly debated.

Models and controls

Global warming and its anthropogenic drive are no longer in doubt. What is remarkable is that human activity alone has been able to affect, at least marginally, such a complex ocean–climate nexus. Atmospheric levels of carbon dioxide have increased by 31 percent since 1750, following widespread industrialization and an exponential increase in the burning of fossil fuels. Such large changes are relatively easy to measure, but the task of computing small variations in global mean temperature, set against the noise of seasonal and decadal cycles, is really quite challenging, especially for the oceans. Major research efforts are at present directed toward the simultaneous, widespread measurement of ocean temperature and other critical properties.

The immense complexity of the natural system, and uncertainty as to the ultimate capacity of its ocean regulator, make it enormously difficult to predict the Earth's continued response to increased greenhouse-gas emissions. It has only become possible very recently to make sophisticated computer models that realistically couple the atmosphere and ocean. Slowly we are beginning to be able to predict future climate change.

Yet even if global warming is now an accepted fact, we still have much to learn about how to respond. A reduction in greenhouse-gas emissions, especially by the industrialized world, seems a logical first step. For some nations, however, a warmer climate would bring benefits for agricultural production and energy consumption. Global warming is one of the largest environmental issues facing humankind; it is not surprising, therefore, that there is still little unanimity about how to tackle it.

Ocean regulator

The control exerted by the oceans is an essential part of the global climate system. The immense amounts of heat and moisture stored within them serve both to moderate change and also to prolong it once it commences. The oceans are some of the Earth's main longterm stores for carbon and carbon dioxide and so act to regulate greenhouse gases in the atmosphere, although with an efficacy we cannot yet predict.

A crucial piece of the climate jigsaw, therefore, is the way in which greenhouse gases are transferred between air and sea. The oceans resemble a gigantic sponge, holding 50 times more carbon dioxide than the atmosphere, and they are thought to absorb 30–40 percent of the carbon dioxide produced by human activity. But the sea surface acts as a two-way control valve for gas transfer, opening and closing in response to two key properties—gas concentration and ocean stirring.

In the first place, the difference in gas concentration between the air and sea controls the direction and rate of gas exchange. Gas concentration in water depends on temperature—cold water dissolves more carbon dioxide than warm water—and on biological processes such as photosynthesis and respiration by the microscopic plants, animals, and bacteria that make up the plankton. Secondly, the strength of the ocean stirring process, caused by wind and foaming waves, affects the ease with which gases are absorbed at the surface. More gas is absorbed during stormy weather and, once dissolved, is quickly mixed downward by strong water turbulence.

A NEW GLOBAL AWARENESS

Human beings have lived on Earth for at least 2 million years, for the most part in harmony with nature. Recently, however, our presence has begun to affect the environment in a potentially disastrous way. Counterbalancing these destructive tendencies is a new awareness of environmental issues and the ecological balances that sustain them. The 21st century will see a struggle between these two trends that may decide the future of humankind.

It is really only in the last 200 years that we have begun to adversely affect the planet, having developed an unprecedented capacity to effect global change. But the situation is by no means all one of doom and gloom. The past half-century has seen the rise of a new environmental awareness, embracing both individual concern and international cooperation. The United Nations Environment Program (UNEP) was founded in 1945, and its birth was closely followed by the opening of governmental environmental departments around the world. Greenpeace, one of the best known environmental movements, was first launched in 1973, paving the way for many other "green" campaign networks focusing on a range of different issues from the plight of rainforests to saving the whale. These and many other environmental issues are now hotly debated at many levels—by children in school classrooms and by the global media machine, by local planners, national politicians, and multinational corporations.

Earth Summit

An increasing pressure for action has emerged out of this new consciousness. However, the threat of global ecological crisis requires global solutions. At the end of a war-ravaged century marked by a growing disparity between rich and poor nations, few expected the world to be capable of embracing the environmental challenge. But that is what it did in 1992, when more than 100 global leaders and around 30,000 other participants met in Rio de Janeiro for the UN Conference on Environment and Development, popularly known as the Earth Summit. Its purpose was to achieve a global commitment from all countries to look beyond their national interests and to come to some kind of agreement over the future management of the planet.

The differences between delegates were almost as great as the complexity of the problems that face the world. Rich nations were reluctant to put at risk their comfortable lifestyle, founded though it is on wasteful consumerism and an inappropriate share of resources. Poor nations were set firm on catch-up mode, paying scant attention to the environmental degradation and pollution that must result. Scientists and environmentalists often could not agree on the validity and interpretation of global

trends, let alone on the most appropriate remedial action. So, as the conference drew to an end, there were, unsurprisingly, diverse views as to whether it had all been worthwhile.

The answer must be that the Summit was a qualified success. The world leaders present adopted a charter for sustainable development and established a new UN body to supervise its implementation. Two global conventions were inaugurated— one on biodiversity, signed by 152 countries, and another on climate change, signed by 150. Although the oceans were recognized as a vital part of the global ecosystem, an opportunity was missed to draw up a separate agreement for their management, which is subject instead to the recently ratified Law of the Sea treaty. However, the Summit's *Principle 25* remains particularly germane to a holistic view of the earth-ocean system: it states that "Peace, development, and environmental protection are interdependent and indivisible."

Protecting the oceans

If real progress is to be made, protection of the marine ecosystems must be at the heart of any global environmental convention. Their conservation involves maintaining natural habitats and ensuring that the rich variety of plant and animal life is neither overexploited nor needlessly destroyed. There are several reasons why this is important. Firstly, the natural balances that ecosystems have achieved are of practical significance to all species including humans—for example, the ocean phytoplankton provide over half the planet's breathable air, removing carbon dioxide from and releasing oxygen into the atmosphere; saltmarshes, estuaries, and mangroves are important nursery grounds for marine life, as well as giant filtration plants for dissolved matter; and biological uptake and long-term storage of carbon provides a vital buffer to rapid climate change.

Secondly, ocean life provides a plentiful supply of food for the global diet, and a wide spectrum of other useful substances. It is particularly important as a storehouse of medicines, including anticancer drugs, powerful new antibiotics, and muscle relaxants, as well as numerous other chemicals, ranging from natural fertilizers to insecticides. Many of these substances have only recently been discovered, and many more no doubt remain to be found provided that the host organisms are not first hastened to extinction by profligate human intervention.

Thirdly, and more philosophically, it is not only unwise but also immoral for humans to reduce the ocean's biodiversity. Everyone is diminished when we destroy the richness of life on our planet. Future generations will hold us to account for gross negligence in allowing even a small part of the vitality and aesthetic beauty of the ocean world to be lost forever at a time when protective action was required. Of course, species will continue to come and go as they always have, with a reason and timescale of their own, but it seems reckless to accelerate that change when we do not yet understand the consequences.

The ticking clock

Many marine species have already vanished under the human onslaught, and time is fast running out for others. The World Conservation Union currently lists as vulnerable or endangered more than 200 species of marine vertebrates, ranging from delicate seahorses to whales. Crisis or change is fast approaching for countless more invertebrate species as their habitats come under extreme pressures.

Coral-reef ecosystems display an astonishing abundance of life. They are often likened in this respect to tropical rainforests, but the likeness also has depressing connotations, as they too appear to be under stress and on the decline. A recent study of

Caribbean reef colonies, for example, has shown that coral coverage has dropped by as much as 80 percent over the past three decades. The causes include natural factors, such as hurricanes, predators, and disease, as well as anthropogenic ones including overfishing, tourism, pollution, and increased sedimentation from the deforestation of surrounding land areas. The consequences are likely to be dire and irreversible—a greatly reduced biodiversity, the collapse of reef-associated fisheries, dwindling tourism, and greater coastal vulnerability to storm damage.

The plight of whales remains of grave concern despite a worldwide ban on whaling introduced progressively through the 1980s. Some whale stocks have become so low, and breeding cycles are so long, that recovery may never be achieved. Pollution, inbreeding, and entanglement with fishing nets all contribute. Dolphins and porpoises continue to be taken in huge numbers as by-catch during commercial fishing operations. Some are illegally hunted, but mostly they drown. The Vaquita porpoise from the Gulf of California is now on the endangered list as a result.

More controversially, noise pollution in the oceans may be proving lethal to whales and dolphins in some areas. The high-intensity sonar systems favored by navies and three-dimensional acoustic surveys carried out by exploration companies, can divert cetaceans from their normal migration pathways to become stranded in shallows and on beaches, to exhibit various types of distressed behavior, and to incur severe acoustic trauma.

Our common future

We all have a stake in the future of our planet and a responsibility to future generations. Although most of today's decision-makers will be dead before the planet suffers the full consequences of their inaction, most of today's children will be alive. Our prime task in the first part of the 21st century must be to take full responsibility for the environment, and for enacting policies that protect rather than destroy it. There needs to be a forced marriage between economy and ecology; a price tag must be put on the environment, coupled with a tight rein on unfettered economic advance.

Central to this process is the protection and understanding of the oceans, first by unreserved and immediate conservation of their living and non-living resources, and second by greatly enhanced scientific investigation. We cannot effectively preserve or manage what we do not fully comprehend, and much about the ocean remains unknown. But the scientists who work in the oceans and other environmental areas cannot simply carry out their research and then sit on the sidelines. They must also explain their results and become active in advising and implementing marine policy.

OUR COMMON HERITAGE

Although the world's oceans may seem limitless, we now know that appearances in this regard are deceptive. Their resources are finite, and cannot be the panacea for all humankind's problems that some have hoped. Yet their potential remains enormous, and managing their riches responsibly presents a daunting but vital challenge affecting the future of the entire planet.

To the energy sector the oceans represent oil and gas in abundance, the promise of frozen methane trapped just below the seafloor, and a plethora of renewable sources awaiting technological advance and the commitment to develop them. To the metals industry they mean nodules and black smokers as well as beach sands of gold, diamonds, and precious gems. To fisheries they stand for a catch of 100 million metric tons a year, new, rich harvests of krill and myriad smaller planktonic species, and gigantic farms for mariculture spread out across the continental shelves. To tourism their broad expanses offer visions of golden beaches and luxury cruises, silent fjords and island tranquility—or else a concrete jungle of hotels, timeshare developments, and coastal apartments. Politicians and the military think in terms of sheltered harbors and missile bases, trade routes and waste-disposal sites, supremacy at sea and wealth for the future. But for ordinary people the oceans must always remain a breath of fresh air to clear our lungs and stir our minds and a free playground for our children to enjoy.

Many dilemmas cloud the issue of the good management of this most precious resource. How is it possible to judge the relative merits of energy supply as against chemicals extraction, tourism against metals, fisheries, or waste disposal? How can we reconcile the multifarious claims at stake? Not only are there competing special interests involved in the use of the oceans, but individual nations with greatly varying degrees of coastal access, wealth, and technological capability make different demands on them and have conflicting expectations of what they can offer.

Law of the Sea

From the earliest times there has been conflict over the right to use the oceans. At first they were ruled by sheer might and by nautical ability. Then, in the early 17th century, a Dutch jurist and statesman, Hugo Grotius, formulated the legal doctrine of *mare liberum* (the freedom of the high seas), which has remained the foundation of all modern international maritime law. About a century later, the concept of territorial seas adjacent to land was recognized, and the notion of national rights started to creep beyond the shoreline. The extent of territoriality was set at 3

nautical miles (approximately 5.5km), which was at the time the maximum range of a cannonball fired from shore.

The two conflicting traditions, one of freedom of the seas and the other of territorial waters under national jurisdiction, coexisted largely unchanged until the 1950s, by which time some fishing stocks were becoming depleted and offshore exploration had begun to reveal substantial mineral wealth on and under the continental shelves. Some countries with long coastlines and significant offshore interests then chose arbitrarily to extend their territorial limits. Others with small coastlines and a limited ability to exploit the promised wealth beyond felt growing unease and a sense of unequal opportunity, as did those that were completely landlocked.

The beginnings of a common policy emerged when the United Nations Organization called the First Conference on the Law of the Sea in New York in 1958. The issues were as complex as the divergence of interest was extreme, and it took two further conferences and 24 years of effort for a draft convention to be adopted. The Law of the Sea Treaty was finally approved in 1982 by a vote of 130 to 4, with 17 abstentions. The United States voted against but, like most of the earlier abstainers, has since been persuaded to join the rest of the world in what is potentially one of the most significant milestones taking international ocean policy forward into the third millennium.

The treaty became legally binding in 1994 although, in practice, individual signatories have chosen to selectively respect or ignore certain provisions. The basic features, however, are now generally accepted: territorial waters extending 12 miles (18.2km) from shore and an Exclusive Economic Zone (EEZ) that extends for 200 nautical miles (370km). Individual nations hold sovereignty over the resources within their EEZs, as well as over all economic activity and environmental protection issues.

All ocean areas beyond the EEZs are considered the common heritage of humankind, and an International Seabed Authority has been set up to oversee any future extraction of deep-sea mineral resources. The treaty also endorsed the values of protecting the ocean environment and of preventing marine pollution, and encouraged freedom of scientific research in the ocean.

The limits of management

Although securing any international agreement on the oceans has been a remarkable achievement, it is far from certain that the provisions the treaty put in place can ensure the sound management of the seas that future generations will demand. The regime that has emerged places about 40 percent of ocean space in the EEZs of coastal states, including the entire areas of such major marginal bodies as the North Sea, the Mediterranean, the Gulf of Mexico, and the Caribbean. The United States, for one, has gained responsibility for a region equal to about one-third of its land area. But fisheries, mineral resources, and pollution do not respect artificial boundaries. The treaty provides no guarantee that resource conservation and environmental management will be pursued rather than more short-sighted goals.

The world's ever-growing population not only needs adequate and sustainable food, energy, and material supplies but also generates huge amounts of waste. Globally, we produce some 6 billion tons of domestic waste and 10 billion tons of sewage annually. Industry and agriculture spew toxic chemicals and gases into the atmosphere and hydrosphere; and at least 4 million tons of waste oil finds its way into the oceans each year. The seas are neither so large nor marine ecosystems so robust that they can withstand these pressures without change. In addition, the pressures on our coastline and marginal seas from urban sprawl, industrial development, and recreational demand are without precedent, and yet it is precisely these regions that now fall entirely under national jurisdiction.

Many of the world's mangrove forests, coral reefs, temperate estuaries, and salt marshes have either been destroyed or severely harmed, and much of the damage has been done in the 20 years since the Law of the Sea convention was first adopted. With fish nurseries and traditional fishing grounds overstressed, the global fishing industry becomes increasingly dependent on aquaculture, which now accounts for 30 percent of all fish consumed and is the fastest-growing sector of the world food economy. But inherent problems of competition and disease spread well beyond the fish farms, eating into the profits of more traditional fisheries. It is estimated that for every 1kg (2.2lb) of farmed salmon or other large carnivorous fishes, 3kg (6.6lb) of wild fish are consumed—a flagrantly uneconomic use of the limited living resources of the ocean.

Toward sustainable development

There can be little doubt that sustainable development for the Earth depends on significant advances in our management of the oceans. At the same time, increasing world attention on the seas has the potential to direct the agenda for care of the whole planet. The principal issues involved are many and diverse. The oceans' fundamental unity requires effective global management regimes, which could then be extended to other areas of international cooperation. The essential indivisibility of regional seas makes agreement on regional management mandatory, perhaps paving the way for similar action on land. The threat of pollution to the ocean ecosystem comes from activities on land, so that effective national actions against it must be based on international cooperation. We also need to ensure nuclear-free oceans.

Both living and non-living ocean resources must be respected and managed by common accord—by controlled fishing quotas, by environmentally sound aquaculture, by well-regulated mineral and energy extraction, and by promoting renewable energy and conservation practices. Almost certainly, the oceans' riches provide scope for redressing the current gross imbalance in global resource distribution and for ensuring that the entire world has an adequate provision of food, energy, and shelter.

Progress, however, will require significant changes in global priorities. In a small way the process has already begun. Some 1,300 Marine Protected Areas have already been designated around the world in an effort to relieve threatened species. The Antarctic Treaty, first signed in December 1959, has provided a model of good practice for an environmental management regime under international agreement extending to the seas around the continent. Agreements on pollution control and coastal environmental standards have also been hammered out on a regional basis, for example by the European Community. The United Nations Environmental Program continues to seek ways of effectively monitoring the oceans so as to address problems of pollution and overfishing.

Very slowly, it seems, the world's nations are starting to take their collective responsibility seriously. Relatively few countries have yet formulated a national oceans policy, or have designated bodies to advise and regulate on maritime issues, but the ones that have must lead the way. It is vital that scientific research should move hand in hand with government and industry in forging the right policies and pathways. We must all become involved in our future and realize that what we do and say really can make a difference to the health and well-being of the planet.

The overarching themes of this book have been the lure of the sea, its incredible beauty and vitality, and its enormous potential to help humankind in our struggle for survival. The oceans offer not just a challenge but also hope for the future.

Glossary

A

ABYSSAL PLAIN The flat, SEDIMENT-covered area forming the greater part of the deep-sea floor between the continental rise and the MID-OCEAN RIDGE at a depth of 3,700–5,000m (12,000–16,500ft).

ABYSSAL STORM Stormlike event caused by higher than normal bottom-current activity at the deep-sea floor.

ABYSSAL ZONE A subdivision of the BENTHIC ocean province from 4,000 to 6,000m (13,000–19,500ft).

ACID A substance that releases hydrogen IONS (H+) in solution, having a pH of less than 7.

ACID RAIN Rain that has become weakly acidic due to the presence of dissolved compounds such as sulfur dioxide and carbon dioxide.

ADAPTIVE RADIATION The effect of natural selection whereby the GENES best suited for survival in a gene pool become more common within a species population.

ADSORPTION The process by which IONS adhere to the surface of an object.

AGGREGATE A mixture of minerals or grains.

ALGAE Simple aquatic plants, unicellular and multicellular, that lack roots, stems, and leaves—for example DIATOMS or seaweeds.

ALKALI A substance whose hydroxyl ion (OH-) concentration exceeds its hydrogen ion (H+) concentration, having a pH of more than 7.

AMINO ACID Organic compound that, in combination with others, forms proteins within organisms.

AMPHIBIAN CLASS of vertebrate spending time both in water and on land, returning to water to breed.

ANAEROBIC Lacking oxygen. Also refers to organisms that do not depend on oxygen for respiration.

ANDESITE A common volcanic rock found in the volcanic arcs above SUBDUCTION ZONES.

ANOXIC Having no free oxygen (O2).

ANTARCTIC BOTTOM WATER Very cold and dense seawater that forms off Antarctica, sinks to the ocean floor, and then flows northward toward the equator.

ANTARCTIC CIRCUMPOLAR CURRENT A deep, eastward-flowing current that completely encircles Antarctica. The largest ocean current on Earth.

ANTARCTIC CONVERGENCE Zone of mixing between the cold waters of the ANTARCTIC CIRCUMPOLAR CURRENT and warmer water to the north in the Antarctic Ocean, between latitudes 50° and 60°S.

APHOTIC ZONE The dark region of the ocean lying beneath the sunlit surface waters.

ARAGONITE A carbonate mineral identical in composition to calcite, but differing in its crystalline form; it is precipitated from ocean water by marine organisms.

ARCHAEA One of the three principal domains or six KINGDOMS of life, made up of single-celled PROKARYOTES. Among the oldest forms of life on Earth, found abundantly in extreme environments.

ARTHROPOD An animal belonging to the PHYLUM Arthropoda, characterized by jointed appendages and a hard exterior covering (exoskeleton). All insects and crustaceans are arthropods.

ASEXUAL REPRODUCTION The process by which offspring are produced from a single parent without the fusion of sex cells.

ASTHENOSPHERE The hot, soft region of the upper MANTLE that lies directly below the LITHOSPHERE.

ATMOSPHERE The vast and complex system of gases and suspended particles that encircles the Earth.

ATOLL A circular coral reef with a central lagoon.

ATOM The smallest component of an ELEMENT, comprising protons and neutrons in a central nucleus, surrounded by electrons.

AUTOTROPH An organism that is capable of producing its own food—for example, plants and other photosynthesizers.

B

BACK-ARC BASIN An isolated marine basin lying between a continent and a volcanic arc.

BACK REEF The area behind the reef front, sheltered from the open sea or ocean.

BACTERIUM (pl. bacteria) Any of a diverse group of microscopic, single-celled PROKARYOTIC organisms; one of the first and most successful lifeforms on Earth.

BALEEN A comb-like structure, used to strain food from the water, that takes the place of teeth in baleen whales.

BARRIER REEF A reef separated by a lagoon from the landmass with which it is associated.

BASALT A dark, fine-grained, glassy, extrusive IGNEOUS ROCK, high in magnesium and iron, and low in SILICATES; it typifies the upper layers of the OCEANIC CRUST.

BASE PAIR The fundamental building block of DNA; the complementary pairings of adenine with thymine and guanine with cytosine that are the basis for genetic code.

BATHOLITH A discordant, irregularly shaped pluton (large body of IGNEOUS ROCK), with a surface area of at least 100 sq km (40 sq mi).

BATHYAL ZONE The portion of the ocean bottom that extends from the edge of the CONTINENTAL SHELF to a depth of 4,000m (13,000ft).

BATHYMETRY The measurement of depth below sea level in the ocean in order to delineate the submarine topography.

BEACH An aggregation of unconsolidated SEDIMENT, usually sand or gravel, that covers the shore.

BENTHIC Pertaining to the ocean bottom.

BERM The flat accumulation of sand on a beach above the high-tide line.

BIG BANG THEORY A model for the origin of the universe in which an incredibly dense, hot state was followed by a sudden, explosive expansion and cooling.

BIOACCUMULATION The process by which chemical substance builds up in the cell or tissue of an organism.

BIODEGRADATION Decomposition by biological agents, especially BACTERIA and ARCHAEA.

BIODIVERSITY The variability among organisms within a specified geographical space.

BIOGENIC SEDIMENT Sediments formed from the remains of living organisms—mostly an accumulation of their hard parts.

BIOLUMINESCENCE The production of visible light by organisms.

BIOMAGNIFICATION The concentration of pollutants or toxins in higher TROPHIC LEVELS of the food chain.

BIOMASS The quantity of living matter, expressed metrically as grams per unit area or unit volume.

BLACK SHALE SEDIMENTARY ROCK rich in organocarbons, typically formed in an ANOXIC environment.

BLACK SMOKER Deep-sea hydrothermal vent, usually occurring at a MID-OCEAN RIDGE.

BLADE A large, flat, leaflike structure typical of green and brown ALGAE.

BOTTOM WATER A general term applied to dense water masses that sink to the floor of ocean basins.

BUTTRESS ZONE The seaward-sloping area of a coral reef, consisting of alternating ridges and furrows.

BYCATCH The non-commercial animals killed during fishing for commercial species.

C

CALCAREOUS Composed of calcium carbonate ($CaCO^3$).

CAP ROCK A relatively impermeable rock that forms a seal above and around a HYDROCARBON reservoir.

CARBONATE COMPENSATION DEPTH (CCD) The depth in the ocean below which material composed of calcium carbonate is dissolved, and therefore does not accumulate on the sea floor.

CARBON SINK An area of stored carbon—for example, the oceans or carbonate SEDIMENTS.

CARNIVORE An animal that feeds on other animals.

CAUDAL FIN The tailfin of a fish.

CELL The basic structural and functional unit of living organisms. All organisms (except viruses) consist of one or more cells, some of which may become specialized into particular functions.

CENOZOIC Of, belonging to, or designating the most recent era of geological time, from 65 million years ago to the present.

CENTRIFUGAL FORCE An apparent force exerted outward from a rotating object.

CEPHALOPOD An animal belonging to the molluskan CLASS Cephalopoda, which includes squids, octopods, cuttlefishes, and nautiloids.

CHANNEL-LEVEE SYSTEM System comprising deep-sea channels and their sediment-mounded banks.

CHEMICAL SINK An area of stored chemicals, such as SEDIMENTARY ROCKS.

CHEMOGENIC SEDIMENT Sediment that forms from the direct PRECIPITATION of chemical compounds from water, or from ionic exchange between existing sediment and water.

CHLOROFLUOROCARBON (CFC) Artificially produced substance that has become concentrated in the Earth's atmosphere, acting as a strong greenhouse gas.

CHLOROPHYLL Green pigment found in plants that is essential for conducting PHOTOSYNTHESIS.

CHLOROPHYTA Green ALGAE, characterized by the presence of chlorophyll and other pigments.

CHORDATE An animal belonging to the PHYLUM Chordata. All have, at some stage of their lifecycle, a supporting cartilage rod (notochord or backbone), a dorsal, hollow nerve cord, and gill slits.

CHROMOSOME A long, coiled molecule of DNA found in the cell nucleus of EUKARYOTES that carries the genetic information. (*See* GENES.)

CILIA Short, hairlike features on many organisms that are used for locomotion, the generation of a minute current of water, or for FILTER-FEEDING.

CIRCULATION Circular movement of a substance, especially water in the oceans.

CLASS A type of taxon used in classifying organisms; contains one or more orders and is within a phylum (*see* TAXONOMY).

CLATHRATE A substance in which guest molecules occupy a space within the crystal lattice of another substance.

CLONE An organism that is directly descended from and genetically identical to a single common ancestor.

CNIDARIAN An animal belonging to the PHYLUM Cnidaria, characterized by a radially symmetrical body with a sacklike internal cavity. Includes jellyfishes, sea anemones, and corals.

COASTAL UPWELLING An upward flow of cold, nutrient-rich water, usually induced by EKMAN TRANSPORT.

COCCOLITH A tiny calcitic platelet that helps protect certain marine PHYTOPLANKTON. Coccoliths are the dominant component of certain limestones, such as chalk.

COCCOLITHOPHORE Photosynthetic PROTIST, covered in calcitic plates, belonging to the PHYLUM Chrysophyta.

COMMENSALISM A symbiotic relationship in which one organism benefits while the other is neither harmed nor benefited. (*See* SYMBIOSIS.)

COMMON WATER Cold, dense, bottom-water mass of the Indian and Pacific Oceans, combining ANTARCTIC BOTTOM WATER and NORTH ATLANTIC DEEP WATER.

COMMUNITY An assembly of populations of different SPECIES that occupy the same habitat at the same time.

COMPETITION Interaction in which two or more organisms in the same space require the same resource.

COMPOUND EYE The eye of most insects and some crustaceans, which is composed of many light-sensitive elements, each having its own refractive system and each forming a portion of an image.

CONGLOMERATE Coarse-grained SEDIMENTARY ROCK composed mainly of rounded rock fragments, finer-grained matrix, and a mineral cement.

CONSUMER An organism that relies on another organism for food.

CONTINENTAL DRIFT The movement of continental masses as a result of seafloor spreading.

CONTINENTAL RISE Gently sloping region toward the deep ocean, lying at the base of the steeper CONTINENTAL SLOPE that surrounds the continents.

CONTINENTAL SHELF Submerged platform region extending from the edge of a continental landmass, with a water depth generally less than 200m (650ft).

CONTINENTAL SLOPE The sloping seafloor region between the edge of the CONTINENTAL SHELF (around 100–200m/300–650ft water depth) and the CONTINENTAL RISE or deep seafloor.

CONVECTION CELL Convection current completing a circular motion, distinct from others around it—for example, the movement of molten material within the ASTHENOSPHERE, or air movement in the atmosphere.

CONVERGENCE ZONE Region where two horizontally flowing water (or air) masses meet, and are forced to change direction, typically flowing vertically.

COPEPOD Member of an ORDER of Crustacea that is shrimplike in appearance and that feeds voraciously on PHYTOPLANKTON.

CORALLIVORE An organism that feeds primarily on coral.

CORAL REEF A marine ridge or mound composed predominantly of compacted coral, together with algal material and biochemically deposited magnesium and calcium carbonates.

CORE The innermost region of the Earth, beginning at a depth of around 2,900km (1,800mi). Thought to be mainly composed of iron and nickel, it is divided into a liquid outer and solid inner core.

CORIOLIS EFFECT Apparent deflection of the path of winds and ocean currents (or other free-moving bodies) that results from the rotation of the Earth.

COUNTERSHADING Protective coloring of an organism by which areas exposed to light are colored darker and areas normally shaded are colored lighter.

CRUST The outermost, thinnest, and coolest layer of the Earth. Consists mainly of either GRANITE (continental crust) or BASALT (oceanic crust). Its thickness varies from 5–70 km (3–45mi).

CRUSTAL DIFFERENTIATION Layering of the CRUST that results from a number of different physical processes during crystallization.

CRYPTIC COLORATION The use by organisms of mirrored or colored surfaces as camouflage.

CRYPTOZOIC Early period of time more or less synonymous with Pre-Cambrian era, from the first beginnings of life on Earth to the start of the Cambrian period. The word means "hidden life," as few FOSSILS are preserved.

CYANOBACTERIA A group of photosynthetic PROKARYOTES that contain chlorophyll and that release oxygen as a by-product of their PHOTOSYNTHESIS. Formerly referred to as blue-green algae; now simply as "blue greens."

CYCLONE Area of low pressure in the atmosphere that displays a circular inward movement of air.

D

DDT An abbreviation of dichloro-diphenyl-trichloroethane, a pesticide belonging to the hydrocarbon chemical group.

DEBRIS FLOW The rapid downslope flow of unconsolidated material, typically comprising an unsorted mix of large boulders, small pebbles, and sand supported in a mud matrix.

DECOMPOSER An organism that breaks down the tissue of other dead organisms and animal wastes.

DEGASSING The emission of gases from within the Earth into the atmosphere.

DELTA A sedimentary deposit formed where a river flows into a lake or the sea.

DENSITY The mass per unit volume of a substance.

DENSITY STRATIFICATION The layering of substances due to differences in their DENSITIES.

DENUDATION The exposure and removal of rock by the action of flowing water.

DEPOSIT FEEDER Animal that feeds on bottom SEDIMENTS, extracting useable organic material and discarding inorganic debris.

DEPRESSION An area of atmospheric low pressure; a cyclone.

DESALINATION The process of removing salt from seawater.

DETRITUS Either organic matter, such as animal wastes and bits of decaying tissue, or inorganic sedimentary material. (*See* SEDIMENT.)

DIATOM Photosynthetic, unicellular PROTIST belonging to the PHYLUM Chrysophyta. Diatoms possess a glassy covering composed of silica.

DIKE A tabular or sheetlike intrusion of IGNEOUS ROCK that cuts across bedding and other structures in the surrounding rocks.

DINOFLAGELLATE Photosynthetic, usually single-celled PROTIST, belonging to the PHYLUM Pyrrophyta, that possesses two FLAGELLA.

DISPHOTIC ZONE The dimly lit zone of the ocean where PHOTOSYNTHESIS can no longer be carried out, but vestiges of light remain. Also known as the twilight zone.

DISRUPTIVE COLORATION A type of coloration found in some animals in which the background color of the body is broken up by lines that frequently run in a vertical direction.

DISSOCIATION The loss of one substance from another; for example, compound gases in the atmosphere, or minerals dissolving in seawater.

DIURNAL TIDE Having only one high tide and one low tide per day.

DNA Short for deoxyribonucleic acid: A complex molecule, made up of two helical chains of nucleic acid connected by complementary BASE PAIRS, that contains an organism's chemically coded genetic information.

DOLDRUMS A belt of variable, very light winds occurring 10–15° to either side of the equator.

DOPPLER EFFECT A change in the observed frequency of a wave that occurs when the source and the observer are in motion relative to one another.

DORSAL FIN A fin found on the upper surface of a fish.

DRIFT NET Very large net used in fishing that is suspended in the water vertically and that may stretch for as much as 60km (40mi). Drift nets entangle fish and other marine animals that swim or drift into them.

E

ECHINODERM An animal belonging to the PHYLUM Echinodermata, including sea stars, brittle stars, sea urchins, sea cucumbers, and crinoids.

ECHOLOCATION Method of determining distance using the speed of a sound wave emitted and reflected off objects. Marine mammals use echolocation expertly.

ECOLOGY The study of the interaction of organisms with one another and with their physical and chemical environments.

ECOSYSTEM A discrete ecological unit consisting of organisms and their environment.

ECOTOURISM A form of tourism that seeks to impose minimal damage on the surrounding environment.

ECTOTHERM An animal that obtains most of its bodily heat from its surroundings.

EDDY A circular flow of water (or wind) generally found at the edge of the main current.

EKMAN SPIRAL The theoretical effect of a steady wind blowing over a water body of unlimited breadth and depth, and uniform viscosity. The resultant surface flow would travel at a 45° angle to the wind, due to the CORIOLIS EFFECT, with further deflection and slower velocity at increased depth, until at around 100m (330ft) it would be traveling in the opposite direction to the wind, thus creating a spiral of water movement.

EKMAN TRANSPORT The net flow of water; 90° to the right of the wind in the northern hemisphere and to the left of the wind in the southern hemisphere.

ELEMENT A naturally occurring substance, composed of all the same ATOMS, that cannot be reduced to a simpler substance by normal chemical means.

EL NIÑO Warm surface waters that usually appear around Christmas off the coasts of Ecuador and Peru. Approximately every 3–5 years, the El Niño effect is intensified, bringing extreme weather conditions to many parts of the globe, especially the tropics.

ENDOTHERM An animal that maintains constant body temperature by generating heat internally

EPIPELAGIC ZONE A subdivision of the PELAGIC province extending from the surface to a depth of around 200m (650ft).

ESTUARY A partially enclosed, tidally influenced body of water where freshwater is mixed with saltwater; typically funnel-shaped and open to the sea.

EUKARYOTE Organisms whose cells contain a clearly defined nucleus containing DNA, and other ORGANELLES. Eukaryotes include all living things except bacteria and archea, which are PROKARYOTES.

EUPHOTIC ZONE The surface layers of the ocean, receiving enough light to sustain PHOTOSYNTHESIS.

EUTROPHICATION A process by which water becomes ANOXIC because of an excess of decomposing organic matter.

EVAPORITE A type of SEDIMENT precipitated from a concentrated aqueous solution, usually by the evaporation of water from a basin with restricted circulation. Includes halite, gypsum, and anhydrite.

EVAPOTRANSPIRATION The loss of water to the atmosphere from plants via the combined processes of evaporation and transpiration.

EVOLUTION The process by which populations of organisms change over time.

EXOTIC TERRANE A small piece of LITHOSPHERE that has been separated and removed from its place of origin by tectonic forces, and subsequently welded onto a different TECTONIC PLATE.

EXTINCTION The disappearance of a SPECIES from its entire geographical range.

F

FAMILY A type of taxon, more exact than an order, comprising several or many different genera. (See TAXONOMY.)

FAULT A natural break or rupture between rocks along which movement takes place. Such movement is one of the principal causes of earthquakes.

FAULT SCARP The often precipitous plane created by the movement of rocks on either side of a FAULT.

FERTILIZATION Initiation of the reproductive process by insemination or pollination; the bringing together of male and female GAMETES.

FETCH The distance the wind blows over a continuous water surface, thereby creating waves.

FILTER FEEDER An organism that filters its food from the water.

FJORD A deep valley cut into the coastline by a GLACIER and subsequently filled with a mixture of fresh and salt water as the glacier retreats.

FLAGELLUM (pl. flagella) A whiplike projection used by some animals, such as DINOFLAGELLATES, for propulsion or for generating tiny currents in the surrounding water.

FLOOD TIDE A rising tide.

FOOD CHAIN A sequence of feeding relationships among a group of organisms that begins with producers and continues in a linear fashion to higher-level CONSUMERS.

FOOD WEB A representation of the complex feeding networks, comprising many food chains or part-chains, that exist in an ECOSYSTEM.

FORAMINIFERAN Amoeba-like PROTOZOAN. Many foraminiferans produce an elaborate shell of calcium carbonate.

FORE REEF The seaward side of a coral reef.

FORM DRAG The resistance to an organism's movement through a fluid medium caused by its shape. Also, the resistance the topography of the ocean floor causes to current flow, typically resulting in EDDY mixing.

FOSSIL Remains or traces of prehistoric organisms preserved within rocks.

FOSSIL FUEL Fuel resource, including coal, oil, and natural gas, formed from the remains of plants and microorganisms that lived millions of years ago.

FOSSIL SUCCESSION The progressive changes in organisms and their fossil remains through geological time.

FRACTURE ZONE A linear zone of highly irregular, faulted topography that is generally oriented perpendicularly to ocean SPREADING CENTERS; also known as a TRANSFORM FAULT. The term can also be used for a region of major faulting on land.

FRICTION DRAG The retarding force associated with the surface of a body moving through a liquid or gas, or a liquid or gas passing over a body.

FRINGING REEF A reef growing at the edge of a landmass without an intervening lagoon.

G

GABBRO A dark, speckled, coarse-grained, intrusive IGNEOUS ROCK similar to BASALT in mineral composition. Formed at depth, it makes up a significant percentage of the OCEANIC CRUST.

GAMETE A sex cell, such as a sperm, egg, or pollen grain.

GASTROPOD A member of the molluskan CLASS Gastropoda, including snails, limpets, and abalones.

GENE A basic, identifiable unit of hereditary information within a chromosome that codes for a particular feature of an organism.

GENUS (pl. genera) A type of taxon, one step above species level. (See TAXONOMY.)

GEOSTROPHIC CURRENT An oceanic current controlled by a balance between the pressure–gradient force and the CORIOLIS deflection.

GEOTHERMAL ENERGY Energy that is derived from natural heat within rocks. Geothermal energy may be released as steam and hot water in volcanic areas or tapped by drilling into hot water reservoirs at depth.

GLACIER A thick mass of ice on land that moves by plastic deformation and basal slip.

GLOBAL WARMING Warming of the Earth's average global temperature due to increasing concentrations of GREENHOUSE GASES.

GONDWANA One of the principal PALEOZOIC continents, composed of the present day landmasses of South America, Africa, Antarctica, Australia, and India, as well as parts of other landmasses including southern Europe, Arabia, and Florida.

GRANITE A light-coloured, coarse-grained, intrusive IGNEOUS ROCK formed deep within the Earth's crust, composed mainly of quartz and feldspar minerals; the dominant rock type of continental crust.

GRAZER An animal that crops low-growing vegetation as a source of food.

GREENHOUSE EFFECT The warming of the Earth's atmosphere due to buildup of GREENHOUSE GASES.

GREENHOUSE GAS Gas in the atmosphere that absorb infrared wavelengths emitted from the Earth's surface. Greenhouse gases include carbon dioxide, water vapor, nitrous oxide, ozone, and CHLOROFLUOROCARBONS.

GULF STREAM Warm ocean current that originates in and around the Caribbean and flows across the North Atlantic to northwest Europe.

GYRE A large water circulation system of GEOSTROPHIC CURRENTS, rotating clockwise in the northern hemisphere and anticlockwise in the southern hemisphere.

H

HABITAT The specific place in the environment where a particular plant or animal lives.

HADAL ZONE A subdivision of the both the PELAGIC and BENTHIC zones, in which the water depth exceeds 6,000m (20,000ft). Also known as the hadalpelagic zone in the open ocean.

HALOCLINE A transition zone in the ocean characterized by a rapid change in SALINITY with increasing depth.

HEAVY METAL Inorganic metallic element with a high atomic mass, such as mercury, lead, arsenic, tin, cadmium, cobalt, selenium, zinc, manganese, or copper. Heavy metals may be highly toxic when concentrated in the environment or in organisms.

HERBIVORE An animal that eats only plants and ALGAE.

HERMAPHRODITE Possessing both male and female functional reproductive organs.

HETEROTROPH An organism that relies on other organisms for its food; also known as a CONSUMER.

HOLDFAST Rootlike structure that secures seaweeds to the SUBSTRATE.

HOMEOSTASIS The internal balance that organisms must maintain in order to survive.

HORSE LATITUDES An area of light and variable wind between 30° and 35°, marked by the vertical descent of air masses.

HOTSPOT Localized zone of melting and UPWELLING in the ASTHENOSPHERE/LITHOSPHERE above which volcanic activity is abundant.

HURRICANE Tropical cyclone in which wind velocities exceed 120 km/h (75mph). The word is used to describe such storms in the North Atlantic Ocean, the eastern North Pacific Ocean, the Caribbean Sea, and the Gulf of Mexico; in the western Pacific such storms are termed typhoons, and in the Indian Ocean they are known as cyclones.

HYDROCARBON Organic compound composed of hydrogen, carbon, and oxygen. Hydrocarbons are the main components of petroleum (both oil and gas).

HYDROELECTRIC POWER Electricity produced by harnessing the power of flowing water.

HYDROGEN BOND A chemical bond consisting of a hydrogen ATOM between two electronegative atoms, especially oxygen, nitrogen, and fluorine.

HYDROLOGICAL CYCLE The continuous recycling of water from the oceans through the atmosphere to the continents and then back to the oceans.

HYDROTHERMAL VENT An opening from which hot, metal-enriched water emanates; commonly found at MID-OCEAN RIDGE sites.

I

ICE AGE Period of time when ice sheets dominate the surface of the Earth, and global mean temperatures are lowered.

ICEBERG A large fragment of drifting ice that has broken off from the terminus of a GLACIER or calved from an ice shelf.

ICEHOUSE EFFECT The cooling of the Earth's average global temperature occurring during an ICE AGE, commonly due to an increase in albedo.

IGNEOUS ROCK Any rock formed by the cooling and crystallization of MAGMA; one of the three main groups of rock. (*See also* METAMORPHIC ROCK, SEDIMENTARY ROCK.)

INFAUNA BENTHIC organisms that live buried in the seafloor SEDIMENTS.

INTERGLACIAL PERIOD Period of time during an ICE AGE when temperatures increase, leading to GLACIER retreat.

INTERNAL WAVE A subsurface wave propagating along a DENSITY discontinuity within the ocean.

INTERTIDAL ZONE The area of the shore that lies between the highest normal high tide and the lowest normal low tide.

ION An electrically charged ATOM produced by the loss or gain of electrons.

IONIC BOND A bond that results from the attraction between negatively and positively charged IONS.

ISLAND ARC A chain of volcanic islands associated with oceanic SUBDUCTION ZONES, lying on the continent side of deep-sea TRENCHES. Formed by the partial melting of the LITHOSPHERE as a plate is subducted.

ISOSTATIC REBOUND The unloading of the Earth's CRUST, typically as a result of GLACIER retreat and ice melting, causing it to rise upward until equilibrium is again attained.

ISOTOPE Different form of the same ELEMENT related to variations in the number of neutrons in the nucleus.

J

JET PROPULSION In marine animals, movement derived from the rearward expulsion of fluid.

JET STREAM Relatively fast, uniform winds concentrated in a narrow band in the upper ATMOSPHERE.

K

KELP Large BENTHIC species of brown ALGAE.

KINGDOM The largest or most general taxon used in classifying organisms; contains one or more phyla.

KRILL PELAGIC, shrimplike creatures belonging to the ARTHROPOD order Euphausiacia.

L

LARVA (pl. larvae) An embryo living on its own before it assumes the characteristics of the adult of the species, generally with a different lifestyle and appearance.

LATERAL LINE A system of canals running the length of a fish's body and over its head that functions in sensing movement in the water by detecting very subtle changes in pressure.

LAURASIA A late PALEOZOIC continent in the northern hemisphere, composed of the present-day landmasses of North America, Greenland, Europe, and Asia.

LAVA MAGMA extruded at the Earth's surface. The word is used to describe both the still molten form and magma hardened into volcanic rock.

LAW OF SUPERPOSITION States that STRATA (SEDIMENT layers) lower in a sequence must be older than those higher up, provided that major earth movements have not overturned the succession.

LEACHING Process by which water removes and transports humus and/or inorganic nutrients and other chemicals in soil.

LITHIFICATION The process by which loose, unconsolidated SEDIMENT is compacted and cemented into a SEDIMENTARY ROCK.

LITHOSPHERE The relatively cool, brittle outer shell of the Earth, including the CRUST and upper MANTLE. Extends to a depth of around 100km (60mi).

LITTORAL ZONE A subdivision of the BENTHIC province between the high and low tide marks, equivalent to the INTERTIDAL ZONE.

LONGSHORE BAR A submarine sand ridge in the nearshore zone that is parallel or subparallel to the shoreline.

LONGSHORE CURRENT A current in the surf zone that runs parallel to the shore and is typically generated by waves striking the shoreline obliquely.

M

MAGMA Fluid rock material, formed at depth, from which IGNEOUS ROCKS crystallize. When it erupts at the surface it is known as LAVA.

MAGMA CHAMBER A reservoir of molten rock within Earth's otherwise solid LITHOSPHERE.

MAGNETIC REVERSAL A complete reversal of the north and south magnetic poles. Such reversals have occurred at irregular intervals throughout geological time.

MAMMAL A member of the CLASS Mammalia, including humans. Mammals are characterized by a covering of hair on the skin, and generally give birth to live young which are suckled by the female.

MANGANESE NODULE Concretionary lump, typically potato-shaped and -sized, containing oxides of iron, manganese, copper, nickel, and various other metals found scattered over the deep ocean floor.

MANGROVE A large, treelike, salt-tolerant plant with intricate networks of partially exposed roots. Dense forests of mangroves dominate the intertidal environments of the tropics and subtropics, forming mangrove swamps.

MANTLE The thick layer of the Earth's interior between the CRUST and the CORE, composed of ferromagnesian (containing iron and magnesium) SILICATES.

MANTLE PLUME A column of MAGMA originating deep within the mantle and rising to the Earth's surface to form volcanoes or BASALT plateaus. Mantle plumes underlie HOTSPOTS.

MARGINAL SEA A semi-enclosed body of water adjacent to and surrounded by continent that is floored either by submerged continental CRUST or OCEANIC CRUST, or by a mixture of both.

MARICULTURE The use of agricultural techniques to breed and raise marine organisms.

MARINE SNOW Translucent, drifting, organic particles and AGGREGATES found in the water column, slowly settling to the seafloor.

MASS EXTINCTION A catastrophic, widespread perturbation in which major groups of SPECIES become extinct in a relatively short period of geological time.

MEDITERRANEAN OUTFLOW WATER Moderately dense, warm, saline water flowing from the Mediterranean Basin into the Atlantic Ocean through the Gibraltar gateway. (*See* OCEANIC GATEWAY.)

MEIOFAUNA Tiny organisms that are adapted to living in the spaces between SEDIMENT particles.

MÉLANGE A mixture, especially of rock types.

MELON An oval mass of fatty, waxy material that is located between the blowhole and the end of the head in cetaceans (whales and dolphins) capable of ECHOLOCATION. The melon serves to direct and focus the sound waves produced by the animal.

MESOZOIC Of, belonging to, or designating the era of time including the Triassic, Jurassic, and Cretaceous periods, between 248 and 65 million years ago.

METABOLISM The sum of all chemical reactions that occur within living cells and organisms, especially the processes by which nutrients are built into living matter.

METAMORPHIC ROCK Any rock altered by high temperature or pressure, and the chemical activities of fluids; one of the three main classes of rocks. (*See also* IGNEOUS ROCK, SEDIMENTARY ROCK)

METAZOAN A member of the Kingdom Metazoa, comprising multicelled animals.

MID-OCEAN RIDGE A long mountain range that forms along cracks in the ocean floor where magma breaks through the Earth's CRUST; the site where new ocean crust is formed and SEAFLOOR SPREADING occurs as two TECTONIC PLATES move apart. Also known as a divergent plate margin.

MILANKOVITCH CYCLE Cyclic variations in climate, with regular periodicities of around 20,000, 40,000, and 100,000 years, as a result of irregularities in the Earth's rotation and orbit.

MINERAL A naturally occurring, inorganic, crystalline solid having characteristic physical properties and a narrowly defined chemical composition.

MITOCHONDRIA (sg. mitochondrion) ORGANELLES found within the cells of EUKARYOTES, containing enzymes responsible for energy production during aerobic RESPIRATION.

MIXED TIDE A tide with two unequal high waters and two unequal low waters each day.

MOHOROVIČIC DISCONTINUITY A compositional and density discontinuity marking the interface between the rocks of the Earth's CRUST and the MANTLE; also known as the Moho.

MONSOON A regional wind system that changes direction predictably with the passing of the seasons; mostly found in south and southeast Asia. Summer monsoons are often accompanied by heavy rain.

MUTATION Any change in the hereditary information in the GENES of organisms.

MUTUALISM A symbiotic relationship in which both organisms benefit. (*See* SYMBIOSIS.)

N

NEAP TIDE The tide with the smallest range between high and low tides. Neap tides occur when the Sun and the Moon are at right angles to one another.

NEKTONIC Describing actively swimming organisms.

NEMATOCYST The stinging ORGANELLE found within the stinging cell of CNIDARIANS.

NERITIC One of the oceanic zones, referring to water lying over the CONTINENTAL SHELF to a depth of around 200m (650ft).

NICHE An organism's role in its environment.

NORTH ATLANTIC DEEP WATER Deep oceanic water of the North Atlantic, most of which originates in the Norwegian–Greenland Sea, with some contribution from the Labrador Sea.

NUCLEAR FISSION A nuclear reaction in which a massive atomic nucleus splits into smaller nuclei, simultaneously releasing energy.

NUCLEAR FUSION Nuclear reaction where small atomic nuclei combine to form a larger nucleus, simultaneously releasing energy.

NUCLEIC ACID A complex organic molecule—either DNA or RNA—made up of a sugar, a phosphate group, and one of four nitrogenous bases, present in the nucleus of the living cell.

O

OCEANIC CRUST The outermost shell of the Earth, around 5–10km (3–6mi) thick, that underlies oceans, composed of the basic IGNEOUS ROCKS—BASALT, dolerite, and GABBRO—and commonly overlain by sedimentary layers.

OCEANIC GATEWAY Partial topographic barrier between ocean basins, through which a deeper, narrow zone or channel allows the interchange of water masses.

OCEAN THERMAL-ENERGY CONVERSION (OTEC) The use of ocean thermal energy to produce power.

OIL TRAP A structural or stratigraphic feature within SEDIMENTARY ROCKS of the Earth's CRUST that confines oil to a certain area.

OMMATIDIUM One of the structural elements, resembling a single simplified eye, that make up the compound eye of insects and arthropods.

OPHIOLITE A distinct sequence of IGNEOUS ROCKS, thought to be a fragment of oceanic LITHOSPHERE now emplaced onto the continent. Composed of a layer of GABBRO overlain by sheeted dikes (interconnected, tabular igneous intrusions) and PILLOW LAVAS.

ORBITAL MOTION The circular motion of water particles that passes energy through a wave.

ORDER A type of taxon containing one or more families that is part of a class. (*See* TAXONOMY.)

ORGANELLE A discrete sub-cellular structure that performs a specific role within a CELL.

ORGANOHALOGEN Complex organic compound that includes one or several of the halogen elements (e.g. chlorine, fluorine, bromine) in its structure

OSMOREGULATION The regulation of salt content in the body fluids of an organism.

OSMOSIS The movement of water through a selectively permeable membrane across a concentration gradient.

OUTGASSING The process whereby gases derived from the Earth's interior are released into the atmosphere and hydrosphere during volcanic and hydrothermal activity.

OVIDUCT The tube responsible for transporting female GAMETES to an area where they can be readily fertilized.

OVIPAROUS Producing eggs that hatch outside the body.

OXBOW SEGMENT A river meander that has been entirely cut off from the main channel, forming a crescent-shaped pool.

OXYGEN MINIMUM LAYER A layer within the water column (typically at a depth of 200–1,000m/650–3,300ft) in which dissolved oxygen concentrations are lower than in the water above or below.

OZONE LAYER Atmospheric concentration of ozone (a form of oxygen with three atoms per molecule) found 10–50km (6–30mi) above the Earth's surface.

P

PALEOCEANOGRAPHY The study of past oceans, their location and characteristics.

PALEOZOIC Of, belonging to, or designating the era of time including the Cambrian, Ordovician, Silurian, Devonian, Carboniferous, and Permian periods, between 545 and 248 million years ago.

PANGAEA The single supercontinent of the late PALEOZOIC and early MESOZOIC eras that comprised all of the present-day continents.

PANTHALASSA The worldwide ocean that surrounded the ancient supercontinent PANGAEA.

PARASITISM A symbiotic relationship in which one organism benefits at the expense of the other. (*See* SYMBIOSIS.)

PARTHENOGENESIS The process by which an egg develops without the introduction of a sperm.

PATCH REEF Small patch of reef located in a lagoon associated with an ATOLL or BARRIER REEF.

PATHOGEN A disease-causing agent, such as a virus or BACTERIUM.

PCB One of many highly toxic and durable synthetic organic compounds that accumulate in the tissue of organisms.

PECTORAL FIN One of the paired fins of a fish, found at the shoulder girdle.

PELAGIC Of the ocean; most commonly used of the open ocean away from the shoreline.

PHAEOPHYTA Brown ALGAE—mostly marine and littoral EUKARYOTIC algae.

PHANEROZOIC Of, belonging to, or designating the eon (period of time) including the PALEOZOIC, MESOZOIC, and CENOZOIC eras, extending from 545 million years ago to the present day.

PHEROMONE A chemical that is used by animals for communication.

PHOSPHORITE A sedimentary deposit consisting of nodules or crusts of phosphorus oxide and hydroxide minerals.

PHOTIC ZONE The part of the ocean, near the surface, that receives enough light to sustain PHOTOSYNTHESIS.

PHOTOPHORE A specialized organ found in some organisms that functions in producing BIOLUMINESCENCE.

PHOTOSYNTHESIS The process by which some organisms use the energy of sunlight to produce organic molecules, usually from carbon dioxide and water.

PHYLUM (pl. phyla) A type of taxon, more exact than a kingdom, and less than a class. (*See* TAXONOMY.)

PHYTOPLANKTON Tiny photosynthetic organisms that float near the ocean surface.

PILLOW LAVA Bulbous masses of BASALT, somewhat resembling pillows, formed when basalt is rapidly chilled under water.

PLACER DEPOSIT Area in a river or on the seafloor where heavy minerals (generally metals) settle out after being transported by river or ocean currents.

PLACODERM One of the "plate-skinned" fishes, of late Silurian to Permian age, characterized by jaws and bony armor, especially in the head and shoulder region.

PLANETESIMAL One of many small bodies thought to have orbited the Sun at the time of the formation of the planets, and from which the planets coalesced.

PLANKTON Animals and plants that float passively in the ocean.

PLANKTON BLOOM The sudden and rapid multiplication of PLANKTON that results in dense concentrations of planktonic organisms.

PLANULA (pl. planulae) The flat, free-swimming, ciliated LARVAE of coelenterates (including corals, sea anemones, and jellyfishes).

PLATE TECTONICS The movement of large segments (lithospheric plates) of the Earth's outer CRUST and MANTLE relative to one another—the new paradigm of the earth sciences developed through the 1960s and 1970s.

POLYP A generally BENTHIC form of CNIDARIAN, characterized by a cylindrical body with an opening at one end that is usually surrounded by tentacles.

POPULATION A group of individuals of the same SPECIES that occupies a specified area.

PREDATION The act of hunting and killing for food.

PREDATOR An animal that preys on other organisms.

PRECIPITATION (CHEMICAL) The process by which dissolved minerals come out of solution.

PRECIPITATION (METEOROLOGICAL) The falling of water from the ATMOSPHERE to the Earth's surface in the form of rain, snow, hail, or sleet.

PRIMARY PRODUCER One of the organisms in a food chain, such as green plants or photosynthesizing or chemosynthesizing BACTERIA, upon which all other members of the chain depend directly or indirectly, but which are not themselves dependent on an external source of nutrients. Also known as AUTOTROPHS.

PRIMARY PRODUCTIVITY The quantity of organic matter that is synthesized from inorganic materials by AUTOTROPHS.

PROKARYOTE A unicellular organism lacking a nucleus and such ORGANELLES as plastids and MITOCHONDRIA; cells of ARCHEA, BACTERIA, and CYANOBACTERIA.

PROTEIN Complex organic material consisting of a particular sequence of AMINO ACIDS found in organisms.

PROTIST A member of the Kingdom Protista, which contains all EUKARYOTIC organisms that are not plants, animals, or fungi.

PROTOZOAN Belonging to the PHYLUM Protozoa, which comprises unicellular organisms that exhibit animal-like characteristics.

PURSE SEINE NET A vast fishing net with a bottom that can be closed off in a manner similar to the pulling of a purse's drawstring.

PYCNOCLINE A transition zone in the ocean that is characterized by rapid change in DENSITY with depth.

PYROCLASTIC Describing fragmented volcanic materials, such as ash ejected from a volcano.

R

RADIOACTIVE DECAY The spontaneous decay of an ATOM to an atom of a different ELEMENT by emission of a particle from its nucleus, or by electron capture.

RADIOLARIAN A type of PROTOZOAN that has an intricate shell made of silica and that uses pseudopods (foot-like projections) to capture prey.

RADULA A ribbon of tissue that contains teeth, unique to mollusks.

RED BED A sedimentary sequence of reddish-colored, iron-stained rocks deposited in continental environments.

RED TIDE A condition that occurs as the result of a population explosion of certain DINOFLAGELLATES (or other PROTISTS) that imparts a red color to the water.

REEF CREST The highest point on a coral reef.

REGRESSION A relative fall in sea level resulting in the deposition of terrestrial STRATA over marine strata.

REPTILE A member of the CLASS Reptilia, including snakes, lizards, crocodiles, and turtles.

RESPIRATION The sequence of biochemical reactions in living cells that break down organic substances and release energy.

RESERVOIR ROCK Sedimentary stratum having porous and permeable properties and acting as a reservoir for HYDROCARBONS.

RHABDOM A transparent rod forming part of a compound eye in an insect or ARTHROPOD.

RHODOPHYTA Lower division of plants, mostly marine and littoral EUKARYOTIC algae.

RIA Drowned river valley transformed by sea-level rise into a high-sided ESTUARY.

RIFT VALLEY The fault-bounded valley found along the crests of many ocean ridges, created by tensional stresses that accompany the process of SEAFLOOR SPREADING. Also occurs on continents.

RIP CURRENT A narrow, swift, seaward-flowing current along the shore that drains water from the surf zone.

RNA (RIBONUCLEIC ACID) A nucleic acid similar to DNA, and a vital component of protein synthesis, in which it reads the genetic blueprint found on DNA molecules.

ROGUE WAVE An unusually large and dangerous wave, usually short-lived and occurring in the open ocean, that is created by constructive WAVE INTERFERENCE.

S

SALINITY A measure of the total concentration of dissolved solids in seawater, generally expressed in parts per thousand.

SALT DOME A columnar intrusion of salt through a sedimentary succession, commonly deforming the STRATA upwards into an anticlinal fold.

SALT MARSH A vegetated intertidal mudflat or sandflat.

SALT PAN A shallow basin in the ground, typical of the coastal zone in arid regions, where salt water is evaporated by the heat of the sun.

SALT WEDGE ESTUARY A type of ESTUARY in which river inflow dominates tidal mixing, producing a highly stratified water column with a sharp HALOCLINE.

SCHOOLING The tendency of many species of fishes and marine mammals to organize themselves into groups.

SEA ARCH An opening through a headland caused by wave erosion, developing from caves.

SEAFLOOR SPREADING The process by which OCEANIC CRUST is created at the crest of MID-OCEAN RIDGES, and by which TECTONIC PLATES diverge.

SEAGRASS A type of marine plant most similar in structure to land plants.

SEA ICE Frozen seawater.

SEAMOUNT A submarine mound, usually of volcanic origin, that rises sharply from the sea floor.

SEA STACK An erosive feature consisting of a small, rocky island a short distance from the shore, left stranded by progressive erosion of a headland.

SEDIMENT Grains or particles of either inorganic or organic origin deposited by air, water, or ice.

SEDIMENTARY BASIN Thick accumulation of sediments, typically 1–15km (0.6–9mi) deep, occurring below both the continents and the sea floor.

SEDIMENTARY CYCLE The process by which SEDIMENTS are formed, transported, deposited, and then undergo LITHIFICATION into sedimentary rocks, before being uplifted by earth movements and once more subjected to weathering and erosion.

SEDIMENTARY ROCK A rock formed from the compaction and cementation of SEDIMENT; one of the three principal rock types. (*See also* IGNEOUS ROCK, METAMORPHIC ROCK.)

SEDIMENT LOAD The amount of sediment carried by air, water, or ice.

SEDIMENT WAVE A series of sediment layers arranged in large-scale, regular undulations, with a typical WAVELENGTH of around 1km (0.6mi). The waves are formed by the deposition of fine, muddy sediment from TURBIDITY CURRENTS and bottom currents.

SEISMIC Pertaining to a naturally occurring or artificial earthquake or earth vibration.

SEISMIC SURVEYING The use of SONAR and explosive devices and measuring equipment to study the nature of natural earthquakes and artificial vibrations.

SEMIDIURNAL TIDE A tide characterized by two equal high tides and two equal low tides per day.

SEWAGE Human or animal body wastes disposed of via the drainage system. One of the main marine pollutants, particularly in coastal waters.

SEXUAL REPRODUCTION The process by which two parent organisms produce an offspring by the fusion of sex cells produced by each parent. (*See also* GAMETE.)

SHELF SEA The area of ocean found at the edge of the CONTINENTAL SHELF, before the CONTINENTAL SLOPE.

SHOCKED QUARTZ Quartz that has been stressed by intense and sudden pressure, as in a meteor impact.

SILICATE A mineral containing silica (SiO_2) as the principal component.

SILICEOUS Adjective describing material whose composition is silica (SiO_2), generally used to referr to biogenically produced siliceous material (from DIATOMS, RADIOLARIANS etc).

SLIDE A type of mass wasting involving movement of material along one or more surfaces of failure; can be both subaerial and subaqueous in occurrence.

SLUMP A type of mass wasting that takes place along a curved surface of failure and results in backward rotation of the slump mass; more contorted than a SLIDE.

SOFAR CHANNEL A zone in the water column where sound waves attain a minimal speed, so that refraction focuses and traps them in a channel.

SOLAR POWER Power created by harnessing solar energy.

SOLUTE A substance dissolved in a solution.

SOLVENT A liquid used to dissolve another substance.

SONAR An acronym for Sound Navigation And Ranging; an instrument used to locate objects underwater by reflecting sound waves.

SPECIES The lowest taxonomic level; a population of similar individuals that can reproduce and produce fertile offspring. (*See* TAXONOMY.)

SPECIFIC HEAT The quantity of heat needed to raise the temperature of 1g (0.035oz) of a substance by 1°C (1.8°F).

SPLASH ZONE The uppermost area of a rocky shore, which is covered by only the highest tides and is usually only dampened by splashing waves and blown spray.

SPREADING CENTER MID-OCEANIC RIDGE where SEAFLOOR SPREADING occurs.

SPRING TIDE A tide that exhibits the greatest difference in height between the high and low tide marks. Spring tides occur when the Sun and the Moon are aligned.

STIPE A stemlike structure found in brown ALGAE.

STRATA Layers or beds of SEDIMENTARY ROCK.

STRATIGRAPHY The branch of geology that studies the age relationship and significance of layered SEDIMENTARY ROCKS and the sequence of FOSSILS they contain. Also known as historical geology.

STRATOSPHERE The lowest part of the middle ATMOSPHERE, 12–45km (7.5–30mi) above the Earth's surface.

STROMATOLITE Laminated CALCAREOUS sedimentary formation produced by lime-secreting CYANOBACTERIA.

SUBDUCTION The movement of one lithospheric plate underneath another so that the descending plate is consumed into the MANTLE.

SUBDUCTION ZONE An area where the LITHOSPHERE is subducted back into the Earth's MANTLE; generally a deep oceanic TRENCH.

SUBLITTORAL ZONE A subdivision of the BENTHIC province extending from the shoreline to a depth of 200m (650ft).

SUBMARINE CANYON Deeply incised, steep-walled valley, commonly V-shaped in profile, that cuts into the rocks and sediments of the outer CONTINENTAL SHELF and the CONTINENTAL SLOPE.

SUBMARINE FAN A cone–shaped sedimentary deposit that accumulates on the CONTINENTAL SLOPE and RISE; generally fed from a distinct point or line-source of sediment such as a major river or DELTA.

SUBMARINE WATERFALL A cascade of water downslope through a narrow underwater channel or gateway. (*See* OCEANIC GATEWAY)

SUBSIDENCE The sinking of large portions of the Earth's CRUST under the influence of major tectonic forces.

SUBSTRATE A general term describing the surface on or within which organisms live or onto which SEDIMENT is deposited.

SUBTIDAL ZONE The region of the shore that is covered by water, even during low tide.

SUSPENSION CASCADING A temporary obstruction of the fall of particulate material through the water column at the interface of water masses of different density.

SWIM BLADDER A gas-filled pouch in many bony fish that is used to attain neutral buoyancy by regulating the amount of gas contained.

SYMBIOSIS An intimate living relationship between two different organisms. (See also COMMENSALISM, MUTUALISM, PARASITISM.)

T

TALUS SLOPE A slope in a subaerial or submarine setting where weathered, eroded, and fragmented material collects.

TAXONOMY The science of naming and classifying organisms according to shared characteristics. A taxon (pl. taxa) is a group such as a SPECIES, GENUS, FAMILY, ORDER, CLASS, or PHYLUM; these represent a hierarchy of decreasing levels of similarity.

TECTONIC PLATE Slab of the Earth's LITHOSPHERE that "floats" on the molten portion of the MANTLE and is transported by convection currents within it.

TECTONIC UPLIFT A rise in topographic height due to crustal movement caused by major tectonic forces.

TERRIGENOUS SEDIMENT SEDIMENTS derived from the weathering and erosion of pre-existing rocks.

TEST The skeleton or shell of some invertebrates.

TETHYS SEA An immense seaway that separated GONDWANA from LAURASIA during the early MESOZOIC .

THERMOCLINE A sharp temperature gradient that marks a transition zone between water masses having markedly different temperatures.

THERMOHALINE CIRCULATION The vertical movement of water masses that results from differences in DENSITY caused by differences in SALINITY and/or temperature.

TIDAL BARRAGE A barrier built across a river to prevent flooding or to be used for the generation of tidal power.

TIDAL BORE A steep wave that moves upriver during the FLOOD TIDE; restricted in occurrence to certain narrow, funnel-shaped ESTUARIES and rivers with especially high tidal range.

TIDAL BULGE Relative high and low points on the Earth's ocean surfaces caused by the gravitational pull of the Moon and Sun.

TIDAL RANGE The vertical difference separating the water level between successive high and low tides.

TRACE ELEMENTS Elements found in small quantities, typically less than 1 part per million.

TRANSFORM FAULT A steep boundary separating two lithospheric plates along which there is lateral slippage.

TRANSGRESSION The incursion of the sea over land areas, resulting in the deposition of marine over terrestrial sediments.

TRAWL NET A large fishing net that is dragged along the seabed or in mid-water by trawlers.

TRENCH Long, narrow, and deep topographic depression, associated with a volcanic arc. Together they mark a collision or SUBDUCTION ZONE where one lithospheric plate is overriding another.

TROPHIC LEVEL A functional or process category for types of feeding by organisms.

TROPOSPHERE The lowest part of the ATMOSPHERE, extending 10–15km (6–9mi) from the Earth's surface.

TSUNAMI A destructive sea wave that is usually produced by an earthquake, but can also be caused by submarine landslides or volcanic eruptions.

TUBE FOOT Tubular structure found in ECHINODERMS that functions in locomotion and feeding.

TURBIDITE Sediment layer, typically with graded bedding, that is deposited by a TURBIDITY CURRENT.

TURBIDITY CURRENT A density-driven current of sediment-laden water that flows swiftly downslope, in some cases traveling many hundreds of kilometers from the edge of the CONTINENTAL SHELF to the ABYSSAL PLAIN.

TURBINE A motor driven by a wheel that is moved by flowing water.

TURBULENCE DRAG Irregular, chaotic flow of a fluid generated by a hydrodynamically inefficient body design that causes considerable drag on an object moving through water.

U

UPWELLING The slow, upward transport of water to the surface from depth, generally recycling nutrient elements and organic material in its course.

V

VENTRAL FIN A fin on the belly of a fish.

VERTEBRATE Any animal that has a backbone.

VISIBLE SPECTRUM The portion of the radiation spectrum visible to the eye.

VIVIPAROUS A type of reproduction in which the young develop from embryos that are attached internally to the mother and that are born alive.

VOLCANICLASTIC SEDIMENT or SEDIMENTARY ROCK containing fragmented volcanic material.Also known as PYROCLASTIC.

VOLCANISM A number of processes associated with the release of MAGMA at the Earth's surface, generally accompanied by hot water, steam, and other gases.

W

WATER COLUMN A vertical section through the ocean.

WATER MASS A body of water identifiable from its temperature, SALINITY, or DENSITY.

WAVE-CUT CLIFF A cliff produced by wave erosion cutting landward.

WAVE INTERFERENCE The overlapping of separate wave trains, either in phase (constructive, producing larger waves) or out of phase (destructive, producing smaller waves.)

WAVELENGTH The horizontal distance between two corresponding points on successive waves.

WAVE PERIOD The elapsed time between the passing of two successive wave crests past a fixed point.

WAVE TRAIN A series of waves coming from the same direction and having a distinctive pattern of wave height and period.

WESTERLIES Air masses moving away from the subtropical high-pressure belts toward higher latitudes. They flow southwesterly in the northern hemisphere and northwesterly in the southern hemisphere.

WESTERN INTENSIFICATION A term describing the nature of the western currents of circulation GYRES, which are swift, deep, and narrow, in contrast to their eastern currents, which are weak, shallow, and broad.

WESTERN BOUNDARY CURRENT Warm currents on the western side of all subtropical GYRES. Such currents flow toward the poles.

WESTERN BOUNDARY UNDERCURRENT A bottom current that flows along the base of the CONTINENTAL SLOPE, eroding sediment from it and redepositing the sediment on the CONTINENTAL RISE. It is confined to the western boundary of deep-ocean basins.

WETLANDS Areas bordering ESTUARIES and other sheltered coastal regions.

WIND FARM A collection of wind TURBINES producing electricity.

Z

ZOOPLANKTON Animal PLANKTON such as FORAMINIFERANS and RADIOLARIANS.

ZOOXANTHELLAE Symbiotic DINOFLAGELLATES that live in the tissue of corals and other reef-building organisms.

ZYGOTE The cell that is formed when a male and female GAMETE fuse, containing a set of CHROMOSOMES from each parent.

Further reading

GENERAL OCEANOGRAPHY & EARTH SCIENCE

Brower, K. and Doubilet, D. *Realms of the Sea*. National Geographic Society, 1991.
Couper, A. (ed.) *The Times Atlas and Encyclopedia of the Sea*. Times Books, 1989.
Deacon, M., Rice, T., Summerhayes, C. (eds.) *Understanding the Oceans*. UCL Press, 2001.
Dixon, D., Jenkins, I., Moody, R., Zhuravlev, A. *Cassell's Atlas of Evolution*. Cassell & Co., 2001.
Earle, S.A. *Atlas of the Ocean*, National Geographic Books, 2001.
Garrison, T. *Oceanography: An Invitation to Marine Science* (4th edition). Brookes Cole, 2002.
Grant Gross, M. and Gross, E. *Oceanography: A View of Earth*. Prentice Hall, 1996.
Kunzig, R. *Mapping the Deep: The Extraordinary Story of Ocean Science*. W. W. Norton, 2000.
Leier, M. and Monahan, D. *World Atlas of the Ocean*. Firefly Books, 2001.
Luhr, J.F. (ed.) *Earth*. DK, 2003.
Pernetta, J. *Philip's Atlas of the Oceans*, Philip's, 1995.
Pinet, P.R. *Invitation to Oceanography*. West Publishing Co., 1996.
Press, F. and Siever, R. *Understanding Earth* (3rd edition). W. H. Freeman, 2001.
Rice, A.L. and Rice, T. *The Deep Ocean*. Smithsonian Institution Press, 2000.
Redfern, R. *Origins: The Evolution of Continents, Oceans and Life*. Cassell & Co., 2000.
Summerhayes, C.P. and Thorpe, S.A. (eds.) *Oceanography: An Illustrated Guide*. Manson Publishing, 1996.
Tarbuck, E.J. and Lutgens, F.K. *Earth: An Introduction to Physical Geology* (5th edition). Prentice-Hall, 1996.
Thurman, H.V. and Trujillo, A.P. *Essentials of Oceanography* (6th edition). Prentice Hall, 1999.
Van Andel, T.H. *New Views on an Old Planet: A History of Global Change*. Cambridge University Press, 1994.

OCEAN LIFE & ENVIRONMENT

Byatt, A., Fothergill A., Holmes, M. *The Blue Planet: A Natural History of the Oceans*. BBC/DK, 2002.
Clark, R.B. *Marine Pollution* (4th edition). Oxford University Press, 1997.
Dipper, F. and Tait, R.V. *Elements of Marine Ecology* (4th edition). Butterworth-Heinemann, 1998.

Ellis, R. *The Empty Ocean: Plundering the World's Marine Life*. Shearwater Books, 2003.
Gibson, R., Hextall, B., Rogers, A. *Photographic Guide to Sea and Shore Life of Britain and Northwest Europe*. Oxford University Press, 2001.
Hall, M., Hall, H., Benchley, P. *Secrets of the Ocean Realm*. Carroll & Graf, 1997.
Hinrichsen, D. *Coastal Waters of the World: Trends, Threats, and Strategies*. Island Press, 1998.
Holland, H.D. and Petersen, U. *Living Dangerously: The Earth, its Resources and the Environment*. Princeton University Press, 1995.
Lean, G., Hinrichsen, D., Markham, A. (eds.) *Atlas of the Environment* (2nd edition). HarperCollins, 1994.
Levinton, J.S. *Marine Biology: Function, Biodiversity, Ecology*. Oxford University Press, 2001.
Morgan, S. and Lalor, P. *Ocean Life*. PRC Publishing, 2001.
Middleton, N. *The Global Casino: An Introduction to Environmental Issues*. Arnold Publishers, 1999.
Meinkoth, N.A. *The National Audubon Society Field Guide to North American Seashore Creatures*. Knopf, 1981.
Nybakken, J.W. *Marine Biology: An Ecological Approach*. Benjamin Cummings, 2001.
Perrin, W.F., Wursig, B., Thewissen, G.M. (eds.) *Encyclopedia of Marine Mammals*. Academic Press, 2002.
Pickering, K.T. and Owen, L. A. *An Introduction to Global Environmental Issues* (2nd edition). Routledge, 1997.
Tackett, D.N. and Tackett, L. *Reef Life: Natural History and Behaviors of Marine Fishes and Invertebrates*. Microcosm, 2002.
Tudge, C. *The Variety of Life: A Survey and a Celebration of All the Creatures That Have Ever Lived*. Oxford University Press, 2000.

USEFUL WEBSITES

FAO Fisheries Department (United Nations)—**www.fao.org/fi/default_all.asp**
IUCN (World Conservation Union)—**www.iucn.org**
Marine Biological Laboratory (Woods Hole)—**www.mbl.edu**
National Oceanic and Atmospheric Administration—**www.noaa.gov**
Scripps Institution of Oceanography—**www.sio.ucsd.edu**
Southampton Oceanographic Centre—**www.soc.soton.ac.uk**
Woods Hole Oceanographic Institution—**www.whoi.edu**
World Wildlife Federation—**www.panda.org**

Picture credits

A	Ardea	BC	Bruce Coleman Collection
C	Corbis	OSF	Oxford Scientific Films
DS	Dorrik Stow	SPL	Science Photo Library

1, 2–3 Royalty-Free/C, 4, 6 Gary Braasch/C, 8 Private Collection, Lauros/Giraudon/Bridgeman Art Library, 11 Staatliche Antikensammlung und Glyptothek, Munich, Germany/Bridgeman Art Library, 12–13 Mary Evans Picture Library, 14 Mary Evans Picture Library, 16 George Bernard/SPL, 18–19 Douglas Peebles/C, 20–21 Roger Ressmeyer/C, 22–23 NASA/Roger Ressmeyer/C, 23 US Geological Survey/SPL, Pekka Parviainen/SPL, 26 Paul A. Souders/C, 26 NASA/C, 26–27 Galen Rowell/C, 28 Charles O'Rear/C, 29 Tom Bean/C, 30 Ralph White/C, 32t Dr. Ken Macdonald/SPL, 32b Ralph White/C, 33 Bernhard Edmaier/SPL, 35 NASA/C, 36t David Parker/SPL, 36b Michael S. Yamashita/C, 37 NASA/C, 39 G. Marcoaldi/Panda Photo/FLPA, 40 Galen Rowell/C, 41 NASA/SPL, 42–43 Robert Gill/Papilio/C, 44 Bernhard Edmaier/SPL, 46 Joel Creed/Ecoscene/C, 47 NASA/C, 48 CNES/SPL, 50 Jean-Paul Ferrero/A, 51t MartinWendler/NHPA, 51b Michael Fogden/OSF, 52 Liz Bomford/A, 53 David Dixon/A, 54 John Clegg/A, 55 DS, 56 Yann Arthus-Bertrand/C, 56–57 Hans Georg Roth/C, 58 DS, 59 NASA/SPL, 60 Yann Arthus-Bertrand/C, 61 Yann Arthus-Bertrand/C, 62 Hans Georg Roth/C, 63 Jonathan Blair/C, 64–65 Hanan Isachar/C, 66 ANT Photo Library/NHPA, 66–67 Yann Arthus-Bertrand/C, 69 NASA/SPL, 70 European Space Agency, 70–71 FLPA, 72 Bernhard Edmaier/SPL, 73 NASA/C, 74 Earth Satellite Corporation/SPL, 76 Yann Arthus-Bertrand/C, 77 Peter Johnson/C, 78 Paul A Souders/C, 82 Roger Garwood & Trish Ainslie/C, 84 Bettmann/C, 84–85 David Pu'u/C, 85 NASA/SPL, 86 Craig Tuttle/C, 89 SPL, 90 Robert Garvey/C, 91 Joel W. Rogers/C, 92–93 Galen Rowell/C, 94 NASA/C, 95 DS, 96–97 Darrel Gulin/C, 98 Ben Spencer/Eye Ubiquitous/C, 99 C, 100t NASA/SPL, 100b Stuart Westmoreland/C, 101 NOAA/SPL, 102–103 Danny Lehman/C, 104 Steve Lindridge/Eye Ubiquitous/C, 105 Mike McNamee/SPL, 106l NASA/C, 106r DS, 108 Morten Strange/NHPA, 108–109 Philip Gould/C, 109 Simon Fraser/SPL, 110–111 Roger Ressmeyer/C, 111 Martin Bond/SPL, 112 Yann Arthus-Bertrand/C, 113 Tony Craddock/SPL, 114t Woods Hole Institute, 114b Charles D. Winters/SPL, 115 Anthony Bannister/Gallo Images/C, 116 C, 117 Bernhard Edmaier/SPL, 118–119 W. Haxby, Lamont-Doherty Earth Observatory/SPL, 134–135 Jeffrey L. Rotman/C, 136–137 Peter Scoones/SPL, 138 Jonathan Blair/C, 139 Wolfgang Bauermeister/SPL, 140t Georgette Douwma/SPL, 140b A.B. Dowsett/SPL, 144t, Jens Rydell/BC, 144c Jane Burton/BC, 145 Norbert Wu/NHPA, 146t Jonathan Blair/C, 146b Anthony Bannister/NHPA, 148–149 Peter Johnson/C, 149 Geological Survey of Canada/SPL,

150–151 Pete Atkinson/NHPA, 153 Kurt Amsler/A, 154 T. Stevens & P. McKinley/Pacific Northwest Laboratory/SPL, 155 R. Cranston/Panda Photo/FLPA, 156 William Gray/OSF, 157 Stephen Frank/C, 158 Flip Nicklin/Minden Pictures/FLPA, 159 John Clegg/A, 160–161 Jan Hinsch/SPL, 161t D.P. Wilson/FLPA, 161b MI Walker/SPL, 163c SPL, 164t OSF, 164b Image Quest Marine/NHPA, 165 Peter Scoones/SPL, 166 Trevor McDonald/NHPA, 166–167 Paul A. Souders/C, 167 Mark Spencer/A, 168 Sanford/Agliolo/C, 169 Richard Herrmann/OSF, 170–171 Peter Scoones/SPL, 172 Jean–Paul Ferrero/A, 173t Ken Lucas/A, 173b Colin Marshall/FLPA, 175 Konrad Wothe/OSF, 176 David Fleetham/OSF, 177t Pam Kemp/OSF, 177b Howard Hall/OSF, 178 B. Borrell Casals/FLPA, 179 Ken Lucas/A, 180 D. Fleetham/Silvestris/FLPA, 181 Brandon D. Cole/C, 182 D. Fleetham/Silvestris/FLPA, 182–183 Francois Gohier/A, 183 Fred McConnaughey, 183, 184 Richard Herrmann/OSF, 185t Paulo De Oliveira/OSF, 185b Norbert Wu/NHPA, 186 D. Fleetham/Silvestris/FLPA, 187 Mary Beth Angelo/SPL, 188 Gordon Bull/OSF, 188–189 Georgette Douwma/SPL, 189 Clive Bromhall/OSF, 190 Douglas Faulkner/SPL, F. Bavendeam/Minden Pictures/FLPA, 192–193 Pat O'Hara/C, 194 Larry Williams/C, 196 Anthony Bannister/Gallo Images/C, 197 Bob Gibbons/A, 200–201 Bob Gibbons/A, 202t Peter Johnson/C, 202b Brian Kenney/OSF, 203 David Woodfall/NHPA, 204 Douglas Faulkner/SPL, 206t Georgette Douwma/SPL, 206b B. Jones & M. Shimlock/NHPA, 207 Kathie Atkinson/OSF, 208t Dr. Gene Feldman/NASA GSFC/SPL, 208b Pete Atkinson/NHPA, 209 F.S. Westmorland/SPL, 210 Peter Herring/Image Quest 3–D, 212t Ralph White/C, 212b Woods Hole Institute, 213 Norbert Wu/NHPA, 214–215 B. & C. Alexander/NHPA, 216–217 David Cayless/OSF, 217 Kim Westerskov/OSF, 218tDoug Allan/OSF, 218b Dr. Gene Feldman/NASA GSFC/SPL, 219 Dave G. Houser/C, 220t Eyal Bartov/OSF, 220b Ehud Spanier, 221 Jean-Paul Ferrero/A, 222 Michael Pitts/Nature Picture Library, 223 Ed Kashi/C, 224 Najlah Feanny/C, 225 ANT Photo Library/NHPA, 226 Lynda Richardson/C, 227t Simon Fraser/SPL, 227b Simon Fraser/SPL, 228–229 Yann Arthus-Bertrand/C, 229 Paul Franklin/OSF, 230t D.Fleetham/Silvestris/FLPA, 230b Georgette Douwma/SPL, 231 Pete Atkinson/NHPA, 232 Rick Price/SAL/OSF, 233 Francois Gohier/A, 234–235 Victor De Schwanberg/SPL, 236–237 Peter Johnson/C, 238–239 R. Visser/Greenpeace/Corbis Sygma, 240 Stuart Westmorland/C

ARTISTS

29, 79b, 146–147 Julian Baker; 152–153, 156 Rob & Rhoda Burns; 141, 169, 198–199, 200–201, 204–205, 210–211 Brin Edwards/Wildlife Art; 174, 175, 206 Alan Hancocks; 152–153, 158 Karen Hiscock; 28, 30, 34, 40, 41, 44–45, 46, 50, 54, 57, 68, 73, 76, 82–83, 92, 94, 97,105, 107, 158, 159, 194 Maltings Partnership; 143, 144–145 Denys Ovenden; 71, 72, 74, 79t, 82, 83, 86, 87, 89, 90, 91, 139, 148 David Russell; 183 Peter Visscher

Index

Index